Also by the Author
The Baby Bonding Book for Dads
Why Babies Do That
Toddler

THE BUSINESS OF BABY

What Doctors Don't Tell You,
What Corporations Try to Sell You,
and How to Put Your Pregnancy, Childbirth,
and Baby Before Their Bottom Line

JENNIFER MARGULIS

Scribner
New York   London   Toronto   Sydney   New Delhi

This publication contains the opinions and ideas of its author. It is intended to provide helpful and informative material on the subjects addressed in the publication. It is sold with the understanding that the author and publisher are not engaged in rendering medical, health, or any other kind of personal professional services in the book. The reader should consult his or her medical, health, or other competent professional before adopting any of the suggestions in this book or drawing inferences from it.

The author and publisher specifically disclaim all responsibility for any liability, loss, or risk, personal or otherwise, which is incurred as a consequence, directly or indirectly, of the use and application of any of the contents in this book.

Some of the material in chapters 4, 8, and 9 was previously published in a different form in *Mothering* magazine.

SCRIBNER
A Division of Simon & Schuster, Inc.
1230 Avenue of the Americas
New York, NY 10020

First Scribner hardcover edition April 2013

SCRIBNER and design are registered trademarks of The Gale Group, Inc., used under license by Simon & Schuster, Inc., the publisher of this work.

For information about special discounts for bulk purchases,
please contact Simon & Schuster Special Sales at 1-866-506-1949
or business@simonandschuster.com.

The Simon & Schuster Speakers Bureau can bring authors to your
live event. For more information or to book an event contact the
Simon & Schuster Speakers Bureau at 1-866-248-3049 or visit our website at
www.simonspeakers.com.

Book Design by Maura Fadden Rosenthal

Manufactured in the United States of America

10 9 8 7 6 5 4 3 2 1

Library of Congress Control Number: 2012031245

ISBN 978-1-4516-3608-6
ISBN 978-1-4516-3610-9 (ebook)

For my mother, Lynn Margulis

Contents

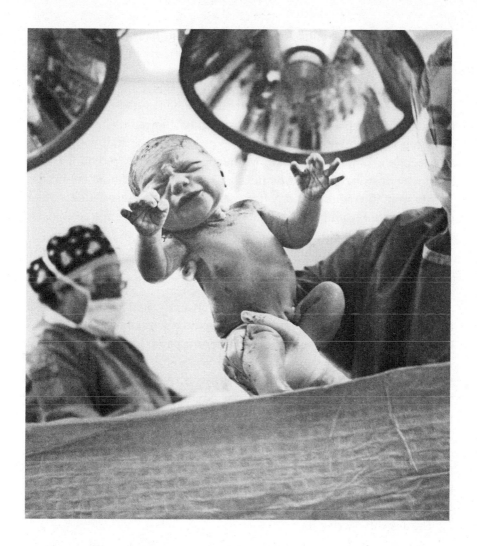

Introduction

Twenty-nine years old and pregnant for the first time, Marijana Picton noticed her nausea only went away when she took long walks and ate *stapci,* Serbian-style salty pretzel sticks. It was 2009. Marijana and her husband, Richard, had been living in England but they moved back to Šipovo, the small town in Bosnia and Herzegovina where Marijana grew up, when she was seven months along. Šipovo's bars were filled with unemployed men and two of the town's four factories had not fully reopened for business—tangible signs of the war that once tore the former Yugoslavia apart.

After taking a birthing class in England, Marijana and Richard had a long list of questions for the staff at the Mrkonjić Grad clinic where they would have their baby: Would her husband be allowed to stay with her? Most husbands in Serbia don't, they were told, but the staff could make an exception. What kind of pain medication would they provide? None, the doctor answered, unless you need a C-section. What about epidurals? "If you want it, you have to buy it yourself," the doctor responded. "Most people get them from Italy. And then you have to find the anesthesiologist to give you the injection."

Marijana's water broke that November on her birthday. She called the clinic to tell them she and Richard were coming so they could turn on the heat in the labor room.

Two years earlier in Oaklyn, New Jersey, twenty-eight-year-old Melissa Farah, a special education teacher at Avon Elementary School, was pregnant for the first time. Melissa and her husband, Dan, were planners: They had been married for almost two years and had begun trying to start a family on their first wedding anniversary. Melissa felt especially lucky because a close girlfriend, Valerie Scythes, was pregnant too. Both women planned to have their babies at the same hospital in Woodbury, New Jersey.

Here's the question: Which young woman would be better off, the one in a small Balkan country still recovering from a brutal civil war, or the mom in the richest and most powerful country in the world with state-of-the-art medical equipment and know-how?

The answer: Marijana.

According to the most recent reports, the likelihood of a mom like Melissa dying due to pregnancy or childbirth in the United States is more than four times higher than in Bosnia and Herzegovina and seven times higher than in Italy or Ireland; the likelihood of her dying as a result of childbirth is five times greater than in Germany and Spain, and fifteen times greater than in Greece.

The United States also lags behind most industrialized countries when it comes to the health and well-being of infants. Eight American children per 1,000 live births will not live to age five.

Fact: A child in the United States is more than twice as likely as a child in Finland, Iceland, Sweden, or Singapore to die before her fifth birthday.

We feel great sadness and shock when we hear about a baby dying:

Avery Cornett of Lebanon, Missouri, who was ten days old when he died on December 18, 2011, of a bacterial infection thought to be contracted from contaminated infant formula; a two-week-old unnamed baby boy who died in September 2011 from complications due to an out-of-hospital circumcision; six-week-old Ian Larsen Gromowski, who died of a severe reaction to the birth dose of the hepatitis B shot on August 10, 2007. We stop in horror, our hearts in our throats for the grieving parents. But we consider these deaths isolated incidents, rare occurrences that garner our sympathy, sure, but that certainly won't happen to us.

Fact: The United States has one of the highest infant death rates of the industrialized world. It is safer to be born in forty-eight countries than in the United States.

Fact: Of the some 4.3 million babies born in America each year, more than 25,000 will die in their first year.

Fact: The maternal mortality rate in the United States is among the highest in the industrialized world.

After taking that birthing class in England, Marijana had been scared of giving birth and was sure she wanted an epidural. But the clinic in Mrkonjić Grad was the only hospital she and Richard could find in Bosnia and Herzegovina where Richard would be allowed to stay with her. So Marijana ended up having no pain medication and no fetal monitoring during labor. When she was fully dilated and climbed onto the ratty operating table, her contractions slowed. The doctor and midwife heaved her upright and told her to walk around the room some more. Her son was born vaginally about three hours after her water broke. The doctor joked you couldn't find another baby like him in all of Bosnia. Richard is half English and half Nepali and the baby, whom they named Stanley, looked to Marijana "like a little Chinaman."

Twenty-eight years old and in good health when she went into the hospital, Melissa Farah gave birth to a healthy baby girl via C-section in April 2007. But the birth did not go as expected: Melissa was transferred to another hospital due to complications from the operation. Doctors were not able to stabilize her and Melissa bled to death the next day. She is not alone: More than seven hundred American women die in childbirth every year, though most of these deaths go unnoted. The vast majority of maternal deaths in the United States are never investigated in any true sense: In 2007 only slightly more than half the deaths related to pregnancy or childbirth were autopsied, and there's evidence that autopsy rates in hospitals are declining; a review of the death is almost always conducted behind closed doors by a committee comprised only of hospital staff, and the information garnered is not released either to the family or to the public; journalists and other outside investigators are often hindered from accessing information because (they are told) of patient privacy concerns. When there is obvious wrongdoing, hospital lawyers work tirelessly to cover it up, negotiating financial settlements that include gag orders so the details of what happened cannot legally be made public. Only twenty-four states require hospitals to report adverse maternal outcomes to the state government. And only a handful of these states—including Ohio, New York, and California—require that this information be available to the public. But even when the reporting is mandatory and the numbers are submitted to the state (noncompliance is an issue), most states have no system in place to investigate maternal deaths. A 2010 investigation revealed that twenty-nine states and the District of Columbia have "no maternal mortality review process at all." Melissa's case became national news because it was the second death at Underwood-Memorial Hospital in Woodbury, New Jersey, in two weeks. Her good friend Valerie Scythes died two weeks earlier after having a planned C-section. Her husband told the BBC that doctors scheduled the operation because Valerie was thirty-five and had had ovarian cysts in college.

How is it possible that a country as wealthy and medically advanced as the United States has a higher maternal mortality rate than a much less affluent country like Bosnia and Herzegovina? Are the high maternal and infant death rates in America really isolated events, or are they mounting evidence that something in our country is going terribly wrong?

Many obstetricians in America throw up their hands and say that our high rates of maternal and infant deaths are either the patient's fault or "an act of God." They argue that our increasing infant mortality rates are due to the rising use of fertility drugs and a greater number of older women having first babies, both of which lead to an increase in twins and premature births. They also point out that more American women are overweight or obese when they get pregnant, making labor more dangerous. While maternal age, obesity, and multiple births can contribute to higher maternal mortality, these factors are only a small piece of the puzzle. If we look at the best evidence, and if we compare American practices to countries where moms and babies enjoy safer outcomes, we find that the science tells a different story.

This book is about how what happened to Melissa points to a larger problem with the way we are treating women and their babies in the United States. "Obstetrics is an ugly business and it's our most primitive medicine," says Stefan Topolski, M.D., assistant professor of Family and Community Medicine at the University of Massachusetts in Worcester. "We say it in meetings in our department all the time. It's the least evidence-based discipline." As this book will show you, time and time again corporate profits and private interests trump what is best for moms and babies. The science is consistently ignored, and practices proven to be harmful are continued. Doctors—even though most have the best possible intentions—often unwittingly go along with a broken and sometimes dangerous system.

Few American parents could imagine deciding what car to buy just based on the ads they see during the Super Bowl or on the first package the car dealer offers. Instead, we do our homework, talk to friends about options, research safety history and gas mileage on the Internet, and read *Consumer Reports*. Ultimately we choose the car that's best for our family after carefully weighing (and usually declining) all the little things the salesman tries to slip in.

But most of us are much more naïve when it comes to parenthood. While it is acceptable to haggle over cars, it's almost unthinkable in the United States to go against what is considered routine when it comes to pregnancy, childbirth, and raising an infant. We defer to the person we believe is the expert, wearing a white coat with a stethoscope around her neck. We are conditioned to trust doctors, to accept that what they tell us is true, and to believe that they only have our best interests in mind.

Like most American children in the 1970s and 1980s, my brothers and I grew up eating Froot Loops, Apple Jacks, and Count Chocula for breakfast. Zach and I sprinted home from school every day to watch ABC After School specials and sing along to Jell-O commercials while our teenage brothers blared Janis Joplin upstairs. We attended public school and received the recommended vaccinations on the recommended schedule.

My parents may have been unusual in that they both had Ph.D.s (my mother was a microbiologist, my father a chemist), but our family was conventional. My parents basically did what everybody else did, fed their children what everybody else fed theirs, and conformed to the values and standards around them. I never thought to question American culture. I always assumed that Wisk *could* solve ring around the collar, that doctors knew better than I what would keep me healthy, and that government officials always had my best interests in mind.

But during my first pregnancy, when I was twenty-nine years old, I found myself sobbing in the car in the parking lot after every prenatal visit. I felt I was being bullied rather than cared for. My husband, who accompanied me on these prenatal visits, would let me cry in his arms and try to hide his worry from me. We were both graduate students and we felt we couldn't change providers because the one we were using was the only practice in Atlanta, Georgia, that our insurance company would pay for. We knew we would not choose to abort, so my husband and I tried to forgo some routine prenatal testing. We weren't trying to be difficult or rebellious—we were just seeking to avoid unnecessary stress. But our health providers did their best to scare us into compliance. One hospital midwife in Atlanta told me I would "buy" myself a C-section if I refused the test she insisted on.

We switched from the hospital midwives to the doctors because, ironically, they seemed less rigid. But toward the end of my pregnancy a doctor ordered an "emergency" ultrasound because she believed I was measuring small. She turned to go to her next client before I could talk to her about it, muttering that she suspected "intrauterine growth retardation." We sat in the waiting room, flooded with anxiety. The scan showed the baby was fine. It wasn't until years later when I started researching and writing about pregnancy that I learned that ultrasound scans have not been shown to be any more effective in predicting intrauterine growth restriction (doctors these days try to avoid using the word *retardation*) than palpation of the pregnant woman's abdomen by an experienced clinician. The same summer my daughter was born, Marsden Wagner, an obstetrician and scientist, and former director of Women's and Children's Health at the World Health Organization, wrote: "There is no justification for clinicians using routine ultrasound during pregnancy for the management of IUGR."

Mine was a low-risk pregnancy. I was young, strong, and healthy. When six months of nausea finally abated it was like someone washed the windows. I exercised daily, zooming past other cyclists on the bike path on Atlanta's East Side. I had been heavy when I first got pregnant, so I only gained a total of twenty pounds. I should have had a straightforward labor and delivery. But my husband and I knew very little about birth; we did not have a great relationship with our health care providers; and we did not have the support we needed in the delivery room. After hours of being left alone while in active painful labor, I was accosted by a brusque nurse who burst through the door, put on a glove, ordered me on my back, and stuck her fingers roughly in my vagina to assess the dilation of my cervix. "Nothing! Not even a dimple," she scolded before rushing out again.

When I vomited during labor and my husband started panicking, nobody reassured us that vomiting was a good sign, an indication that my hormones were kicking into gear and that my body was cleaning itself out to make room for the baby. Instead, the staff acted disgusted that they had to clean it up. I have a family history of low blood sugar, and as the contractions continued hour after hour I felt myself getting weaker. I begged for something healthy to eat. The nurses refused. The doctor on call when our daughter was born was a floater in the practice, the only man, and the only one we had never met before. Knowing I didn't want an epidural or Pitocin (a synthetic hormone that stimulates the uterus to contract), he chastised me for selfishly putting my family through "so much waiting," and told me, while I was having an intense contraction, that I should stop thinking only about myself. I ended up giving birth on my back with Pitocin and an epidural, needing stitches for a badly torn perineum, and having side effects from the anesthesia (one of my legs went numb) that lasted for months.

After our baby was born the nurse bustled into the room with a tray. "Time for the hep B vaccine!" she announced. I knew enough to know that hepatitis B is a sexually transmitted disease; I felt totally protective of the skinny frog-legged baby whose life was my responsibility. We told the nurse we wanted more information. Instead of explaining the rationale behind vaccinating an hours-old baby for a sexually transmitted disease my husband and I had both tested negative for, the nurse slit her eyes in anger.

Two weeks later a pediatrician applauded our decision and told us that a fax on her desk warned that hepatitis B should not be given to newborns. I would find out ten years after that what happened: Since vaccines are tested individually and not in combination, scientists at the CDC had rather shockingly and embarrassingly discovered that they had overlooked the fact that the mercury load in the infant vaccine schedule might be damagingly high. In July 1999, the American Academy of Pediatrics (AAP) and the U.S. Public Health Service issued a joint statement asking for the mercury-based preservative thimerosal to be removed from infant vaccinations, and that the birth dose of hepatitis B be withheld from newborns whose mothers tested negative. Our stand against the hepatitis B vaccine had been more than warranted. It was a pivotal moment of disillusionment; suddenly I knew that it was up to me to educate myself in order to make the best-informed health decisions for my baby.

In the weeks after my daughter was born, a slim envelope arrived in the mail. It was a hospital bill for more than $6,000. The insurance company denied the claim because they considered the pregnancy a preexisting condition, even though they had been my insurer for more than four years. As graduate students, my husband and I each made an annual stipend of about $11,000. I sat down to nurse my newborn, gazing at her tiny ears that stuck straight out like my husband's, inhaling the warm smell of her scalp, and fretted about the unpaid bills.

Though the insurance company eventually acknowledged their mistake, my confidence was shaken. I felt like a trapped gerbil: I used an OB practice that I did not like because it was the only one my insurance would pay for; I was frightened into having a birth with expensive intervention that I did not want or need; and then I ended up with a bill for procedures I had not wanted that I could not afford to pay.

Unfortunately, the experiences I had during the birth of my first child are not unique. The difficulties that started with prenatal care and continued well into my daughter's childhood are part of a larger system that I expose in this book, a system that puts large companies and corporate interests ahead of the best interests of the mother and her baby. You may be surprised to read that the baby wash used in American hospitals contains formaldehyde, and that the same company that makes it changed their formula for the European Union. You may be dismayed to learn that the pediatric "expert" who promoted what we now know is a mistaken idea of delayed potty training was a paid spokesperson for Pampers. Though some are no longer even American owned, the businesses that are hurting babies in order to increase profits—like Gerber and Pampers and Johnson & Johnson—are such an ingrained part of American culture and so aggressively promote their public image that many of their unconscionable practices are not questioned even by the most educated and savvy caregivers.

When we were little my brother Zach, just eighteen months older than I, and I used to ask our father which of us he loved best.

"The one who likes ice cream," my father would invariably answer.

Zach and I would throw our heads back and laugh. I knew my dad loved *me* best because I liked ice cream. My brother knew my dad loved *him* best because he liked ice cream.

My father never said, "I love you both the same," because he, in fact, didn't love us the same. None of us loves our children equally, we love them specially, according to their needs and our needs, their age and our age, their temperament and our temperament, and a host of other factors. This book is not an advice book and it does not offer One Right Way to parent. Because every baby is different, every parent is different, and every circumstance is unique, each triad of parents and baby comes to figure out their own way in their own time.

What it does offer, however, is valuable information to help all of us in America stop blindly following the status quo and start doing our own research. In order to most effectively protect and raise our children we have to figure out the best way to do things ourselves. Parents, not for-profit companies or health care professionals, are actually the real experts when it comes to gestating, birthing, and raising our babies. Though we need support and advice, we do not need to be bullied or intimidated. We know more about what works best for our families and for our children than anyone else. As Benjamin Spock famously told the generation of nervous parents at the beginning of the baby boom in 1946, we can trust our own instincts; we know more than we think.

Still, it's not easy to tell a well-intentioned doctor, who acts like he knows better than you, that you question his reasoning, or that you want a second opinion before proceeding. What if you're wrong? What if they're wrong?

The Business of Baby will show why the practice of self-education, standing up to the system, voicing your concerns, and acting on them is more crucial than ever to the well-being of our babies. It is my hope that the stories of the men and women who kindly agreed to be included in this book, as well as the testimony from doctors, nurses, and other health care practitioners who are deeply concerned about the failings in our system, combined with a more detailed knowledge of both the science and of who is benefiting from the ways things are being done today, will inspire readers to seek out their own path and find a healthier and happier way to raise America's children.

Author's Note

This book begins in pregnancy and ends at baby's first birthday, following in rough chronological order many of the situations and decisions parents face as they gestate, give birth to, and care for their new babies. An entire book could be written in the place of each chapter (and many excellent books already have), so I've included recommendations for further reading on each topic in a resource section in the appendix.

When describing medical procedures, I use medical terminology that may be unfamiliar. Because I did not want to interrupt the flow of the story, I left many of these medical terms undefined. For this reason, a glossary of terms and abbreviations is included at the end of the book.

While I use the real first and last name of the people whose stories I share the vast majority of the time, some names and identifying details have been changed. In the interest of readability, once I introduce a person, I refer to that person by first name. However, I have decided to follow contemporary American social convention and use last names when referring to doctors and scientific researchers. I'm not altogether comfortable with this hierarchy but I couldn't find a way around it that felt respectful to my sources (some of whom will be offended that I do not always refer to them as "Dr." every time their name appears) and to parents. If a doctor is sharing a story primarily as a parent and not a professional, or if I'm describing a setting where everyone is on a first-name basis, I use first names.

I have similarly deferred to Icelandic social convention. Doctors in Iceland are called by their first names because last names are actually patronymics (so siblings of different genders do not share the same last name: *Jónsdóttir* means "John's daughter," *Jónsson* means "John's son").

CHAPTER 1

GESTATION MATTERS:
The Problem with Prenatal Care

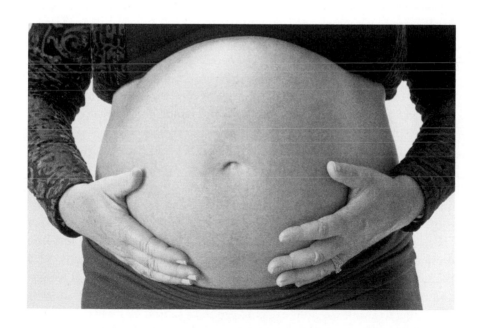

Like many middle-class American women, Elizabeth Goodman-Logelin, at thirty, had put childbearing on hold to establish a career. In her twenties she used her smarts, drive, and organizational skills to become a high-powered management consultant. When she and her husband, Matt, decided it was time to get pregnant, the couple from Minnesota felt as if they had it all: a spacious new home in L.A., a happy marriage, and a baby on the way.

But pregnancy was harder than Liz expected. A petite woman, Liz struggled to gain weight. As Matt details in his memoir, *Two Kisses for Maddy*, Liz had debilitating morning sickness and vomited frequently. Just to be safe, when she was about twenty-eight weeks along, her obstetrician referred her to a high-risk specialist in Pasadena, Greggory DeVore, M.D. DeVore performed an ultrasound and announced that Liz's amniotic fluid levels were low, the baby was small for gestational age, and the cord was wrapped around her neck.

Then he prescribed three weeks of bed rest.

When a subsequent ultrasound indicated no improvement, DeVore insisted—though the baby's due date was still nine weeks away—that Liz spend the rest of her pregnancy in the hospital. Terrified first-time parents, Liz and her husband did not question this advice.

After two weeks in the hospital, the doctors told her and Matt that it was time for the baby to come out, believing she would be safer outside the womb. Madeline Logelin was born seven weeks premature via C-section on March 24, 2008. She was rushed to the neonatal intensive care unit (NICU), placed in a plastic incubator, and given supplemental oxygen "as a precaution" although her father was told she was breathing just fine on her own. It turned out that Maddy was a perfectly healthy baby. At 3 pounds 13.5 ounces and 17.25 inches long, she was in the normal weight and height range for her gestational age. Her only issue was that she was too young to swallow—an iatrogenic problem caused by early delivery.

Twenty-seven hours later, getting ready to finally hold her daughter for the first time, Liz mumbled that she felt light-headed, fell backward into the wheelchair waiting to take her to the NICU, and then slumped forward onto the floor.

She was dead.

The doctors told Matt that Liz most likely died of a pulmonary embolism—a blood clot or other blockage that impedes blood flow to the lungs. Matt believed "shitty luck" killed the woman who had been the love of his life.

Expectant parents such as Matt and Liz are rarely told that putting a patient on bed rest dramatically increases the risk of pulmonary embolism (one study showed that embolism is almost twenty times more likely for pregnant women on bed rest). "I was put on strict bed rest at three different hospitals and never told there was a risk," says one mother of four from Pennsylvania. Most expectant parents are also unaware that several studies have shown that it is difficult to access the amount of amniotic fluid in utero accurately. When researchers at the University of Mississippi Medical Center analyzed three techniques used for testing amniotic fluid level, all three were found to be "moderately accurate," giving correct estimates only between 59 and 67 percent of the time.

In other words, at least one third of a doctor's diagnoses from ultrasound that a woman's body is generating too much or too little amniotic fluid are wrong. These misdiagnoses lead to more pregnancy interventions, including C-sections. As Sarah Buckley, an Australian family physician and obstetrician, birth advocate, mother of four, and author of *Gentle Birth, Gentle Mothering: A Doctor's Guide to Natural Childbirth and Gentle Early Parenting Choices,* explains, using ultrasound to detect low amniotic fluid levels "has been shown to lead to overdiagnosis of problems resulting in high rates of induction for healthy babies." The science on this subject is abundant: One double-blind randomized study of more than fifteen hundred pregnant women concluded that measuring amniotic fluid levels for women who were at forty weeks gestation was not significantly correlated with better fetal outcomes. It was, however, a predictor for more unnecessary intervention. The study concludes: "Routine use [of amniotic fluid index, that is, an approximate estimate of amniotic fluid levels from ultrasound] is likely to lead to increased obstetric intervention without improvement in perinatal outcomes."

According to her husband's account, the doctor spent only a few minutes looking at an ultrasound before prescribing Liz bed rest. Yet more than half a dozen studies have determined that bed rest is either of no proven benefit or that there is simply not enough available evidence to support or refute that it works. While there is not sufficient evidence showing bed rest is of benefit, there *is* evidence that it causes harm. Researchers have found that it not only dramatically increases the likelihood of getting blood clots, but it can also lead to significant bone loss in pregnant women.

"I don't put pregnant women on bed rest anymore," asserts Reynir Tómas Geirsson, M.D., the chair of the Department of Obstetrics and Gynecology at the Landspítali University Hospital in Reykjavik, Iceland. Reynir has been practicing obstetrics for thirty-six years and has attended more than three thousand deliveries. "I sometimes ask women at risk for preterm labor—and occasionally those with high blood pressure—to stop work, to rest at home, or occasionally in the hospital, and I direct them on what exertion to avoid, and when and how during the day to take rest periods. But enforced strict bed rest has never been proven of use."

Since Baby Maddy turned out to be of normal height and weight for her gestational age, it's possible that *nothing* was wrong with Liz's pregnancy in the first place. Liz's tragic story begs the question: Would this young mom be alive today if the doctors had done less testing, paid attention to the existing scientific literature instead of prescribing a course of action that is of no proven benefit, and taken into consideration that she was a petite woman likely to have a smaller-than-average baby?

TESTING 1, 2, 3

"The twenty-first century has skewed every aspect of modern life," argues Michael Klaper, who has forty years of experience practicing medicine, and has done postgraduate training in obstetrics at the University of California–San Francisco. Dr. Klaper and I talk for more than an hour via Skype teleconference. "We've made pregnancy—which is normally this joyous, wonderful process—become a perilous and often tragedy-filled event." Take, for example, the astonishing number of routine prenatal tests that almost every pregnant woman is subjected to in the United States today.

This dizzying array of testing can make prenatal care an unpleasant, uncomfortable, time-consuming, and expensive experience for women and their partners. "You go in and it's like, 'Let's see what horrible new problem you have today that you've never heard of before,' " says one dad from Buffalo, New York, who accompanied his wife to every appointment. "Between the false negatives and the false positives, and the endless stream of tests, when are you ever going to feel reassured?"

As a mother of four, I know firsthand how stressful it can be to have prenatal testing (or to refuse it and incur the wrath of a provider), but I wanted to hear an obstetrician's point of view. Which is how I find myself on a bitterly cold December evening walking past a bicycle wrapped in Christmas lights on the front lawn of a home in Arlington, Massachusetts.

Stephanie Koontz, an obstetrician at Mount Auburn Hospital in Cambridge, grew up with me about eight miles from here, in Newton Centre. Though we haven't been in touch in years, she has graciously agreed to gather colleagues to talk about prenatal care.

There's apple cider warming on the stove as Mary Baker, a certified nurse midwife with a ready laugh and a thick Boston accent, tells me she feels her practice overwhelms patients by having to offer so much testing when they come in for a first prenatal visit: serial sequential testing, AFP4, CVS, nuchal translucency screening, glucose tolerance test, eighteen-week ultrasound, amniocentesis, cystic fibrosis testing regardless of genetic risk. Stephanie says she feels like she needs to give her patients a crash course in statistics and risk assessment to help them understand that most of this early prenatal testing gives you only a percentage of risk, and that a higher risk estimate on a test helps guide patients in considering the option for further (and usually more invasive) testing, not proof that anything is wrong.

The two obstetricians and two hospital midwives I'm interviewing don't hesitate to admit that we are overtesting moms. They're as frustrated by it as some of their patients, but they also feel their hands are tied. Hospital doctors and midwives are required to follow state guidelines, answer to their colleagues, and be hypervigilant that they have offered and performed every test required in order to satisfy both health and malpractice insurance providers in the event of a bad outcome.

Mary and the other certified nurse midwife, Phyllis Gorman, both agree they would rather spend their time talking, *really talking*, to expectant moms: asking them what they've been eating, how they are feeling about having a baby, how they are preparing to welcome that baby into the world, how they can work together to be a provider-parent team and make shared decisions.

Stephanie's colleague Brian Price, M.D., associate clinical director of Harvard Vanguard Medical Associates, interrupts to interject that in the current model of obstetric care in the United States, when you've got dozens of items to cover and very little time per appointment, those kinds of conversations are impossible.

Brian practiced for four years in Brooklyn, New York, in the late 1980s and early 1990s at the Brooklyn Hospital Center, where there was virtually no prenatal testing available. They did not do alpha-feto-protein testing, routine fetal surveys, or ultrasound scans to check fetal growth. And yet now that we are doing more prenatal testing, he points out, the outcomes have not improved.

Brian is articulate, opinionated, and incisive. Bald and broad-featured, he sits up straighter to emphasize his point: "We have *not* decreased morbidity or mortality," he says, talking loudly to be heard over his colleagues, who are clamoring to agree. "All we've done is increase the cost of care, and increase the anxiety."

GLUCOSE TOLERANCE TESTING

As any pregnant woman who has had it can attest: One of the most unpleasant prenatal tests is the screening for diabetes. According to the American Diabetes Association, about 7 percent of pregnancies become complicated by gestational diabetes. Gestational diabetes arises when too much sugar builds up in the bloodstream, either because the body is not producing enough insulin (the hormone that clears sugar from the blood and takes it into the cells to use as energy) or pregnancy hormones block the effects of insulin. Many women have gestational diabetes that is not a cause for concern. It is usually a mild condition that develops in the third trimester, causes no symptoms, and clears up after the pregnancy is over. But pregnant women with elevated blood sugars can have larger-than-normal babies and be at more risk for high blood pressure. For this reason, care providers order a diagnostic gestational diabetes screening test. A woman is told to drink an unnaturally sweet and horrible-tasting syrupy beverage that has a high glucose content. One brand, GLUTOLE, has 75 grams of glucose. Its ingredients: glucose syrup, maltodextrin (a creamy-white slightly sweet starch derived from processing corn or wheat), purified water, acidity control compound E330 (a pH control agent), preservative E211 (also called sodium benzoate, used as an antifungal and antibacterial; one study found that when paired with artificial food additives it causes hyperactivity), cola aroma, foodstuff color E150 (a caramel coloring made from sucrose), and carbonic acid (a weak acid with a tart taste found in many sodas and also in champagne).

The beverage made Angela Decker of San Mateo, California, when she was thirty-four years old and pregnant for the first time, so sick that she vomited it up in the waiting room. But the real problem with this test—in addition to having an already queasy pregnant woman drink an unappealing nonfood beverage laden with artificial colors, preservatives, and additives—is that there are no international standards for the amount of glucose in the test for pregnant women (50 grams, 75 grams, or 100 grams) and there is no clear consensus on what glucose response is elevated enough to be cause for concern. A positive result leads to further screening and, often, aggressive intervention, including the use of insulin during pregnancy, a scheduled early induction, or even C-section because of a fear the baby will grow too big.

If you are having a healthy pregnancy, your baby is measuring normally, and you have no risk factors for diabetes, should you even have this test? Although the American College of Obstetricians and Gynecologists (ACOG) recommends that every woman be screened for gestational diabetes regardless of risk factors, the U.S. Preventive Services Task Force, which does systematic reviews of all the available scientific evidence, has concluded that "the evidence is insufficient to recommend for or against routine screening for gestational diabetes," and the American Diabetes Association concludes that "low-risk status requires no glucose testing."

AVOIDING GESTATIONAL DIABETES THROUGH LIFESTYLE CHANGES

Kristen Boyle of Denver, Colorado, was surprised to be diagnosed with gestational diabetes during her first pregnancy seven years ago. Thirty years old, slender, and fit, Kristen had no diabetes in her personal or family history and she thought she ate well. But her diabetes got so bad that she had to take insulin during the last trimester. At every visit she had fetal monitoring and an ultrasound. At thirty-eight weeks her blood pressure was high.

"Your amniotic fluid is low," she was told after an ultrasound. "We need to induce you today." She was given a Pitocin drip and then an epidural. Hoping for a natural childbirth, Kristen ended up birthing her seven-pound daughter Sofia via C-section after thirty hours of unproductive labor.

"My body wasn't ready to have her," Kristen remembers sadly when we talk. "Baby wasn't ready to come out."

For her second pregnancy, Kristen decided to do things differently. She went to a midwife instead of a doctor and she and her midwife devised a proactive diet to keep her from developing diabetes in the first place. It was then that she realized that what she had been eating during her first pregnancy had not been healthy. Instead of being advised to avoid processed foods and eat whole fresh vegetables and fruits, high-quality protein and fats, and whole grains, Kristen had been told to eat granola bars and take glucose tablets when her blood sugar was too low. This time Kristen added high-quality protein to her diet, eating hemp seeds, eggs, chicken, and the occasional steak. She cut out all refined sugar (though she still ate foods sweetened with agave or maple syrup). She made sure to eat a lot of fiber. Like in her first pregnancy, she practiced yoga, took prenatal swimming classes, and walked two and a half miles around Sloan's Lake almost every day. Kristen monitored her own blood sugar levels by pricking her finger. She had no problems and gave birth at home, vaginally, to a healthy nine-pound baby boy.

In one survey from August 2012, 62 percent of women who were expectant or had given birth reported that their providers did not talk to them about how to care for their health during pregnancy. Doctors often don't tell women that eating foods that are low in nutritional content but high in sugar and starch—like white bread and bagels, white pasta, ice cream, cake, candy, and soft drinks—can induce or worsen gestational diabetes. These foods also add to excessive weight gain during pregnancy, which can lead to high blood pressure. The best way for a pregnant woman to avoid developing gestational diabetes is to eat a healthy diet that contains high-quality protein and no added sugars and no refined grains. If elevated blood sugar becomes a problem, pregnant women can start removing naturally occurring sugars from their diet, like ripe bananas and other fruits.

But even women who continue to eat a high-sugar diet can help their bodies process the sugar by exercising. This is why most homebirth midwives have women keep a food log over three or four days to record what they eat and spend a good deal of time during prenatal visits strategizing about how to improve their diets and how to find time to walk, swim, run, bike, take yoga classes, or do other daily exercise.

"When you're under the hospital model they're just looking at numbers, not the whole picture," Kristen says. She now suspects the anomalous high blood pressure reading that led to an induction and then a C-section in her first pregnancy was because she was excited about having her first baby. "If my blood pressure had been high at thirty-eight weeks, my homebirth midwife would have said, 'drink some water, relax, meditate, go for an easy walk, and we'll check you again.' First-time moms are very suggestible. We look at doctors as being the experts. When they say something, we say okay. I never questioned it. In hindsight, when they said induction, I should have asked, 'Can I come back and you can monitor me? Do I really need this? Is it really a medical emergency?' "

NUTRITIONALLY DEFICIENT DOCTORING

"'What do you eat? What do you have for breakfast, lunch, and dinner?' When doctors start asking those questions of pregnant women, then we'll see a change in prenatal care," says Dr. Klaper, who looks younger than his sixty-four years and directs the nonprofit Institute of Nutrition Education and Research in Manhattan Beach, California. "In the early days and weeks of pregnancy, just after conception, when a woman doesn't even know she's pregnant, the embryo is the most vulnerable—just a cluster of a few cells. *This* is the time the mother's diet is key." Klaper, talking rapidly and convincingly, continues, "So she goes to a fast-food restaurant for 'lunch,' and she's ingesting a witches' brew of chemicals, flavorings, colorants, and stabilizers. All these molecules are infusing the baby in this very vulnerable time. Who knows what this does to the fetus?"

When I ask Klaper why obstetricians don't emphasize the importance of nutrition during pregnancy, he chuckles. "No one tells us it's important!" he exclaims, shaking his head. "We go to medical school to learn how to work in the body repair shop—which is what hospitals are. If you break your body, go to the hospital. They'll fix it. But then get out of there. No one is going to mention nutrition to you before, during, or after, because no one mentions it to us."

Doctors actually denigrate nutrition, believing it is irrelevant and unimportant, Klaper says. "There is an inherent contempt for nutrition built into Western medicine. Nutrition is a sissy sport among physicians." One look in a doctor's own refrigerator will show you that he is eating the same junk food that most Americans eat, Klaper points out. I think of a friend's husband, a doctor who eats vanilla cupcakes for breakfast, and a pediatrician I know who has been trying to eat more healthy food but who still feeds the family sugar-laden processed cereals, bread full of additives like calcium peroxide (a bleaching agent), white pasta, granola bars, and conventional milk (because organic is too expensive). "Real doctors work in the operating room. Real doctors deliver babies," Klaper says. "The sad irony is then they go back to their offices and see a waiting room full of fat, unhealthy women who are sick because of what they're eating."

DO PRENATAL VITAMINS MAKE
PREGNANT WOMEN SICK?

At first Jenna Nichols thought she was so sick because she was pregnant during a heat wave in Philadelphia. A petite woman with fair skin and hazel eyes, Jenna, twenty-four, looked and felt green. But reaching the first trimester milestone, when morning sickness is supposed to abate, didn't help. Every day, shortly after Jenna took a one-a-day prenatal vitamin, her hands would sweat, she would feel clammy all over, and get so nauseous she could barely breathe. She lay as still as she could on the couch, air-conditioning at full blast, feeling more motion sick than she had on the dragon ride at the carnival when she was six years old.

One night about halfway through her pregnancy, Jenna unscrewed the cap on her vitamins to check if she needed more. The smell triggered a fresh wave of nausea and Jenna realized it might be the vitamins themselves that were making her sick. Because she was vegan before she got pregnant, Jenna was worried her baby might be malnourished. She kept a detailed food log of everything she ate and started trying different vitamin brands: Target, Trader Joe's, Whole Foods. Nothing changed. Her symptoms only went away when she stopped taking vitamins.

Sarah Jane Nelson Millan, a mother of two in Los Angeles, broke out into mouth sores every time she took her conventional prenatal vitamin. Every time Katherine Womack, a Las Vegas–based mom of a two-year-old, took her vitamins she threw up twenty minutes later. Another young woman experienced unpleasant nausea in five pregnancies. Because she had undergone cancer treatment and been told by the doctors she would stop menstruating, she did not know she was pregnant for the sixth time until the second trimester. She took no prenatal vitamin and for the first time in six pregnancies had no morning sickness.

Though some pregnant women are not aware of experiencing any adverse reactions to prenatal vitamins, others report painful constipation, horrible stomach pains, and dizziness. Since bad reactions to prenatal vitamins are not unlike pregnancy symptoms, most women don't realize when their body is reacting badly to the vitamins.

During pregnancy a woman's body undergoes enormous changes: first the fertilized egg implants into the lining of the uterus and then, before a woman even knows she is pregnant, the egg (called a blastocyst) starts separating into cells that will become the placenta and cells that will become the baby. Just two weeks after fertilization, the layers of the embryo itself actually start to differentiate into specialized parts. The outer layer will become the hair, skin, eyes, and nervous system; the middle layer will become the heart, reproductive organs, bones, muscles, and kidneys; and the inner layer the baby's liver, lungs, and digestive system. As a pregnant woman's body grows an entirely new organ (the placenta) to sustain the baby, her blood increases in volume and her heart enlarges. Because of the demands of pregnancy, it is thought that women need more of certain nutrients, especially iron, calcium, and folic acid. Women are told that prenatal vitamins—be they over-the-counter or prescription—will help ensure they are getting what they need nutritionally to grow a healthy baby. They are also told that folic acid around conception and in the first trimester is especially important. This has been the recommendation since 1999 after a large study of pregnant women and infants in China found that taking 400 µg of folic acid a day reduced the likelihood of neural tube defects, which occur in early gestation when openings in the spinal cord or the brain do not properly close. Prenatal vitamins are considered a "nutritional insurance policy" to protect a growing fetus.

But what most pregnant women and their partners (and even doctors) don't realize is that many prenatal vitamins contain extra nonvitamin ingredients that are known to be harmful. There are no standard guidelines or any clear consensus on what amount of which vitamins should be included in prenatal vitamins. For example, some doctors now believe that to have a positive effect pregnant women need more than twice as much folic acid—as much as 1,000 µg—an amount available only by prescription.

Women also don't realize that just because they are taking vitamins it does not mean the body is able to absorb and use them. In their promotion of prenatal vitamins, the medical community overlooks this problem, as well as that the studies of the importance of folic acid supplementation have never included a group of pregnant women who receive folic acid from whole-food sources (lentils, kidney beans, broccoli, spinach, kale, and citrus fruits are high in folic acid).

Despite its name, there is nothing natural about Sundown Naturals, an inexpensive over-the-counter prenatal vitamin. In addition to synthetic vitamins, Sundown "Naturals" contain vegetable cellulose, vegetable stearic acid, calcium silicate (anticaking agent), vegetable magnesium stearate, titanium dioxide color, FD&C Yellow No. 6 Aluminum Lake, FD&C Red No. 40 Lake (CI 16035), FD&C Blue No. 1 Lake (CI 42090). Other brands, some of which are much higher in price, have identical ingredients.

This "natural" vitamin contains four artificial colorants. There is no good reason for any artificial colors to be added to prenatal vitamins, but there are ample reasons to avoid them. Titanium dioxide, also found in paint and sunscreen, has been shown to cause neurological damage, cell injury, mutation, and ultimately respiratory tract cancer in rodent experiments, and is now believed to be carcinogenic in some forms to humans. It has also been shown to harm marine animals and has been linked to autoimmune disorders. Yellow 6, Red 40, and Blue 1 are petroleum-based dyes that in industry-sponsored animal studies have all been found to provoke allergic reactions as well as nerve-cell damage (Blue 1) and possibly tumors (Red 40 and Yellow 6). Red 40 and Yellow 6 have also been found to be contaminated with carcinogens.

Stuart Prenatal Multivitamin/Multimineral supplement tablets cost $29.99, more than *four times* as much as Sundown Naturals. Yet the list of unpronounceable ingredients you would never feed directly to an infant—and probably wouldn't want in contact with a developing embryo—is twice as long. It includes sodium aluminosilicate, mixed glycerides, sodium benzoate, polysorbate 80, polyethylene glycol, as well as cornstarch and sugar, to name just a few. How can we assume that it is safe or beneficial for a pregnant women to swallow a pill loaded with nonfood substances our bodies have not evolved to ingest?

When Jenna finally realized her prenatal vitamins were making her sick, she was told to switch to Flintstones children's chewables. This is common advice: Doctors and midwives tell women who cannot stomach the prenatal vitamins to take a children's multi. But it left Jenna perplexed. She did not want to eat a sugary product (the children she used to nanny thought their vitamins were candy), and she knew the amounts in vitamins designed for children were not comparable with those for pregnant adults.

Indeed, it is impossible to know if the amounts listed on the label of any vitamins are actually what's inside the vitamin. In 2004, testing by ConsumerLab.com, a White Plains, New York–based company that independently evaluates health and nutrition products, revealed that one prenatal brand tested could not fully disintegrate, suggesting it would not deliver its nutrients to the body. It also contained twice the amount of folic acid listed on the label. That same report found that a children's gummy vitamin was contaminated with high amounts of lead. More recent testing of multivitamins showed thirteen out of thirty-eight brands did not contain the amount of nutrients as listed in the ingredients.

Since 1994, when the Dietary Supplement Health and Education Act was signed into law by President Bill Clinton, dietary supplements, including prenatal vitamins, fall under a new regulatory framework. While the government plays a strong role in ensuring food safety, it is the company that makes the vitamins that is responsible for determining if they are safe. Manufacturers take full responsibility that the representations or claims made about vitamins are substantiated by adequate evidence that is not false or misleading. Prenatal vitamins do not need approval from the FDA before they are marketed, their contents do not have to be tested by the FDA, and the claims made about their health benefits are not independently verified. The manufacturer does not even have to provide the FDA with the evidence it relies on to substantiate safety or effectiveness before or after it markets its products.

"We find problems all the time," says Tod Cooperman, M.D., president of ConsumerLab.com. "We have found problems with prenatal vitamins in the past. It wouldn't surprise me to find a prenatal that was contaminated, didn't disintegrate, or didn't have all its ingredients. I have three kids. When my wife was pregnant, we only used products we had tested. Just to cover our bases, I had her switch each day between two different products."

According to *Consumer Reports,* American consumers spent $26.7 billion on supplements in 2009. The average prenatal vitamin costs about 30 cents a pill. If the more than four million pregnant women each year in the United States take a prenatal vitamin every day for at least nine months, they collectively spend at least $336 million a year. With so much money at stake it is no wonder that pharmaceutical salesmen peddle their brands directly to American obstetricians and family practitioners, bringing treats to their offices (like doughnuts or catered lunches) and giving free samples to the staff. When doctors distribute these samples to their pregnant patients, they are openly endorsing the brand. Endorsement by a trusted physician is an extremely effective marketing strategy. "We get salespeople in here all the time trying to promote their vitamins and saying why they are better," says Dr. Lester Voutsos, section chief of obstetrics at Providence Hospital in Novi, Michigan. "I hear their presentation and they give us free samples, which I give out. But I don't prescribe them."

Voutsos doesn't think pregnant women should waste their money on expensive prenatal vitamins when they can buy less expensive over-the-counter or generic brands. "Pregnancy is a profound experience," Voutsos says. "When they are pregnant, women are so focused on the pregnancy they are willing to spend $30 or $40 a month on vitamins. Frankly, I don't think it's worth it."

FOOD RULES

When she was six months pregnant, a young woman walked into a bagel shop and asked if they had bagels made with whole wheat flour. The young clerk behind the counter looked perplexed.

"No," she said in a low, almost conspiratorial tone, "none of our products are made with wheat."

The clerk was mistaken. Though there were no *whole-grain* bagels on hand, all of the products in that bakery, even the pumpernickel (which also contained rye flour), were made with wheat flour. Like the bagel clerk, most Americans don't know the difference between processed wheat and whole wheat.

A wheat kernel is made up of three components: the bran, the germ, and the endosperm. The bran, which contains most of the plant's fiber, and the germ, which contains most of the nutrients, are both removed in the process of converting whole flour into white flour. What is left is the starchy endosperm. If your eyes are already glazing over, here's what you need to know: Because so many vital nutrients are taken out during processing, most food companies add chemical nutrients back into white flour, which is why the flour is called "enriched." However, so many nutrients are lost in the refining process that "enriched flour," though it sounds healthy, can never be as nutritious as whole-grain flour.

"Most of the nutrients that were there to begin with are never reinstated," Larry Lindner, former executive editor of the *Tufts University Health & Nutrition Letter* and an expert on nutrition, explained to me. "These include vitamin E, vitamin B_6, pantothenic acid, magnesium, manganese, zinc, potassium, and copper."

Phytochemicals are another vital component of the grain lost during processing. Phytochemicals are substances found in plants that are not vitamins or minerals but that play a part in promoting good health. "Unlike vitamins and minerals, phytochemicals are not put into prenatal supplements," Linder explains, "but they are in whole grains." In fact, researchers are just beginning to isolate and identify these compounds—there are literally thousands of them.

Whole grains (like brown rice, unpearled barley, whole millet, and oats) are high in fiber, which can help alleviate constipation. Since whole grains take longer to digest, pregnant women suffering from low blood sugar report fewer symptoms when they eat whole grains. "Whole grains metabolize more slowly in your body," explains a former obstetric nurse from Brattleboro Memorial Hospital in Vermont and an expert on nutrition. "They reduce problems of high glucose that can come from eating refined grains. They have more fiber, which helps your intestines stay cleaner, and more vitamins and minerals."

White bread, white pasta, and white rice not only fill you up with empty calories, they also spike your blood sugar levels, exacerbate constipation for pregnant women who may find their bowels have become more sluggish, and predispose you to diabetes. A 2012 meta-analysis of available research on white rice found that eating it in large quantities, like white bread, other refined carbohydrates, and sugary food, is associated with a higher risk of developing insulin resistance and diabetes.

A whole foods–based diet that includes plenty of vegetables, high-quality proteins, whole grains, healthy fats, and fresh fruits—and that has little or no processed foods, soft drinks, sweets, or added sugar—is a pregnant woman's best nutritional insurance policy. Yet while there is a billion-dollar industry of vitamins and supplement manufacturers with deep pockets to study how synthetic chemicals they manufacture are good for humans and promote those findings to pregnant women, there is very little, if any, financial incentive to study and promote whole, fresh, healthy foods. You won't see a farmer going to a doctor's office with free kale in the hopes of getting pregnant patients hooked.

DOCTORS HAVE NO FINANCIAL INCENTIVE TO TAKE THEIR TIME

It's quick and easy to scribble a prescription for prenatal vitamins, hand it to a pregnant woman, and move on to the next patient. Educating a woman about nutrition, asking for an inventory of what she's had to eat, and counseling her on preventive medicine take time. Even if they were knowledgeable about nutrition during gestation, most obstetricians would not make the time to talk to their pregnant clients about it.

In today's model of prenatal care, pregnant women find themselves waiting for long stretches in the doctor's office only to be hurried through appointments. Paul Qualtere-Burcher, an obstetrician with twenty-three years of experience who has participated in more than four thousand births, is as frustrated by this as his patients. Qualtere-Burcher says many of his colleagues are frustrated too, which is why they retire early. "They do a high-volume practice and then are tired and cranky and burned out by the time they are in their late forties or early fifties." In his soft-spoken but emphatic way, Qualtere-Burcher explains that there is an enormous financial imperative for obstetricians to see as many women as possible, which is leading to an inadequate level of care. In order to pay for fixed overhead costs (including office staff and medical liability insurance), obstetricians in private practice squeeze as many prenatal appointments into a day as possible, keeping women waiting and then racing through the visits.

Qualtere-Burcher recently chose to take a $150,000-a-year pay cut to move from a practice in Eugene, Oregon, to a medical school in Albany, New York. In his new position he is doing more teaching and supervising. He tells me he can talk more honestly now because he doesn't have to worry about jeopardizing his job or offending his colleagues. He patiently runs through the numbers and the different work models, using words like median productivity (the national standard across the country of how many patients an obstetrician sees on average) and Relative Value Units (RVUs—the more patients you see, the more income you generate). Then Qualtere-Burcher cuts to the chase: Whether in private practice or employed with a steady salary by a health group or HMO, the more women an obstetrician sees in a day, the more money he makes.

"Human behavior responds to incentives, and the incentive for OBs is to spend less time with people," Qualtere-Burcher explains. "Whether you spend five minutes or fifteen, the fee is essentially the same. The financial incentive is to run them through quickly. There is no financial incentive to take your time."

At Qualtere-Burcher's last job for a nonprofit medical group, more incentives came quarterly. If you delivered enough women over median productivity, you could make as much as $48,000 more a year beyond your base salary.

For OBs in private practice the financial incentives are even more pronounced: Once you've paid fixed expenses, everything else you make is yours. Even doctors who don't go into obstetrics for the money find that earning more becomes an irresistible motivator. "Once you've covered your expenses, it's all gravy," Qualtere-Burcher says. "Why take home $200,000 when you can increase your patient volume and make $500,000 instead?"

To boost his bottom line even more, one of Qualtere-Burcher's colleagues refused to provide care for medically complicated pregnancies. If a pregnant woman had any kind of problem, he sent her to another doctor. The doctor bragged to his colleagues that only seeing straightforward pregnancies meant he could see more than forty women a day. "That was his moneymaking conveyer belt," Qualtere-Burcher says, though he is quick to add that this colleague was not typical. "Most OBs like doing all aspects of their specialty, and do enjoy taking care of complicated patients. It reminds us that we have some skills."

But Dr. Edward Linn, chair of obstetrics and gynecology for the Cook County Health and Hospitals System in Chicago, Illinois, who has made time to talk to me in between seeing gynecology patients, says that he has spent many years watching doctors "punt" complicated pregnancies and "cherry-pick" their clients because they do not get higher reimbursement for high-risk women who take up more time. We sit in his spacious fifth-floor office, which has a wide banker's desk piled with academic journals, papers, and files. While many doctors refuse to give any care to women on public assistance, other doctors prey on them, Linn says, maximizing their profits by following a woman during pregnancy and collecting per-visit reimbursement from the state with no concern about the outcome. These doctors "see pregnant patients and carry them along with no intention of ever delivering them," Linn tells me. "They tell them that once they go into labor, they should show up in the emergency room of the hospital of their choice." These doctors increase their income margin by cramming in as many office visits as possible, collecting a per-visit fee for each visit, but refusing to be present during labor and at the birth (which takes time away from the office). "We see these patients coming into our emergency room," Linn continues. "People accept that. They don't understand why, but they trust their doctor. Doctors who are practicing like that lack professional integrity. They're looking for the billable opportunity."

This kind of unethical behavior leads to fragmented care and poor outcomes, Linn says, especially for socioeconomically disadvantaged women who have among the worst outcomes and who most need follow-up and continuity of care. Sharon Rising, a certified nurse midwife with more than thirty years of experience, former faculty member at Yale University's School of Nursing, and founder and CEO of a woman's health care advocacy nonprofit, agrees that the kind of prenatal care we are delivering in America is not working, especially for America's poor.

"We have terrible premature birth rates, which if anything are getting worse. We still have a high percentage of pregnant women who are coming late to prenatal care, lots of women who aren't breastfeeding, and the overuse of triage and emergency rooms. These outcomes are really less than desirable," Rising says. "What's really happening? What kind of a system do we have? We have a care system designed to support the hospital and the clinicians, and is tailored to their convenience. When the outcomes aren't great you need to change the system."

"This is only fixable at a systems level," agrees Qualtere-Burcher. "You want to say this is physician bad behavior, but it really isn't. . . . Nothing is going to change until you take the profit out of medicine. The economics of medicine keep distorting medical decision-making and the doctor-patient relationship, even the relationships between physicians and midwives, because you view a colleague as a potential competitor. It just distorts everything. Until you remove that it is going to be impossible to make substantive changes."

MIDWIFERY: PROACTIVE VERSUS REACTIVE CARE

When Alice Domurat Dreger, Ph.D., a professor of clinical medical humanities and bioethics at Northwestern University Feinberg School of Medicine, asks her medical school students to describe the kind of woman who would give birth with a midwife rather than an obstetrician, they imagine someone who wears long hippie skirts, eats vegetarian food, and drives a VW minibus. Her students are always shocked to learn that when Dreger, a pants-wearing omnivore and self-described "science geek" and her partner—also an academic—became pregnant in 2000 they chose to have their prenatal care and delivery overseen by midwives.

Why? Dreger scoured the scientific literature to learn everything she could about the safest way to have a baby. Her reading revealed she should walk a lot during pregnancy, Dreger explains in an article in the *Atlantic*; get regular checkups of her weight, urine, blood pressure, and belly growth; and avoid vaginal exams. She also found out that she should not agree to any prenatal sonograms in her low-risk pregnancy because doing so would be extremely unlikely to improve the baby's health, but could result in further testing and intervention that increased risk to her and her baby with no benefit.

But her obstetrician and his team were uncomfortable with such an "old-fashioned" approach. So Dreger and her partner quit the practice, instead engaging a midwife who, she says, "was committed to being much more modern."

The modern midwife's approach is to be proactive during pregnancy and childbirth. Instead of aggressively treating gestational problems with the latest medications and the most advanced technology *after* they arise, good midwives (like the one who helped Kristen avoid gestational diabetes by changing her diet) work closely with their pregnant clients to ward off problems before they start.

"The medical model of obstetrics is reactive," explains Stuart Fischbein, an obstetrician who has spent much of his thirty-year career working closely with midwives, when I interview him via Skype teleconferencing at his home in southern California after he's spent a long day seeing patients. "When a woman develops diabetes, they'll fix it. When a woman develops preeclampsia they'll treat it. When a woman develops anemia, they'll deal with it. The midwifery model is preventive. They help build a woman's hemoglobin so she won't become anemic. They educate her about not eating sugar and doing exercise so she doesn't get gestational diabetes. Ten minutes of every hour visit is spent talking about nutrition so a woman won't develop preeclampsia."

Though many hospital midwifery practices run their groups much like obstetricians, racing through appointments and reacting only after problems arise, midwives usually offer a higher level of individualized prenatal care than obstetricians. Qualtere-Burcher thinks one reason for this is that midwives don't face as many financial pressures as doctors. They make less money, pay a much lower rate of malpractice insurance, and usually don't graduate from school with the kind of education debt that physicians do. When Qualtere-Burcher worked at Olean General Hospital in an academic position for SUNY–Buffalo from 1995 to 2002, his medical malpractice insurance cost about $65,000 a year. The midwives, who supervised just as many pregnancies and delivered just as many babies, paid only $3,500 a year. If the midwives run into a problem they are not usually sued. "If they have a patient who gets into trouble, they are going to consult an ob-gyn, so an ob-gyn will be involved and available to be sued," Qualtere-Burcher explains. "So they will always drop the lower-level professional. I'm told that's pretty typical."

It is partly this constant fear of being sued that leads most obstetricians to see themselves as "managing" the pregnancy, dictating to pregnant women what they should eat, how much weight they should gain, and how many weeks they will "allow" gestation to continue before medically inducing labor. They react strongly and swiftly at the first sign that something might be wrong, whether the issue is a high blood pressure reading or the suspicion of a larger-than-average baby.

Of course obstetricians have a laudable reason for wanting to monitor their clients' behavior: They want their patients to have healthy pregnancies and healthy babies. But they also have financial incentives and professional concerns that have little to do with a pregnant woman's health. No pregnant woman should have to sit for an hour in the waiting room because a doctor has overbooked clients in order to maximize profits. It is wrong for care providers to hand out samples of expensive prenatal vitamins, or to undermine a woman's confidence about breastfeeding before she's even had a baby by giving her formula samples and coupons. This is why pharmaceutical salesmen are not allowed into the doctors' offices, and women who receive prenatal care from Harvard Vanguard Medical Associates, where Brian and Stephanie practice, are not given free samples of branded products.

"We have never allowed 'free' pharmaceutical samples to be given to patients because nothing is ever 'free,' " Brian explains in a follow-up email. "We also do not give out new baby bags with formula samples or allow any product placement in our patient handouts. Several studies have clearly shown that clinicians are very clearly swayed by all of these subtle forms of advertisement and promotion. I have gone so far as to forbid 'drug lunches' and pen distribution."

It is also wrong for doctors to prescribe radical medical intervention before trying less invasive ways to fix a potential problem. Treating gestation as an illness or an accident waiting to happen not only takes the joy out of pregnancy, it also creates the opportunity to intervene unnecessarily and potentially do harm. Expectant moms and dads who question their providers, seek a second opinion, or ask for gentler or less invasive care are often met with scolding, bullying, or genuine surprise ("Why wouldn't you want that test? No other patient has ever refused it!"). During my four pregnancies, I discovered I didn't need stressful prenatal testing and intervention; I needed wholesome food, time to exercise, sunlight, quality sleep, effective ways to destress, and friends and family to listen patiently to my hopes and fears and give me hugs when I cried (pregnancy is an emotional time). The best prenatal care happens when health care providers take the time to really listen to their clients, examine them gently, offer evidence-based advice, and support gestation as a natural, healthy, awe-inspiring, life-changing event.

Dr. Qualtere-Burcher believes that healthy women having normal pregnancies should go to midwives, be they certified nurse midwives who work in a hospital or professional midwives who deliver babies at home. He points out that in European countries such as Ireland and Norway, where the maternal and fetal mortality rates are much lower than in the United States, midwives, not doctors, provide prenatal care. Like Dreger and her partner, Qualtere-Burcher and his wife chose not to see obstetricians. The births of his own three children were supervised by midwives.

"Midwives do a much better job caring for pregnant women than I'm able to," Qualtere-Burcher confesses. "I wish all OBs would work in tandem with midwives. We can always be there for backup if something goes wrong."

American College of Obstetricians and Gynecologists gross receipts for the 2009–2010 fiscal year: $80 million

Average salary of a high-risk obstetrician: $446,886/year

Average salary of a hospital midwife: $100,000

Total cost of prenatal visits with a doctor: $3,940

Total cost of prenatal visits with a homebirth midwife: $1,300

Cost per minute to have pregnancy supervised by a doctor: $15.00

Cost per minute to have pregnancy supervised by a homebirth midwife: $1.67

Ingredients in GLUCOLE (gestational diabetes test): glucose syrup, maltodextrin, purified water, acidity control compound E330, preservative E211, cola aroma, foodstuff color E150, and carbonic acid

Money spent on prenatal vitamins per year: more than $336 million

Nine-month supply of brand-name prenatals: $1,169.55

Nine-month supply of generic prenatals: $134.91

Cost to eat organic kale: $1.99 a bunch

Jennifer Penick: My Midwife Knew Me Better After One Hour Than My OB After Four Babies

Jennifer Penick, thirty-two, a mom of five from La Vista, Nebraska, liked her obstetrician. He was always personable during prenatal appointments. It was true that she often had to wait as long as an hour, and that once she had to reschedule an appointment because he was running behind, but she just assumed that was how things worked. Then she found another option.

I grew up military so when I got out of that life and had to pick my own doctors, I did it pretty randomly. My husband and I didn't know better. I was twenty-two when I got pregnant for the first time. When our OB finally came into the room, he would get through everything as fast as possible, listen to the baby's heartbeat, reassure us that everything was okay, and rush to the next appointment. I honestly didn't think there were any alternatives. I didn't think midwives still existed in the United States.

My first birth didn't go very well. The OB told us that he wanted to induce me because he was concerned about the size of the baby. I was thirty-seven weeks. He told us it would be better if we induced because the smaller the baby the easier it comes out, and there's less chance of needing a C-section. We didn't want to be induced. I had heard from enough people "Don't get induced, because it hurts way worse." But when he told us, "You'll have a better chance of a vaginal birth!" we agreed to the induction. At thirty-eight weeks, one day I was induced with Pitocin, and then the obstetrician broke my water. I had an epidural, but it only worked on one side and I was in excruciating pain. After an hour and a half of pushing, the doctor tried to get the baby out with forceps. When that didn't work he wheeled me in for a C-section.

My son, A.J., weighed 7 pounds 3 ounces. I was completely doped up on pain medication and exhausted. I only have one picture of my son and me after he was born and it always makes me cry. The worst part is I don't have any memory about the birth. The only thing I remember is saying to the doctor, "He was supposed to be bigger, that's why we went through all this." The doctor didn't say anything. He didn't apologize.

I got pregnant again when A.J. was just nine months old and we went back to the same OB because we weren't sure where else to go. He was supportive of us having a VBAC [vaginal birth after Cesarean]. I'm only five-three. I have no torso, and I only gain like twenty pounds in my pregnancies. There's nowhere for the baby to go but out, which is why I look so big. The doctor figured it out this time and wrote "LARGE FOR GESTATIONAL AGE" on the outside of my chart. I forgave him. I take some responsibility for not doing my own research and making my own decision. He was very nice to us no matter how much in a hurry he was. I never remember him being gruff or upset or like I was a bother. But he definitely didn't take his time or get to know us. If I saw him at the grocery store he wouldn't have a clue who I was, even though he's delivered four of my kids.

My fourth birth in the hospital was pretty awful. The doctor barely made it and the nurses kept saying, "Don't push! Don't push! Wait for him to get here!" The resident who ended up delivering the baby didn't know what he was doing. So when a midwife opened a birthing center, The Midwife's Place, in our area last December, I decided to deliver there. She took time to talk about my past medical history and births. My midwife knew me better after one hour than my OB did after four babies. It felt like she cared. She hadn't booked five people for that hour. She made time to sit and talk to me. She explained there are tests but that I could opt out of them if I wanted to and she would sign the paperwork. I always hated bringing my kids to the obstetrician's office because I felt like I had to tell them to keep still and be quiet. At the birthing center there are toys for my kids to play with. My three-year-old was so excited because she could go on the rocking horse.

It was so much less clinical. I didn't feel like I was walking into this scary place where everyone was dressed in scrubs and white coats waiting for something to go wrong. Every time a baby is born at the birth center the midwives add the baby's footprint to the wall outside the birth rooms. Instead of flimsy paper gowns and pink paper draped over your thighs, you wear a fluffy plush white robe any time you have a procedure. I checked my own weight and blood pressure. When

I needed a Pap smear, the midwife warmed up the speculum so it would be more comfortable. It's odd to talk about a nice Pap smear! But they just care about your comfort. I wish we hadn't waited so long to make the switch.

But when I was thirty weeks pregnant, I found out that the birth center would not let me have my baby there, because of my one prior C-section. I was heartbroken. My husband and I toured the hospital that backs up the birth center; I've had friends deliver there who had good experiences and I thought I could make peace with it. But I couldn't. At thirty-two weeks I told my husband I wanted a homebirth. That's not easy in Nebraska. Certified nurse midwives are not legally allowed to attend homebirths in our state. But I had a friend who is a homebirth midwife who agreed to assist us, even though it was last minute and she was due two weeks after me.

Three of my four babies had been induced with Pitocin and I had epidurals for all four births. This time we let the baby decide when she was ready to come. I was pretty miserable the last month. I thought I would be pregnant forever. I got a pedicure, had acupuncture twice, and tried herbs but I was still pregnant. At forty weeks five days, on August 29, 2012, I woke up feeling like my back was a wall of pain. I padded around the house unable to get comfortable. This was it! We filled up a birthing tub in our living room and when the pain got really intense, Adam came in the water with me. He kissed my head and told me I could do it, that I'd done it four other times before, even though I said I couldn't and I wanted to leave.

Our daughter, Unity Dale Penick, shot across the pool at 6:36 a.m. She weighed 8 pounds 11 ounces and was 20.5 inches long. Our biggest baby yet! I only had one tiny tear that didn't need stitching. My midwife stayed for a few hours to be sure everything was okay. After she and my doula and the photographer all left, there was no parade of nurses, doctors, and other staff interrupting us every few minutes to check me and the baby or to take the baby away to the nursery. It was just us, learning how to be a family of seven.

CHAPTER 2

SONIC BOOM:
The Downside of Ultrasound

When she was thirty years old, Karen Bridges went to the doctor for an injured knee and burst into tears when he told her he couldn't prescribe pain medication because she was pregnant. She was crying tears of joy. She and her husband, Josh, had been trying so long to get pregnant they had just about given up. As quickly as she could, Karen made an appointment with an obstetrician who did a vaginal ultrasound to confirm the pregnancy. She remembers having an ultrasound at every monthly visit after that. Starting at seven months, at risk for preterm labor, the doctor ordered an ultrasound every week. It was thrilling to see her baby in black-and-white on the monitor, and the ultrasound techs were always happy to print photos for her to take home. Josh's health insurance, through General Electric, paid for every scan. The excited dad-to-be couldn't wait for their little baseball player to be born.

An obstetric ultrasound, also called a sonogram, works by sending frequencies of sound waves through a pregnant woman's body. Like the echolocation a bat uses to find its way in a dark cave, the sound waves bounce off the fetus's organs, fluids, and tissues, creating an image that is projected on a screen.

Obstetric ultrasound is used to detect conditions that might be harmful to the fetus or the mother. It can also be used to confirm pregnancy, estimate the date of conception, assess the size of the fetus, determine the location of the placenta, and identify fetal abnormalities (like club feet) or confirm fetal demise. First used for obstetrics by a Scottish doctor named Ian Donald around 1957, by the late 1970s—about a decade after I was born—ultrasounds had become a routine part of prenatal care in the United States. In 2001, the most recent year for which we have reliable statistics, 67 percent of pregnant women had at least one ultrasound.

Less than ten years later, in 2009, there is evidence from Canada to suggest that the number of pregnant women in North America who had at least one ultrasound jumped to 99.8 percent, with an average of three ultrasounds per woman. Women who have high-risk pregnancies are subject to ultrasound scans as often as once or even twice every month. Some report having as many as twenty-five ultrasounds per pregnancy.

Ultrasounds are now the standard of care, given as early as twelve weeks gestation to date the pregnancy if a woman is unsure of when she started her last menstrual cycle. They are also used in first-trimester genetic testing to assess a baby's risk for Down syndrome and other chromosomal abnormalities by measuring the translucent space at the back of the baby's neck (babies with heart problems and Down syndrome tend to have more fluid at the back of their necks in the first trimester, making that space appear larger). This nuchal translucency test, or nuchal screening, is done between eleven and thirteen weeks six days gestation. A second trimester ultrasound, usually between sixteen and twenty weeks, is done to assess fetal development as well as to learn the gender of the baby.

Most doctors argue that the benefits of early detection outweigh the possible harm of having an ultrasound. "We recommend an eighteen-week ultrasound with a fetal survey," says Stephanie Koontz, my childhood friend who is an obstetrician at Mount Auburn Hospital in Cambridge, Massachusetts. "I tell my patients that it is something that's often helpful for identifying things in pregnancy. I've never had anyone not want that ultrasound."

Most first-time parents are excited about getting ultrasounds. They can't wait to see the baby in grainy black-and-white, get pictures of their little alien, and share those pictures with family and friends. More than a diagnostic test, a second trimester ultrasound has become a rite of passage—a time to find out the baby's gender, upload images to Facebook, email them to friends, and be reassured that everything is okay.

AN IMPERFECT TEST

But in their enthusiasm for ultrasound scans, most pregnant women and their partners don't realize that the test is not always accurate. Though doctors are increasingly relying on scans for information about everything from the baby's weight to the level of amniotic fluid in a woman's uterus, they readily admit that the scans are problematic.

"Ultrasound is no better than my personal assessment, what I can do with my hands," contends obstetrician Felicia Cohen as she drinks a cup of coffee in the cafeteria of Three Rivers Community Hospital in Grants Pass, Oregon, where she has just delivered a 9-pound 6-ounce baby via C-section. "We all know the estimated fetal weight at term is not accurate. It can be off by as much as a pound either way. So if someone tells you the baby weighs eight pounds, you could have a nine-pound baby or a seven-pound baby. There's a range of reasons for this—the amniotic fluid level, the mom's weight, the position of the baby, the skill of the technician reading the scan."

Sometimes these inaccuracies have droll consequences. Though parents are often eager to discover the gender of their baby, early gender identification can be difficult depending on the fetus's positioning and the technician's experience. One study found that gender identification before fourteen weeks was inaccurate nearly 20 percent of the time. Ask Melanie Plisskin, a thirty-one-year-old first-time mom who was told she was having a girl. Melanie displayed her baby's first gift, an oversized bunny wearing a pink tutu, proudly in the nursery, chose a name, and printed pink baby shower invitations only to find out from the same ultrasound technician when she was seven months along that Eva Grace was actually a boy.

Sometimes inaccurate ultrasound information can lead to unforeseen benefits. In May 2010, when one of her clients was measuring big at the end of her first trimester, Margaret Jones, a homebirth midwife based in Salt Lake City, Utah, recommended an ultrasound. The results came back that everything was normal. But at thirty-four weeks Margaret was puzzled.

"If I didn't know better," Margaret told her assistant, "I would say there are two babies in there."

Margaret sent her client back for a second scan.

This time the images clearly showed that Margaret's client was having twins: a boy and a girl in separate amniotic sacs. Margaret has a policy of not delivering twins that share one amniotic sac—she feels they are at greater risk for becoming entangled during delivery—but based on the ultrasound information Margaret agreed that homebirth was a safe option. At thirty-eight and a half weeks Margaret's client gave birth kneeling by the side of the bed, moaning through contractions, her mom, sister, niece, and husband all there to support her.

Surprise! The babies were identical boys, sharing one amniotic sac. "It was a blessing for that mom. She had a beautiful homebirth," Margaret laughs. "I might not have done it had the ultrasound actually been accurate."

But inaccurate ultrasound readings often have much more serious consequences: causing weeks of stress to a pregnant woman when there is actually no reason to worry, or falsely reassuring parents when something is actually terribly wrong.

STRESS-INDUCING TESTING

Most American women have been led to believe that prenatal testing is necessary for a healthy pregnancy. Yet when a test has a positive outcome—either because something real *is* wrong with the fetus or because the test has high false positive rates—the stress can be overwhelming. "Ultrasound can't promise us a healthy baby," Margaret reminds her clients, "and it sometimes gives us things to worry about that aren't really a problem."

When she was pregnant with her oldest son in 1998, Rachelle Eisenstat was told during the ultrasound that her baby had cysts on his brain and an echogenic bowel. At the amniocentesis that same week, the technician told her the accompanying ultrasound showed the baby had a hernia in his stomach that was developing into his chest cavity. But the three Long Island, New York, doctors examining the screen could not agree on what they saw and left Rachelle and her husband in the examination room so they could argue out of earshot. They finally ordered a fetal echocardiogram. The tests came back normal and an ultrasound at seven months revealed the "cysts" were gone.

"My husband and I liked the tests, we're not antitesting," Rachelle tells me during a phone interview. "Still, it was an incredible amount of stress."

We know that when a pregnant woman's body is flooded with stress hormones, the stress has a negative impact on the fetus. Several scientific studies have shown that exposure to prenatal stress negatively affects the physical development of the infant—it is even correlated with smaller birth weight, smaller head size, and structural malformations. There's a growing body of scientific evidence that this stress can have a lasting effect on an infant's development, evidenced by poor psychomotor performance and more difficult behavior during the first ten years after birth. Not only does stress take the joy out of a pregnancy, it also disrupts a woman's sleep, shakes her confidence, and churns her already emotional state with ongoing worry. "I've had the ultrasound technicians say things like, 'The brain doesn't look quite right.' My client goes back six weeks later and everything looks fine," Margaret says. "But that mom just spent six weeks worrying about the baby. She's not sleeping and the anxiety is flooding her system with stress hormones. How is that good for the baby? How is that good for the mom?"

FALSE ASSURANCES

Louana George, a childbirth educator and nurse midwife in Los Angeles who delivered babies at home for twenty years, believes pregnant women need to be asked about their opinion about abortion before being offered ultrasounds and other testing. "I think it's a psychological lie for women," says Louana. "They say, 'I want my ultrasound to make sure everything's okay. So I'll have one at every visit.' No one says, 'Are you prepared to see a baby that's not perfect?' So they think that ultrasound equals a baby that's all right."

Though Stephanie La Croix Hinkaty, a mother of three from Hastings-on-Hudson, New York, would not have had an abortion, she did have a nuchal screen when she was twelve weeks pregnant because she felt more comfortable knowing in advance if something was wrong. Stephanie, who was thirty-seven, and her husband, Chris, were told the test showed they had a lower-than-average probability of having a baby with Down or another trisomy: a 1 in 667 chance.

To do more invasive testing would have put their baby at risk. An amniocentesis—where a needle is inserted through the abdomen into the uterus to extract about two tablespoons of amniotic fluid—is 99 percent effective in diagnosing chromosomal abnormalities but carries a risk of miscarriage that can be as high as 1 in 100. The other test Stephanie could have had is called a CVS, short for chorionic villus sampling, which is also invasive and carries a miscarriage rate of between 1 and 3 in 100, as well as a small risk of uterine infection or of having a baby with a limb missing. Because the risk of problems generated from more invasive tests exceeded Stephanie's risk of having a baby with a chromosomal abnormality, Stephanie's OB did not recommend them.

Stephanie and Chris concurred. Still, Stephanie had six more ultrasounds as part of her routine care and even consulted with a fetal cardiologist, who thought he saw something amiss with the baby's heart. The couple was then assured that everything looked fine. But when Iris was born, her muscles were loose and floppy, and her eyes were slanted more than would be expected for the child of two Caucasian parents. She needed oxygen and was whisked to the neonatal nursery. Four days later a blood test confirmed what the nurses and neonatologists in the hospital suspected: Iris has Down syndrome.

"I feel like all the testing is not worth anything because it's inconclusive," Stephanie says. Down syndrome detection using the nuchal translucency screening is thought to be accurate approximately 79 to 87 percent of the time (slightly less in younger women and if the test is performed at thirteen instead of eleven weeks) with a false positive rate of 5 percent. As often as 20 percent of the time—like what happened to Stephanie—the test will give false assurances. If a woman chooses to terminate a pregnancy based on an assessment of Down syndrome, she may be aborting a healthy fetus.

A MONEYMAKING SCAN

Dr. Edward Linn talks with his hands. He's a big man—tall and solid. The blinds are open and I can just see the West Polk Street traffic buzzing outside. Linn explains that Cook County was once a neighborhood hospital within walking distance for Chicago's poor, but that public housing projects were razed to build more medical facilities and high-end condominiums.

Linn is a bit rushed—he has gynecology rounds with a group of medical students in half an hour. But he takes time to explain that he believes obstetricians are overusing ultrasounds. He admits that doctors are often eager to sell patients additional ultrasounds partly because the patients enjoy seeing their unborn babies. But, says Linn, there is also often a clear financial motivation on the part of the doctor.

"It depends on how you're reimbursed," he explains. "If your OB fee is capped and you're not getting paid extra for anything you do within that fee, you'll see fewer tests being ordered. But often doctors get paid more based on the extra testing they do. So they'll say, 'Hm, something was a little unclear on the ultrasound, let's repeat it just to be sure,' or, 'Let's double-check if the baby's growth is okay.' It may just be the doctor's own anxiety, but these behaviors really get reinforced when there's also extra revenue attached. So being really cautious by ordering unnecessary ultrasounds actually helps you financially."

ROUTINE ULTRASOUND DOES *NOT* HELP YOU HAVE A HEALTHY BABY

Most women look forward to multiple ultrasounds because they are lulled into the assumption that this technology will catch potentially fatal abnormalities—such as a heart defect—early, so they can be fixed. When doctors tell pregnant women they will only get one or two scans, some are terribly disappointed, feeling that they won't be able to bond as effectively with the baby or worrying that the doctor won't know that the baby is growing normally. But one study of 15,151 pregnant women published in the *New England Journal of Medicine* showed that an ultrasound scan does not improve fetal outcome. The study, which was conducted by a team of six researchers over almost four years, compared pregnant women who received two scans to pregnant women who received scans *only* when some other medical indication suggested an ultrasound was necessary. The results showed no difference in fetal outcomes.

"[T]his practice-based trial demonstrates that among low-risk pregnant women ultrasound screening does not improve perinatal outcome," the authors conclude. Even when the ultrasound technology uncovered fetal abnormalities, the fetal survival or death rate was the same in both groups.

What the authors did find, however, was that routine ultrasounds led to more expensive prenatal care, adding more than $1 billion to the cost of caring for pregnant women in America each year.

Another study, of 2,834 pregnant women, published in the *Lancet,* showed that the babies of the randomly chosen group of 1,415 women who received five ultrasounds (as opposed to the group of 1,419 women who had only one scan at eighteen weeks) were much *more likely* to experience intrauterine growth restriction, a scary combination of words that means the fetus is not developing normally. Ironically, intrauterine growth restriction is one of the conditions that having multiple ultrasounds is supposed to detect.

A DISTURBING HYPOTHESIS: ULTRASOUND EXPOSURE TRIGGERS AUTISM

Though one follow-up study tracking newborns who showed intrauterine growth restriction at birth did not reveal lasting neurological damage, other studies have shown that growth-restricted children experience long-term developmental delays, and there is a growing body of evidence that extended exposure to ultrasound can damage fetal brain tissue.

In 2006, Pasko Rakic, M.D., a neuroscientist at Yale University School of Medicine, found that prenatal exposure to ultrasound waves changed the way the neurons in mice distributed themselves in the brain. Rakic and his team do not fully understand what effect the brain cell migratory alteration might have on brain development and intelligence, but they noticed, rather alarmingly, that a smaller percentage of cells migrated to the upper cortical layers of the mouse brain and a larger percentage to the lower layers and white matter.

At first reluctant to publish these results because they were preliminary and might discourage pregnant women from accepting medically necessary ultrasounds (the mice studies are part of a years-long double-blind experiment that is testing the effects of ultrasound on primate brains), Rakic decided the findings were too significant to ignore and concluded that all non-medical use of ultrasound on pregnant women should be avoided. "We should be using the same care with ultrasound as with X-rays," Rakic said.

Manuel Casanova, M.D., a neurologist who holds an endowed chair at the University of Louisville in Kentucky, contends that Rakic's mice research helps confirm a disturbing hypothesis that he and his colleagues have been testing for the last three years: that ultrasound exposure is an environmental factor directly contributing to the exponential rise in autism.

When Casanova began researching autism fifteen years ago he discovered that neuroscientists had not been able to isolate the differences between an autistic brain and a normal brain, unlike with Parkinson's disease or Alzheimer's, where the damage in the brain has been localized. Casanova realized that in order to understand both the causes and the potential cures for autism, scientists needed first to figure out where in the brain of autistic children damage was occurring.

Since no damage to individual neurons had ever been isolated, Casanova theorized that we might not be examining the brain in the right way. He began looking at the brain as a system instead of isolated parts.

Understanding that the brain is modular is the first key to uncovering what causes mental disorders, as the work of neuroscientist Vernon B. Mountcastle shows. In 1980, Mountcastle was a professor at the Johns Hopkins University School of Medicine when Casanova was a research fellow there. Mountcastle found that when he placed microelectrodes in the cortex of cats' brains in a vertical alignment, straight down through the gray matter, all the neurons in the line responded together in the same way. But when he impaled neurons horizontally, parallel to the brain surface ("tangentially" is the scientific way to describe this), they did not respond together.

Mountcastle gave these vertically oriented groups of neurons that work together a name: "minicolumns."

It is these columns of neurons working together that are responsible for higher cognitive functions like facial recognition, joint attention (if I turn my face and look somewhere, a child will turn and look too. Not because I told the child to look, but because the normal human brain is wired to do so), and much more. Joint attention is one of the many qualities that appear to be abnormal in the brains of autistic children. Inspired by Mountcastle's research, Manuel Casanova recognized the imperative of studying the circuitry within the brains of patients with autism and other psychiatric conditions. He and his colleagues found something surprising: brains of autistic patients have a 10 to 12 percent *higher* number of minicolumns as compared to nonautistic brains.

They also found another anomaly. During the normal formation of the human brain, cells divide in the hollows (ventricles) of the brain and then migrate to the surface (cortex), acquiring a vertical organization into columns. At the same time, other cells migrate tangentially and meet up with the columns. Casanova calls these migrations "a very fine ballet," and explains that the cells that migrate tangentially have an inhibitory role, acting like a container to keep the cells in the minicolumn from spilling into other parts of the brain. Compared with other animals, even primates, the neurons in the human brain have to travel a much longer distance, and during this long migration there is, unfortunately, ample opportunity for things to go wrong.

Casanova explains: "You know that a shower curtain keeps water inside of the bathtub. If you have a defect in the shower curtain, water will spill out of the tub. If the radial migration is not coupled with the tangential migration of inhibitory cells, then the minicolumns will have a faulty shower curtain of inhibition and information will no longer be kept within the core of the minicolumn, it will be able to suffuse to adjacent minicolumns and have an overall amplification affect. Actually the cortex of autistic individuals is hyperexcitable and they suffer from multifocal seizures. One third of autistic individuals have suffered at least two seizures by the time they reach puberty."

Translation: As the "minicolumn" brain cells move outward, if the complementary cells that inhibit them don't keep pace, the information in the minicolumns will suffuse out to surrounding cells, causing a chain reaction that can result in seizures.

Ultrasound waves, Casanova explains, are a form of energy known to deform cell membranes. In fact, in the early 1990s the FDA approved the use of ultrasound to treat bone fractures because ultrasound increases cell division. Some cells in the human body are more sensitive than others.

Among the most sensitive cells? Those stem cells in the brain that divide and migrate.

Casanova's hypothesis: Prolonged or inappropriate ultrasound exposure may actually trigger these cells to divide, migrate, and form too many minicolumns. They divide when they're not supposed to and there are no inhibitory cells to contain them.

"Something impinges on the germinal cells and causes them to divide at a time when they should not," Casanova explains. "Cells migrate to the cortex, but because it is at an anomalous time they are not synchronized with inhibitory cells, so there is an excitatory-inhibitory imbalance."

Casanova points to a large body of corollary evidence that buttresses the theory that ultrasound exposure may be causing autism:

1. High-risk women who receive multiple ultrasound scans are at higher risk for having autistic children.
2. Autism is much more common among educated, upper-middle-class families, who have access to the "best" prenatal care and are more likely to have multiple scans.
3. In Somalia, autism is virtually unheard of (though the low numbers of autism may partly be due to a lack of testing, epidemiological data make it clear that autism is much less prevalent in the developing world), but Somalis who come to the United States and other industrialized countries, where they receive aggressive prenatal care, are experiencing skyrocketing autism rates.
4. People who do not use ultrasound, like the Amish, are at lower risk for autism.

It's easy to dismiss this corroborating evidence. Correlation, we know from Statistics 101, is not causation. Yes, the number of red SUVs on American highways has increased dramatically at the same time that autism has been on the rise, but the two phenomena are obviously unrelated. Casanova himself is quick to dismiss the value of circumstantial evidence. But in the past few decades, autism has increased worldwide. There are more neurologically damaged children in the United States today than ever before. As of 2007, 5.4 million children (the entire population of Finland) have been diagnosed with attention disorders, and today one in every eighty-eight children in America has been diagnosed with an autism spectrum disorder. Japan, Norway, Iceland, Denmark, Australia, France, Germany, Canada, and the United States are among the industrialized nations that are seeing a huge, troubling, and seemingly inexplicable rise in the numbers of autistic children. These countries are geographically and culturally different. Their vaccine schedules are different. The labor and delivery experience is also different: In Scandinavian countries and Japan many more pregnant women tend to choose unmedicated vaginal births. As writer Caroline Rodgers points out in an article in *Midwifery*

Today, these countries are more diverse than they are the same—they do not share the same air, water, clothing, building materials, diets, pesticides, or ancestral history. No single genetic or environmental factor can explain the relentless increase in the neurological damage we are seeing in small children.

But all these countries do have one thing in common: The vast majority of pregnant women are getting regular prenatal care and being exposed to ultrasound in the form of anatomy scans and fetal-heart monitoring. In countries with nationalized health care, where virtually every pregnant woman is exposed to multiple ultrasounds, autism rates are even higher than in the United States.

"I have spent most of my working life in medical research," wrote a commentator in an online article about autism and ultrasound. "I manufactured ultrasonic transducers for twelve years. I have used ultrasonic cleaners to clean surgical instruments (and jewelry). I work with someone who recently told me that she was having bi-weekly ultrasounds for her new pregnancy, but now that she's reached nineteen weeks, she will have an ultrasound session that takes half an hour. This last comment is what made me think. I did not know that someone would have such a long ultrasound session. Ultrasound travels through moisture, and specifically, ultrasonic cleaners work by cavitation . . . they make pits in the surface of what they hit, such that they knock off dirt/tissue/blood by vibration. Perhaps this vibration could knock little weak spots in myelin sheeting of nerves or such, I don't know. Knowing about potential for cavitation, even more so than heat problems or the auditory potential to a fetus, is something which I am concerned about and its possible implication in autism's unbelievable rise."

The ultrasounds done on pregnant women today use sound waves with eight times the intensity used before 1991. This time period roughly coincides with the alarming increase in the incidence of autism within our population. Even more disturbing, the majority of technicians using ultrasound machines (as many as 96 percent) do not understand the safety margins they must adhere to in order to make sure the fetus is not exposed to harm.

As ultrasound equipment gets smaller, less expensive, and more portable, it has also become available—without any regulation—to anyone who knows how to surf the Internet. Want to see or hear your baby? You can buy your own ultrasound machine on Amazon or eBay.

"Most people believe it's just about taking pretty pictures," Manuel Casanova says, his voice thick with regret.

When X-rays became widespread in the United States, people delighted in how this new technology could capture the insides of a human body. The pictures weren't exactly pretty, but they were amazing.

Discovered in 1895, X-ray technology was so celebrated and became so widespread that when my mom was growing up in Chicago in the 1940s, her feet were not measured by stepping on a metal plate with a ruler. Instead, she stuck them in an X-ray machine, pushed the button, and saw her bones on the screen. Great fun for her and her three sisters! And a boon for avant-garde shoe salesmen, who boasted that these machines gave the most accurate shoe size.

Like shoe salesmen, some scientists and doctors were so enthusiastic about this presumably harmless and amazing technology that a 1935 journal article insisted X-rays should be a routine part of obstetrical care. Published in the *Proceedings of the Royal Society of Medicine*, L. N. Reece claimed, "[A]ntenatal work without the routine use of X-rays is no more justifiable than would be the treatment of fractures." Marsden Wagner, author of *Born in the USA: How a Broken Maternity System Must Be Fixed to Put Women and Children First*, points out that early obstetrical textbooks denied that exposure to X-rays could pose any danger to the fetus or the mother. Since no safety studies had been conducted, medical practitioners were quick to insist the use of X-rays posed no evidence of harm. Technically that was true.

As use of obstetric X-rays became more widespread, an industry grew up around it. Obstetric X-ray equipment was needed, as well as training for specialized radiologists or radiographers. Although concern continued to mount, for almost fifty years X-ray technology was used to estimate fetal head size, calculate a pregnant woman's projected due date, and screen for fetal abnormalities. A pregnant woman would be strapped to a table and an enormous machine, beaming X-rays at her abdomen, would take images of her pregnant belly for doctors to analyze. The routine use of prenatal X-rays was justified because the information gathered was thought to help ensure a safe pregnancy outcome.

We know now that obstetricians were terribly mistaken. Today doctors take great care to educate pregnant women on how to avoid radiation exposure. We have discovered that repeated exposure of a baby in the womb to X-rays—long considered benign and harmless—causes damage to fetal cells that can result in everything from miscarriage, mental retardation, and birth defects to leukemia and other kinds of childhood cancer.

What if we are making the same mistake with ultrasound?

A WOMB WITH A VIEW

Yet prenatal ultrasounds have become so attractive to first-time parents that many are delighted to have as many as possible, not to screen for potential problems but just to see the baby. Some doctors offer "bonus" four-dimensional ultrasounds in their offices, which clients pay for out of pocket since these scans are not covered by insurance. Others doctors and ultrasound technicians have found another way to cash in on the appeal: Nationwide, there are more than seventy businesses, often located in malls, that offer three-dimensional and four-dimensional elective ultrasounds for the expectant couple and their entire family. These companies that offer non-medical ultrasounds, which started becoming popular about ten years ago, have cutesy names like Prenatal Peek, Fetal Fotos, and Womb with a View.

"We were super excited when we found out we were having a girl. We did the four-dimensional just to check," says Lisa Nguyen who went to A Baby Visit in San Diego with her mother, mother-in-law, husband, and three-year-old son when she was thirty weeks pregnant. "It was really sweet. We were really happy," she remembers. "We heard the heartbeat, saw her move. We even saw her smile and sucking her thumb." The best thing about the visit for Lisa was how much more pleasant it was than the ultrasounds at the doctor's office. The tech was very personable and made her feel comfortable. "She was just as excited about seeing my baby as I was. It makes you feel they're enjoying what they're doing—sharing their happy moments with you—instead of just doing their job."

At Before the Stork in Anaheim, California, as many family members as desired ("We never limit the number of guests you can bring," boasts their website as The Supremes belt out "Baby Love") can assemble in a large scan room with black sofas and a zebra-striped rug. The mom-to-be lies discreetly on an examination table in the corner while images of her baby are projected on a seventy-five-inch screen for everyone to see. The February 2012 Valentine's Day special costs only $159 for a twenty- to twenty-five-minute viewing of the baby. And you can even get a follow-up appointment for $79 extra.

Fetal Fotos has branches in Oregon, Idaho, and Utah; the Oregon location offers a twenty-five- to thirty-minute session that costs $275 and includes guaranteed gender determination, a "free" four-dimensional DVD set to music, and twelve glossy color photographs (duplicate DVDs are available for purchase).

The Food and Drug Administration has issued several warnings against these non-medical entertainment ultrasounds. In the absence of a valid medical reason, the FDA cautions, fetal ultrasound may do more harm than good. There are several problems: for nonprescription purposes the equipment itself is not regulated; the person performing the scan may not be adequately trained, especially in terms of how to handle a problem with the baby or if a fetus has died; and the fetus is often being exposed to sound waves for much longer than normally occurs for a medically prescribed sonogram simply to create photos and videos for the family. These FDA warnings against entertainment ultrasounds are echoed by several major medical associations, including the American College of Obstetricians and Gynecologists and the American Institute of Ultrasound in Medicine.

"When ultrasound enters the body, it heats the tissues slightly," explains Robert Phillips, Ph.D., a physicist with the FDA's Center for Devices and Radiological Health in an FDA statement. "In some cases, it can also produce small pockets of gas in body fluids or tissues." The long-term effects of tissue heating and of the formation of partial vacuums in a liquid by high-intensity sound waves (cavitation), according to Phillips, are not known.

When Danielle Driscoll accompanied her best friend to get a thirty-six-week four-dimensional ultrasound in southern California, in 2007, she found the experience deeply disturbing. Ten family members piled into the room to watch as the technician failed to get a good picture of the baby's face. So the session, only supposed to last half an hour, went on for an hour and a half.

"That baby looked like it was in pain," Danielle recalls. "The baby actually flipped transverse and buried its face in the placenta and covered its ears with its hands. It was pretty obvious that it was in distress. . . . I wasn't a parent and I didn't feel like I could say anything, but the tech kept making cute remarks about it. She'd say, 'Oh, look at his toes, look at his fingers! Oh, he's really moving around today! Oh, it looks like he's running from us and trying to hide! But don't worry, we'll get him!' "

OPTING OUT

Though the American College of Obstetricians and Gynecologists recommends that obstetricians discuss the advantages and disadvantages of having an ultrasound scan with pregnant patients, ACOG does not explicitly recommend the screening. ACOG explains that ultrasound *may* reduce fetal mortality rates because women who discover they are carrying fetuses that are incompatible with life will often choose abortion, but ACOG also specifies that ultrasound *has not been proven to be effective* for reducing infant mortality in any other way.

Their policy statement continues: "Screening detects multiple gestations, congenital anomalies, and intrauterine growth restriction, but direct health benefits from having this knowledge currently are unproven. The decision ultimately rests with the physician and patient jointly." The authors of the definitive, exhaustive, 1,385-page textbook for obstetricians, *Williams Obstetrics,* take a similarly conservative stance about ultrasound and do not explicitly recommend it for low-risk pregnancies: "Sonography should be performed only with a valid medical indication," the authors write, "and with the lowest possible exposure setting to gain necessary information . . ."

Yet doctors and other birth providers take great exception if low-risk pregnant women refuse to be scanned. In 2004 when Lia Joy Rundle, a mom of three from Mazomanie, Wisconsin, was just a few weeks pregnant with her second child, she changed insurance providers. The new obstetrician reviewed her paperwork. "We might be able to do a quick ultrasound today, if the machine's available," she said. "Then you can take a look at your baby." Though they were planning to have a twenty-week ultrasound, Lia and her husband saw no benefit to doing an early ultrasound and felt there might be some risk. But when they declined the scan, the obstetrician insisted there was no way to get an accurate due date without it. "Look at him, he's fine," she scoffed, pointing at their one-year-old son. "How many ultrasounds did you have with him?"

Though Lia held her ground, women who do not want ultrasounds are often pressured into them anyway. That's what happened to Wendy Scharp, a mom of three in Portland, Oregon. Wendy was nursing her second baby and had not had a period when she got pregnant again, so she had an ultrasound to date the pregnancy. It was an unpleasant experience: The technician was huffy, insisting on using a vaginal probe because Wendy's bladder wasn't full enough. At her twenty-week visit Wendy and her partner, Shawn, told the OB they did not want any more ultrasounds. The obstetrician was astonished—Wendy had never questioned her instructions and she had never had a client refuse a scan. At a subsequent appointment when Shawn wasn't with her the same obstetrician convinced Wendy she needed an ultrasound because the baby was measuring big. Wendy was feeling particularly vulnerable and reluctantly agreed. "After the fact I was so upset. Even if he was big, I was not going to agree to a C-section or induction," she explains. A young mom (Wendy was twenty-one at the time), she felt bullied by the doctor. The ultrasound made no difference to the outcome of her low-risk pregnancy. Her son, Sol, was born vaginally on his due date weighing in at 9 pounds 6 ounces.

After multiple scans during a low-risk pregnancy followed by a difficult hospital birth, Sam Smith and her husband chose to have a homebirth the second time. Their midwives agreed there was no valid medical reason for Sam, who was thirty-one, to have an ultrasound. But when she was five months pregnant, Sam found out her best friend Ellie was carrying an anencephalic fetus (a baby with no brain) who could not live outside the womb. Ellie and her partner chose to have a second trimester abortion. Sam and her husband anguished: What if their baby had an undetected congenital problem? Since abortion was not an option, they decided the risks of exposing the fetus to sound waves outweighed any benefits. If their baby had a condition incompatible with life, they decided, at least they would have the chance to welcome it into the world and say good-bye in person. Their daughter was born at home on a snowy New England afternoon. She cried a lot as a newborn, Sam says, but she was perfectly healthy.

In 2004, twenty-seven-year-old Jennifer Cario had two ultrasounds: one in the first trimester to confirm she was pregnant and a second one, at eighteen weeks, to examine the fetus and find out the gender. During the prenatal visits Jennifer's OB told her he was supportive of her wish for a natural birth. But when Jennifer's water broke at the start of labor and labor was slow to start, one of the nurses at Riverside Memorial Hospital in Columbus, Ohio, "threw a fit" when Jennifer got out a bottle of juice. "She insisted I couldn't have anything by mouth," Jennifer remembers. The on-call doctor pressured her to agree to Pitocin, and the situation made Jennifer so stressed that her body stopped laboring. She ended up having large doses of Pitocin and an epidural, even though she was hoping to avoid both. Elnora was born vaginally after twenty-nine hours of labor. She weighed 8 pounds 4 ounces. Against Jennifer's wishes, the hospital nurses gave the baby a bottle. Made to wait forty-eight hours to see a lactation consultant (there were a lot of babies born that day and not enough staff), Jennifer found herself unable to breastfeed. She pumped exclusively for thirteen months.

Two years later, when she got pregnant again, she decided to do things differently. Through her church, Jennifer met a lay midwife who delivered babies at home for the Amish community. The midwife had delivered some four hundred babies and was also a trained emergency medical technician. Jennifer had no ultrasounds during that pregnancy. When another friend who was seeing the same midwife developed placenta previa, the midwife identified it without the use of ultrasound and referred her to a doctor, who took over her care. That experience confirmed for Jennifer that her midwife would be able to identify potential problems competently, even without the use of ultrasound or other tests.

Jennifer tells me twelve of her friends were pregnant around the same time. "Every other person I know was scheduled for a C-section or an induction," she says when we talk on the phone. "The doctor does a routine ultrasound and says he thinks the baby's too big or doesn't like the way it's lying. Followed by weeks of stress. That would set the stage. A few weeks later they do another ultrasound. Then another. Then they schedule an induction and the baby weighs only six or seven pounds." Jennifer's son Emmitt, who weighed 9 pounds 8 ounces, was born peacefully at home after five hours of labor, with just twenty minutes of pushing. "Ultrasounds have their place, but for low-risk pregnancies I think they just give more opportunities for doctors to insist on more intervention," she says.

Manuel Casanova's grandson has curly black hair and an angular face like his grandfather. But unlike Casanova, an eminent neuroscientist and researcher who holds a prestigious endowed chair at the University of Louisville, trains hundreds of medical school students every year, and is articulate and charming, Bertrand, who is four years old, has never been able to say a word. Most of the time he is unaware of his surroundings, he has difficulty making eye contact, and he is plagued by seizures. Two of Casanova's four daughters have moved to Salt Lake City to help their sister care for Bertrand, who is severely neurologically damaged.

Casanova believes there is mounting evidence to suggest the ultrasound scan the doctor ordered when Bertrand was a ten-week-old fetus smaller than a lemon has something to do with Bertrand's autism.

Karen Bridges's blond-haired, blue-eyed son Brock flaps his hands, speaks indiscriminately to strangers, and has an IQ of 61. He too is autistic. Did prenatal exposure to ultrasound cause Brock's or Bertrand's autism? Since autism is a spectrum disorder that manifests differently in different children and may be triggered in children with genetic susceptibility in different ways, and since it is very difficult to ethically design brain experiments for human fetuses, the jury is still out.

Not every pregnant woman who was X-rayed had a baby who later developed cancer, but once the connection between cancer and X-ray exposure was definitively established, the use of X-rays in obstetrics stopped. Once promoted by Harvard researchers as a drug that could make healthy pregnancies even more robust, the synthetic estrogen diethylstilbestrol (DES) was prescribed to pregnant women for more than thirty years. Even though by 1953 published research showed that DES did not work to prevent miscarriage, it took almost twenty more years before the FDA advised doctors to stop prescribing it. Pregnant women were told that DES was safe because there was no evidence it caused harm. We now know exposing a developing fetus to synthetic estrogen can, years later, cause a devastating form of vaginal and cervical cancer, as well as reproductive tract anomalies, infertility, and benign testicular growth in boys whose mothers took the drug. We know from animal experiments that sound waves cause brain abnormalities. We know there are no proven benefits to using ultrasound in low-risk pregnancies. If we followed best practices, obstetric ultrasound scans, if done at all, would be conducted as infrequently as possible and as quickly as possible. Instead, doctors and hospitals have financial incentives to order more scans. It's fun to see the baby before he is born. But is it worth it?

Salary of an ultrasound technician: $47,000/year–$50,500/year

Salary of a radiologist (a doctor who specializes in reading scans): $486,764

Voluson 730 Pro GE ultrasound system: $29,500 (The Voluson 730 Pro is in the medium price bracket)

Cost of an ultrasound: $200–700

Cost that doing ultrasounds adds to the health care system: $1 billion

Prenatal office visit without ultrasound: $180

Prenatal visit with first-trimester ultrasound: $695

Prenatal visit from homebirth midwife: $50–100

Cost of 4D ultrasound scan with video clip: $140 (Free rescan if the baby hides! Weekday rates are cheaper. Extra prints: $2–30, depending on the size)

Cost to forgo an ultrasound: $0

Louana George: Seduced by Technology

Louana George, sixty, started as a childbirth educator and then became a labor and delivery nurse for three years. She opened her own midwifery practice in 1987 and spent twenty years delivering babies as a homebirth midwife in southern California. During her long career as a birth professional, Louana became a trained ultrasound technician. But she now has deep reservations about the widespread use of these machines.

When I first started helping pregnant and laboring women, there were no ultrasounds. Doctors and midwives would find out everything they needed to know by palpating a woman's belly. Now some doctors do an ultrasound at every prenatal visit. I've watched ultrasound go from nothing to everything.

I took an ultrasound course in 1997 when it was still cutting edge and I happily scanned women. Of course it's fun to see the baby! And know the gender! Then I can know to decorate in pink or blue, what kind of clothes to buy. Moms feel like it helps them relate to the baby. And the dads can feel like they're part of it too. It's very seductive. When I took that class, I couldn't get enough of it. It's TV, isn't it?

Just a few months after that class I went to a professional birth convention in Long Beach, California, and there was a vendor selling ultrasound machines. In those days the machines cost between $50,000 and $100,000.

"Wow, these are expensive," I said to the vendor.

"Yes," he agreed. "But if you routinely scan all of your pregnant women, you can charge the insurance company extra for that, and you can capture back your money in no time."

His ready answer made me start thinking about why we were doing ultrasounds, how necessary they were, and who was really benefiting. Of course he was right—how else is a physician in private practice going to recuperate his money after he buys expensive equipment? I was opening my own midwifery practice so around the same time I also took a workshop for independent providers on how to work with HMOs and PPOs to get adequate reimbursement for your time. The workshop leader mentioned that if you are a physician and you have advanced equipment but you don't use it, and your client has a bad outcome, you will have a very big problem. If the case goes to court, the judge and a jury are going to slap you with a guilty sentence because you didn't use the machine. She also told us how to use a different billing code for doing ultrasounds. The reimbursement is higher than for sitting and counseling the patient. So she advised us to ultrasound as often as possible, both because if you have an ultrasound machine in the office and you don't use it and something is wrong, you will be sued by the patient and because it's an effective way to "capture the money."

Those were the two incidents that really opened my eyes. I started researching what we know about the safety and dangers of ultrasound and I've come to realize that there are many disadvantages to giving women ultrasounds. On the fact sheet we were given for the course, we were told that "there have been independently confirmed significant biological effects in mammalian tissues" that is exposed to certain intensities of ultrasound. We also know that actively dividing cells of the central nervous system of the embryo or fetus are most disturbed by sound waves.

The new machines project a pulsating beam of sound instead of a constant beam of sound, but we do not know if the problems have actually been solved. It's very common to see babies cringing away in utero even though we don't know why. Is it the sound? Is it the heat? If it's so benign what is the baby running away from?

Doctors and other birth professionals are seduced by the technology without even realizing it. It gives them the false sense that they are in control. They think there is no such thing as prenatal care without ultrasound. In the forties and fifties doctors routinely X-rayed women to see if their pelvises were big enough and said it was perfectly safe. Now they say that ultrasound is perfectly safe but there are no long-term studies on this current technology. Should we really take their word for it? Most doctors don't realize that even ACOG does not recommend routine ultrasounds on women who do not have complications. It's astounding to me that doctors are still doing it. As a midwife, I would rather put my hands on a client than take my chances with this unknown.

CHAPTER 3

EMERGING EXPENSES:
The Real Cost of Childbirth

For British journalist Molly Castle, birthing her daughter in 1936 at a small hospital in southern California, labor was a nightmare. She had a bad reaction to the anesthesia and fought so wildly that three attendants—her husband, a nurse, and the doctor—could barely subdue her. For fifteen minutes, her eyes wide with terror, Molly clawed to get the ether mask off her face, screaming in a shrill, terrified way. When she finally went limp the doctor inserted metal forceps (which look like giant-sized salad tongs) into her vagina, used scissors to cut the skin between Molly's vagina and her anus, and pulled out her baby in a gush of blood. Afterward Molly had only a vague and unpleasant memory of what had happened: People in gas masks leaning over her, the terrifying thought that World War II had finally begun.

Though every hospital had different ways of treating women in labor, from as early as 1906 until the 1960s laboring women in American hospitals would be injected with sedatives to dull labor pains. One combination of psychotropic drugs—usually morphine and scopolamine—was supposed to put the woman into *Dämmerschlaf* or "Twilight Sleep," a semiconscious, peaceful state that would make childbearing easy. By 1915, when the drug cocktail had become popular among well-to-do women looking to avoid the pain of childbirth, American women were promised that Twilight Sleep would "take away the terrors of childbirth" and act as "a preserver of youth and beauty through the removal of agony."

The problem was that Twilight Sleep didn't work very well. Some of the first doctors to try this drug cocktail on their patients at the turn of the century became outspoken critics against it. Scopolamine is "one of the most dangerous of all poisons—a poison incalculable in its action," one doctor reported. Another found that extremely large doses of the drugs would have to be administered to some women in order to have any effect at all, while small amounts of the drugs made other patients "wild" instead of calm.

As this kind of childbirth anesthesia became more widely used in the United States, it became increasingly clear that the obstetricians who had originally opposed it were correct: Women in Twilight Sleep would become out of control, hallucinating, screaming themselves hoarse, and thrashing so much they hurt themselves or the hospital staff if they weren't physically restrained. Some hospitals started using metal cuffs or leather straps to secure women to the bed, a disturbing practice that continued into the 1960s. Some used loops of lamb's wool so they wouldn't have to explain the bruises on a woman's ankles and wrists to a worried spouse afterward. Their wives couldn't tell them what happened. At the dosages administered during labor, scopolamine functioned as an amnesiac, leaving women with only a hazy memory, if any, of what had happened.

Companies manufacturing Twilight Sleep drugs, like Boehringer & Son and Hoffmann-La Roche Chemical Company, knew a financial bonanza when they saw one. Hoffmann-La Roche, the Swiss-based company that made scopolamine, made an "educational" film about the tranquillity and ease of Twilight Sleep, using language that would appeal to a country intent on becoming more modern (they called it "science's greatest triumph"), to advertise the practice, and held screenings across the country. A cottage industry of American doctors and pharmacists, working to perfect the shelf life of scopolamine by synthesizing it themselves, also sprang up. Forward-thinking entrepreneurs leased land for private sanatoria to provide Twilight Sleep to wealthy women when hospitals would not. These companies and entrepreneurs anticipated the future of childbirth in America, which was becoming less of a personal family affair and more of a profit-making business with every passing day.

If Twilight Sleep was bad for some women, it was also bad for their babies. A well-known side effect of morphine is respiratory distress, and when the technique was introduced in America, doctors started noticing that babies born to women in Twilight Sleep were often so oxygen deprived they were born blue.

"I see almost every day comments on this," wrote one doctor in 1915, who believed the technique was unsafe, "and the consensus of opinion is that it is not to be used much because a large per cent are blue babies and many die."

Despite poor outcomes and negative repercussions to the newborns and the moms, some upper-middle-class women and obstetricians championed it. Mrs. Francis Carmody, the wife of a well-to-do attorney, traveled to Germany with her obstetrician so she could birth her second baby using this method. The experience was so flawless that Francis became one of Twilight Sleep's most public supporters, opening a Twilight Sleep hospital in Brooklyn, staging rallies, and telling women if they wanted this kind of pain relief they would have to fight for it. In August 1915, Francis died tragically in childbirth with her third baby. Though her death (which may not have been a direct result of the drugs) put a damper on the movement, American obstetricians continued to inject women subcutaneously with Twilight Sleep drugs until as late as 1974.

FROM HOME TO HOSPITAL

In Colonial times and during most of the nineteenth century, the majority of births in America took place at home. Birthing women were usually attended by informally trained midwives who passed on their skills from generation to generation. If the mother-to-be was well-to-do and could afford the extra fee, a family physician might come to the house to be present alongside the laboring woman's friends, neighbors, and the other experienced women who were often there as well. As medicine became increasingly professionalized in the United States, however, more doctors began attending homebirths. By the beginning of the twentieth century, roughly half the births in the United States were attended by midwives and half by doctors. These births took place almost entirely in the home. As the twentieth century progressed, however, that began to change.

In 1929 the first Indiana limestone was laid on the foundation to build the University of Chicago's Lying-In Hospital, which was open for business two years later. An imposing Gothic complex with tall windows, archways, and a cloistered hallway like a nunnery, Chicago's Lying-In quickly became in vogue with wealthier Chicagoans.

Huge bay windows looked out onto a courtyard planted with maple trees and a redbud alive with pink blossoms in the springtime. "In the early days of Lying-In it was very much the place to go to have your children," says John Easton, a senior science writer at the University of Chicago's Medical Center, who is giving me a tour. "It was the carriage trade who came here."

When my mother was born here in 1938, more and more well-to-do women were heeding the call of the hospital's founder, Joseph DeLee, an obstetrician who claimed that all women should give birth in the hospital. Considered by many to be the founder of obstetrics in America, DeLee argued that childbirth was "destructive," "pathogenic," and "pathologic," an event requiring highly skilled hands-on obstetric care. He advocated in favor of systematizing birth to avoid pathology by inducing labor, managing labor with drugs (in particular scopolamine), and using forceps during delivery.

"Midwives still hate him," John tells me.

But by 1940, two years before he died, Joseph DeLee may have been ready to recant. In a Mother's Day address to a lay audience he warned pregnant women to stay away from physicians who tried to streamline the birth process. "Mother Nature's methods of bringing babies are still the best," he told the crowd.

By then it was too late.

The American labor ward had become a place of systematic emotional and physical abuse.

"My mother-in-law was a labor and delivery nurse in the 1940s," says Mary Fauls, who works as a doula giving support to women in labor and runs a pregnancy program at John H. Stroger Jr. Hospital in Chicago. Mary's so tall she has to bend down a little to talk to me. "She says no one has ever made a horror movie that was anything as bad as a labor room back then." But *Ladies' Home Journal* did write about it. In a 1957 investigation, "Cruelty on the Maternity Wards," dozens of women shared how they were treated during labor. "Women are herded like sheep through an obstetrical assembly line, are drugged and strapped on tables while their babies are forceps-delivered," testified a mom of three from Columbus, Ohio. "I was left entirely alone for most of the sixteen hours of labor," says a mom from Haddonfield, New Jersey. "When my baby was ready the delivery room wasn't," attests another. "I was strapped to a table, my legs tied together, so I would 'wait' until a more convenient and 'safer' time to deliver." Other women report being hit in the face ("When my husband saw my bruised neck, face and arms, he questioned the doctor and was told that first mothers knock themselves around"), having steel clamps put over their shoulders and chests ("for the mother's protection"), laughed at, and told by medical personnel, "You've had your fun, now you can suffer!"

We cringe when we read about how women in the past were drugged, left to have psychotic episodes and labor alone, and generally mistreated during childbirth. Today we would never give a laboring woman morphine or scopolamine. We have highly trained obstetricians and state-of-the-art medical facilities. We have open wards where family members can stay with the birthing woman. We have beautiful pastel birthing rooms with framed drawings of flowers on the walls, like in a hotel. We have fetal monitoring equipment to ensure that the baby is tolerating contractions well. We give women IV lines in their hands so that we can react quickly in an emergency. We have a system of birth and delivery that is both proactive—preparing for the worst-case scenario—and safe—ensuring that pregnant women and their babies will emerge from delivery healthy and alive.

Don't we?

What most Americans don't realize is that it is actually *more dangerous* to have a baby in the twenty-first century than it was two or three decades ago.

DEADLY DELIVERY

The United States spends more money on health care than any other country in the world. Hospital charges related to pregnancy, delivery, and infant care are among the top five most expensive conditions requiring hospitalization. Pregnancy-related hospital costs are second only to coronary heart disease. Bills for pregnancy-related and newborn care totaled more than $98 billion in 2008 (the most recent year for which we have reliable statistics), more than for any other area of medicine.

Considering all the money spent to ensure healthy pregnancies, the United States should be one of the safest countries in which to give birth. We would expect that harm to mothers and babies would be isolated events, so rare and shocking that they would make national news. Logically we should be among the industrialized countries with the best birth outcomes.

But we are not.

In fact, the statistics in our country related to maternal and infant mortality lag so far behind other countries that in 2010 Amnesty International called the situation for mothers, especially women of color and low-income women, a "crisis."

Amnesty International's report *Deadly Delivery* reveals that women in the United States have a greater lifetime risk of dying of pregnancy-related complications than women in forty other countries. There's ample evidence that things are getting worse: Data collected by the United Nations shows that while the vast majority of countries reduced their maternal mortality rates (for a global decrease of 34 percent), the maternal mortality rate in the United States *doubled* between 1990 and 2008, from 12 to 24 in 100,000 births.

While our maternal outcomes are much better than in war-torn countries like Afghanistan or impoverished countries like Niger, they are inexcusably high: More than two women die every day in the United States from pregnancy-related causes.

Contrary to our cultural stereotype, it is not just impoverished drug addicts without prenatal care who die in childbirth in America. Well-educated middle-class women, like thirty-two-year-old Diane Rizk McCabe, who died following complications from a Cesarean section at

Albany Medical Center in 2007; thirty-seven-year-old Karen Vasques, who died during a C-section at Beth Israel Deaconess Medical Center in 2008; thirty-five-year-old Hope College professor Jennifer Tait, who died in March 2011 after having a C-section at Holland Hospital in Michigan; and thirty-three-year-old Candice Boyle, a high school teacher who died in February 2011 at Aultman Hospital in Canton, Ohio, after developing a severe form of preeclampsia are also victims.

Because record-keeping is neither centralized nor mandatory in the United States—like it is in many countries where the birth outcomes are better—many health care professionals believe the number of pregnancy- and postpartum-related deaths may actually be much higher.

In fact, epidemiologists from the CDC have stated that the maternal death rate is "grossly underreported." Some birth experts estimate that the death rate in some areas of the United States is as high as 35 deaths per 100,000 births. The New York Academy of Medicine estimated that the maternal death rate for women in New York City was 23.1 deaths per 100,000 births. With a national goal of 4.3 deaths per 100,000 births, the official death rates in some places in America are more than *five* times higher than what is considered acceptable.

THIRTY-FOUR THOUSAND "NEAR MISSES" A YEAR

Almost as troubling as the high maternal death rates are the numbers of American women whose bodies are damaged during childbirth. While other countries report an increase in positive outcomes and a decline in childbirth-related complications, "near misses"—pregnancy-related complications so acute a woman almost dies—are climbing in the United States. Only a fraction of these cases come to the attention of the public, like twenty-nine-year-old Abbie Dorn, who suffered severe hemorrhaging and brain damage after her uterus was nicked during a Cesarean section at Cedars-Sinai Medical Center in 2006.

Nicole Dennis, a genetics counselor in Pleasanton, California, suffered such bad nerve damage after having her baby in the hospital that she could not walk unassisted. Though she had had sciatica previous to pregnancy, Nicole sat immobilized in bed for eleven hours after receiving an epidural and Pitocin. Because the baby's heart rate decelerated when she was turned on her side, the nurses kept her in one position. A hospital neurologist theorized that the constant pressure of her baby's head in her pelvic cavity is what caused the nerve damage.

Though Nicole could not walk unassisted for weeks (four years after her daughter's birth, she continues to have little sensation in her leg), her birth trauma is too mild to even be counted.

Yet between 1998 and 2005, near misses increased by 25 percent in the United States. Severe pregnancy- and childbirth-related complications currently affect at least 34,000 women every year. It's impossible to know how many women suffer from putatively mild childbirth-related complications.

The United States has the highest maternal mortality rate of any country in the industrialized world. And it continues to get worse. How is this possible? Why are so many women injured during labor and birth? Why do other countries have better birth outcomes? What are we doing wrong?

LABORING DOWN

The night shift on the labor and delivery floor at a hospital in upstate New York, despite having almost every room filled, is surprisingly quiet. At this moment there are no women moaning through contractions. No one is walking the halls. The nurses and doctors converse in hushed tones and the only loud noises are of technology—the shrill ring of a telephone, the *deet deet* of a doctor's pager, and the incessant beeping of the centralized screens outputting information from room monitors that are strapped around each laboring woman's abdomen.

The obstetrician I am shadowing is helping two low-risk women give birth tonight. He shows me with pride that the labor and delivery rooms all have Jacuzzi bathtubs and birthing balls. Both women are young and healthy: a first-time mom in her midtwenties, and a thirty-four-year-old who has one small daughter at home. But they both have been given epidurals (spinal pain medication), so they can feel nothing that is happening in their bodies. One looks bored, reclining in bed with her spouse by her side, waiting for something to happen. The other looks anxious, her excited husband running back and forth from the nurses' station to refill his orange juice. Since both women are completely numb from the waist down, the only information they have about their bodies and their babies is what the hospital staff read off monitors.

The OBs call what these women are doing "laboring down," lying flat on their backs, unmoving, after reaching 10 centimeters dilatation. If you've been given an epidural, you have no choice but to labor down. Continuous fetal monitoring also requires a woman to spend much of her time immobile and in bed, as movement interferes with getting an accurate readout.

In other cultures and in nonhospital settings in the United States, women are not expected to birth in bed. They stand, they squat, they get down onto their hands and knees (a position that helps the pelvis be as wide as possible). While it's easier and more convenient for birth attendants to have women labor and give birth on their backs, being supine means that gravity is working against a woman's body and her baby.

"What position did I birth my first in? Flat on my back," writes a woman named Karen. "I was already strapped down with the fetal monitor. They asked me to put my feet in the stirrups, and I said, 'I don't want to be on my back.' My OB said, 'What do you want to do?' I wanted to answer 'All fours,' but just then a contraction hit and I couldn't speak. By the time the contraction was over, they had pushed me into the stirrups and were holding my feet back so I couldn't move. They enlisted my husband to hold me down 'or she won't be able to push the baby out and we'll have to do a C-section.' I felt raped and I hated my husband for being part of it even though he was as scared as I was and felt he had no choice."

If a woman has not been medicated and is able to move around during childbirth, she will respond to the intensity of each contraction by shifting her weight, swaying her hips, leaning on a partner or friend for support, rising on her toes, getting on her hands and knees, or sitting or lying quietly. As she moves her baby is also moving, shifting positions during the journey through the birth canal and out into the world. In response to what is happening inside her body, she may choose to birth in a semi-upright position on the bed, but it is just as likely that she will find herself simply standing up, or in a supported squat, or on her hands and knees. Birth is a dynamic process that benefits from freedom of movement; even as the baby is crowning, a laboring woman will not necessarily stay still.

During my third labor I shocked Kathleen, a doctor who was attending my birth as a friend, by crawling away from the midwife's waiting hands after my son's head had already been born. "Where are you going, Jennifer!?" Kathleen cried. "Just don't drop the baby!" I cried back, twisting myself onto the bed. I was not acting consciously or with any clear purpose, I was responding to what my body told me to do. One push later my son's shoulders and body were born. Afterward, the midwife, who had ten children of her own all born at home, said she thought my movement at that moment was what made the pushing stage so easy. My body knew what to do. I went along for the ride. In most American hospital settings doctors will not allow women to move during the pushing stage of labor and are even uncomfortable if a woman is off the bed. The labor room floor in most hospitals is not a place you would want to be on your hands and knees. In fact, anything but the lithotomy position is so inconvenient for health care providers, and may cause fetal monitoring to give inaccurate readouts, that it is either actively discouraged or just not possible in a hospital setting.

MONITORING MISHAPS

Three years ago when a friend invited me to photograph and support her during labor, the nurse rushed into the bathroom where she and her husband were riding out contractions in the tub, a fetal monitor strapped to her abdomen.

"This just doesn't look good," the nurse said, shaking her head. My friend's doctor appeared for the first time half an hour later to announce that an "emergency" C-section was needed because of nonreassuring heart tones on the fetal monitor. The doctor ducked out to scrub up. The baby, once born, got an Apgar score of 9 out of 10 and showed no signs of having been in distress.

In 2005 ACOG reviewed the scientific literature available on electronic fetal monitoring and concluded that there was solid and consistent scientific evidence to show that constant monitoring results in a high false positive rate for adverse outcomes. In other words, based on monitoring, doctors conclude that something may be wrong with the baby when in fact everything is okay. As a result, fetal monitoring is associated with more interventions, including Cesarean sections, vacuum extraction, and forceps deliveries. Fetal monitoring does *not* reduce vaginal birth–related brain injuries, such as cerebral palsy, by catching when a fetus is in distress, as is commonly thought. Since monitoring became routine, rates of cerebral palsy have stayed the same. It does, however, result in more worry. The false positive rate for predicting cerebral palsy is greater than 99 percent.

LEFT ALONE

It's dinnertime and the obstetrician has gone downstairs to the cafeteria to have dinner with his wife, who has brought their two small children to the hospital. The labor nurse at the upstate–New York hospital sticks her hand up the woman's vagina to feel the "strength" of her contractions. Her shift is almost over and she's impatient.

"These aren't doing anything," the labor nurse mutters, peeling off a blue glove and throwing it in a medical waste bin as she leaves the young couple alone in their room. "I doubt if she's even going to make it. The head hasn't budged." She shakes her head. "It's never going to happen." Half an hour later her shift is over and she goes home.

What most first-time mothers don't realize as they dutifully go to their prenatal appointments is that their doctor—or whichever doctor is on call—will not be in the room with them for most of the time they are in labor. The constant care is left to the labor and delivery nurses. The obstetrician comes in to reassess the patient at various intervals, evaluate a complication, order more medication, or when a woman is pushing. When the labor and delivery nurses are busy, the laboring woman—unless she has hired a birth attendant to stay by her side—will be mostly left alone.

Yet evidence shows that constant labor support helps ensure good outcomes. A 2007 review of the scientific literature, including sixteen trials from eleven countries involving 13,391 laboring women, found that women who received continuous support during labor were more likely to have vaginal births, less likely to need invasive interventions like vacuum-extraction or forceps, less likely to need pain medication, were more satisfied with their birth experience, and even had slightly shorter labors.

THE CASCADE OF INTERVENTION

Though our cultural assumption is that the hospital is a safe place to give birth, and though many women—especially first-time moms—are eager to get there quickly, the problems that arise in labor are often created once the woman arrives at the hospital.

"We often see people going to the hospital who are in strong active labor, but after they get to the hospital their contractions start to space out or even stop," says Dr. Stuart Fischbein, who has been practicing obstetrics in the Los Angeles area for more than thirty years. "Why? Because the hospital is a fear-based environment. A human's body responds to that fear just like any other mammal's. So the doctor says, 'Now we might as well start some Pitocin.' But the Pitocin creates really strong contractions, which bother you. Now you have to stay tethered to the bed, so you can't get up and move around to help with the pain. So now you get an epidural, which lowers your blood pressure. The baby doesn't like that and his heart rate starts decelerating. So they turn off the Pit and you get wheeled in for a C-section. That's the cascade. I've seen it hundreds of times."

Kristy Boone was twenty-six years old in 2004 when she gave birth to her daughter Emy at Huron Valley–Sinai Hospital in Commerce, Michigan. When Kristy and her husband explained to the obstetricians in the group practice they went to that they were hoping for a childbirth with as little intervention as possible, some rolled their eyes, others shrugged and said, "Okay, we'll see." One obstetrician got really huffy.

"She made it seem like I was an idiot, like I didn't know what I was talking about," Kristy tells me over the phone as she drives her four-year-old son home from his sports class. "*She* knew what I needed. And *she* would decide. The fact that I wanted to avoid intervention bothered her."

Kristy was ten days past her due date and an ultrasound revealed the amniotic fluid was low. So after she was admitted the doctors decided to augment her labor with Pitocin.

"They gave me the Pitocin and kept turning it up. I was having crazy Pitocin contractions, one after another. The contractions were unbearable. It was insane," Kristy recalls. "I had my second child at home, he weighed over ten pounds and it was hard, sure, but nothing compared to the pain I felt in the hospital. When they give you that Pitocin there's no break between contractions, just violent pain, and it snowballs and gets worse and worse and worse."

Kristy's experience was not unique. It's common for obstetricians to order nurses to continue increasing the dose of Pitocin to speed up delivery. One labor and delivery nurse explains:

Labor and delivery nurses all over this country (including myself) have been bullied, yelled at, cursed out, and downright humiliated by birth attendants who want you to "keep cranking the Pit" regardless of maternal contraction or fetal heart rate patterns . . .

I once had an obstetrician, while in the patient's room, call me "incompetent" in front of the patient and her entire family because I had not continuously increased the Pitocin every fifteen minutes until I reached "max Pit" and instead, kept the Pitocin at half the maximum dose because increasing it any more caused my patient to scream and cry in pain and her uterus to contract every one minute without a break . . .

Another time I had a physician (who via a program called "OBLink" can watch her patient's monitor strips from her own home or office) call me on the phone from her house to chew me out about not having the Pitocin higher. When I explained that I had to shut the Pitocin off an hour earlier and start back up at a slower rate because the baby started to have repetitive and deep variable decelerations despite position changes, IV fluid bolus, and 10 liters of oxygen via face mask, I was told that the decels "weren't big enough" to warrant such a "drastic measure as shutting off the Pitocin" and I was "wasting her time" because "at the rate [I] was going [her] patient wouldn't deliver until after midnight."

I had yet a third doctor tell me once that he wished that only the "older" nurses on the floor would take care of his patients because they aren't "as timid" and "are not afraid to turn up the Pitocin when a doctor orders them to." That younger nurses like me are "too idealistic" and don't understand "how the world really works."

And yet another time I had a physician tell me that I needed to "crank the Pit to make this baby prove himself either way" and that if I couldn't do "what needed to be done" for his patient, then he would ask the charge nurse to "replace me with a nurse who could."

When I came in the next day and read the birth log, I discovered that 3 out of those 4 patients ended up with Cesarean sections after I had left . . . for "fetal distress."

Too much Pitocin can cause hyperstimulation of the uterus, uterine rupture, and brain damage to the fetus. If a fetus is showing signs of Pitocin-induced distress, shutting off the Pitocin often isn't enough; a doctor has to perform an emergency Cesarean to get the baby out.

Doctors increasing Pitocin levels too quickly is so common that a 2008 nursing manual cautions labor and delivery nurses against following "Pit to Distress" orders:

Even if the oxytocin order calls for a low-dose, a high-dose, active management of labor, or "Pit to distress," the nurse needs to follow a conservative protocol and decrease or discontinue the oxytocin infusion to protect the fetus (Clark et al., 2007; Freeman & Nageotte, 2007). The best plan is to discontinue the infusion to allow the uterus time to rest, and to allow the placenta and fetus time to receive oxygen. In addition, "Pit to distress" is not an acceptable order. If a provider writes "Pit to distress," notify your charge nurse or supervisor . . . At all times, you must practice to prevent harm.

A HOSPITAL IS A BUSINESS

Why use continuous fetal monitoring when we know it does not improve birth outcomes? Why immobilize women in bed? Why expect or even bully women into accepting pain medication? Why discount or ignore the requests of laboring women? Though our cultural bias is to view hospitals as altruistic institutions with the primary objective to help sick people become well, the reality is quite different. More than a thousand hospitals in the United States, about 18 percent, are investor-owned for-profit institutions. These hospitals are businesses with one bottom line: to make money. While the remaining hospitals in America are not-for-profit institutions, some of which are state- and local-government-owned, these hospitals too must generate income. They bill private and public insurance for their services and often have their revenue managed by for-profit contractors. Only 213 hospitals in the United States are federally run institutions (like military, veterans, and prison hospitals) where patients are not charged and insurance is not billed for patient care. These hospitals' operating costs are paid for almost exclusively by tax dollars.

Hospitals in America need to make money. Laboring women are rushed through childbirth because delivering quickly is in the best economic business interests of the hospital. "The rushed atmosphere in labor and delivery is often caused by staff and physicians who schedule or maneuver birthing times for convenience, or to assure that beds are available for the next mother who comes in for labor," says Marsha Walker, a Boston-area-based registered nurse, who has more than thirty years experience as a women's health advocate. "Then there's the fact that the flow of patient admissions and discharge is affected by staffing levels—hospitals save money when they are minimally staffed. So the trend in many hospitals is to reduce or eliminate expert positions, like highly trained lactation consultants, and replace them with lesser-credentialed staff who aren't paid as much. While this may save hospitals money, it makes it very hard for breastfeeding mothers and infants to receive the level of care that they need. Unfortunately, I've seen time and again how a hospital's bottom line can take precedence over the health and safety of patients."

The need to maximize profits and minimize costs helps explain why women having hospital births are often so rushed: It is not in the hospital's financial interest to have the mom take her time.

"In and out is how reimbursement is maximized," explains Richard Anderson, who worked in hospital finances in California as a health care financial consultant as well as an auditor for thirty-seven years. According to Richard, hospitals maximize their profits when women either deliver vaginally as quickly as possible or have surgical births. Here's why: Many insurance companies in California pay the hospital an all-inclusive fixed rate for uncomplicated vaginal deliveries, so hospitals are only reimbursed for a maximum of one day and are not reimbursed for expenses incurred on the day of discharge. This means that if a woman has a long vaginal labor and needs to stay for two days, the hospital potentially loses money, since insurance will only reimburse for one. "Hospitals have an incentive to get 'normal' deliveries out of the facility fast because what they're going to get paid is capped," Richard explains. "They also have an incentive to perform C-sections and get paid an additional two days of reimbursement. The Finance Department always monitored this to make sure the mothers were discharged after one day or three days as appropriate." A database compiled from California's birthing records from their 253 hospitals corroborates Richard's observations: while low-risk women had a 9 percent chance of having a C-section at the nonprofit Kaiser Permanente Redwood City Medical Center, they had a *47 percent* chance of having a C-section at the for-profit Los Angeles Community Hospital. Overall, women delivering at for-profit hospitals were 17 percent more likely to have surgical births.

Another way hospitals make money is to charge the insurance companies or the private individuals as much as they can for each intervention. Though Anna Wilde Mathews was relieved to get an epidural after twenty hours of labor, she was shocked at the price tag. The three-day hospital stay for herself and her newborn, despite having no complications, totaled $36,625. On her itemized bill from Cedars-Sinai Medical Center in Los Angeles, there were fourteen items for the baby and thirty-four items for her, not including doctors' fees. The hospital charged $530.29 just for the tray of sterile equipment used to give her an epidural. The total cost of the epidural was $4,212.84. The hospital charged $2,382.92 for the ninety minutes Anna spent in the birthing room after delivery, *$26.48 a minute*, though nothing further was done to Anna besides a nurse checking her vital signs.

CRATE RATE

At the same time that most hospitals have a financial incentive to do as many interventions as possible and deliver women as quickly as possible, today's obstetricians often have financial incentives to deliver as many women as they can as quickly as they can. Sometimes this is part of the infrastructure of a doctor's contract—obstetricians at some practices get end-of-the-year bonuses on top of their base salary for the more women they deliver. Although each insurance company reimburses differently (and may even pay a different reimbursement rate for the same procedures performed at the practice across the street), insurers pay what some doctors call a "crate rate"—one global fee to the doctor for prenatal care and delivery. The problem: As the cost of malpractice insurance has been steadily climbing, this global fee to doctors to care for pregnant women has been steadily going down.

This situation incenses Dr. Edward Linn. When Linn left private practice to become an administrator in 1987, he paid $55,000 a year for malpractice insurance and was paid, on average, $2,800 by the insurance companies to give prenatal care and help each mother deliver her baby. Today the average cost of malpractice insurance is almost three times higher but obstetricians, especially those who care for women on public assistance, are being paid about half as much as they were. "The average premium for malpractice insurance is closer to $150,000," Linn explains. "But as that premium has tripled, the rate of reimbursement has plummeted. Now, because of HMOs and the government lowering Medicare fees, doctors are getting more like $1,500 or $1,600 per woman. It doesn't matter if the delivery takes two hours or twenty-four hours. If you're a doctor in private practice and the majority of your income is coming from obstetrics, you have to deliver as many women as you can." Another young obstetrician tells me her colleague was under investigation by the practice he joined because a chart review revealed his C-section rate was too low.

HMOs. Insurance premiums. Medical malpractice. The abbreviations abound, the systems are complex, and the numbers head-spinning. But Linn, and many other obstetricians I talked to, tell me the same story: Their ability to care for laboring women with kindness, patience, and compassion is compromised by their need to make a living. Their desire to allow women vaginal birth after Cesarean—or even to be patient and not intervene in the event of a long labor—is mitigated or negated by their fear of litigation (doctors believe they are more likely to be sued for *not* intervening than for intervening unnecessarily, which is another motivation to intervene more). They tell me their beliefs about what they think is right, safe, and healthy are overridden by the dictates of the hospitals they work for and the malpractice providers who insure them.

"Expediency, economics, and the fear of litigation are the three things that control pretty much every decision that's made in medicine today," says Dr. Fischbein, "especially obstetrics."

OVERMEDICALIZED BIRTH

Seventy-year-old Ina May Gaskin has long white hair, wire-rimmed glasses, and an understanding smile. She and her team of direct-entry midwives have been delivering babies for forty years. Direct-entry midwives are midwives who have no formal medical training but who learn to assist birthing women through apprenticeships, a midwifery school, self-study, or a higher education program that is not part of nursing school. Direct-entry midwives like Ina May approach childbirth as a natural process for women, learning the best practices for vaginal birth usually by observing and participating in out-of-hospital births rather than by studying current obstetric practices. Popular among birth advocates and midwives, Ina May has also gained the respect of the more mainstream obstetrical community. A former director at the World Health Organization called Ina May "the most important person in maternity care in North America," and in 2011 she won Sweden's Right Livelihood Award for teaching and advocating safe childbirth.

Ina May is the most famous midwife in America today for two reasons: She is outspoken in her support for laboring women and in her critique of America's birth system, and she has enjoyed excellent outcomes in her own practice. By anyone's standards, the statistics at the Farm Midwifery Center in southern Tennessee are impressive. The vast majority (95 percent) of women who live at the Farm—an intentional community that was started in the early 1970s and has undergone many iterations since—or who come to the Farm for prenatal care and childbirth have safely delivered their babies at the Farm. Of these 2,844 women (including ninety-nine breech births and seventeen sets of twins) fewer than fifty have needed Cesarean sections, giving the Farm a C-section rate of 1.7 percent, nearly *twenty times* lower than the national average of 32.8 percent, despite the fact that Ina May actually takes on what other practitioners would consider high-risk cases. The Farm's infant mortality rate (4 per 1,000) is half the national average, despite the fact that their statistics reflect the outcome of every mother who receives prenatal care at the Farm, even those who "risk out" and deliver in the hospital. Their maternal mortality rate is zero. Ina May and the other midwives at the Farm have never lost a mother during labor or from post-labor complications.

Ina May argues that the reason birth is so much more dangerous in the United States than in other industrialized countries is because of unnecessary, often harmful, medical intervention. Ask most obstetricians why women are dying in childbirth in America and they are likely to blame the woman herself—for being obese, developing gestational diabetes, or not seeking prenatal care. But Ina May argues that the overmedicalization of low-risk birth in America—not a defect on the part of a woman or her body—often causes problems that then can only be rectified by more intervention.

"Those who are used to the birth ways of other mammals know that it is easy to cause complications during labor by disturbing the mother," Ina May explained in an interview published in the *Sun*. "If we put horses, goats, and cows through the restrictions and indignities that most laboring women in U.S. hospitals are routinely subjected to, the animals would surely have as many complications as we do. The astonishing thing to me is that we have come to believe that our human bodies are not as well designed for birth as other mammals' are."

DILATING ON A TIMELINE

Because her pregnancy with twins was considered high-risk, forty-one-year-old Sara Schley of Wendell, Massachusetts, was informed her delivery would have to take place in the operating room in case of emergency. She would be allowed to labor in a normal room, but once she was taken to the OR, policy dictated she could only have her spouse with her. Knowing she would be safer, calmer, and more confident with her women friends present, Sara fought back. She and her husband finally got the head obstetrician to agree in writing to allow two women to stay with her, though this went against hospital protocol.

Early that evening, after Sara had been laboring for nine hours, the head obstetrician arrived to examine her. Shadowed by a nervous young male resident, Dr. Lucy Bayer told Sara she was a hundred percent effaced and one centimeter dilated. "These babies will be here by three a.m.," she announced. Trying not to be discouraged that she was only one centimeter dilated after nine hours of what felt like pretty intense labor, Sara kept working, moaning through contractions, listening to the classical music provided by a friend, and hanging on to her husband. At 10 p.m., Bayer returned and did another vaginal exam. With the manner of a woman with no time to waste, she declared Sara still at one centimeter and ordered the nurse to start Pitocin to speed the contractions. Sara and her husband, Joe, had been adamant they did not want any labor augmentation, but Sara was too exhausted and discouraged to protest.

After four more hours of what felt like agony to Sara, Bayer reappeared. She did a manual pelvic exam and found that Sara was two centimeters open. Sixteen hours of labor, four hours of Pitocin, and she was now told she was "failing to progress."

"You need to prepare yourself now that this birth is not going to go the way you wanted it to go. I know you are someone who likes to get things done, but this is not something you can will to go your way." Bayer spoke firmly, looking down at Sara strapped to the bed. "You're forty-one, you've been on bed rest for seven weeks, you have twins. Twins distend the uterus so it doesn't contract effectively. Your babies are now at risk. I know you want a natural birth, but death is also natural at birth. This is my one-thousandth labor and your first. I know it's disappointing, but everything in my experience tells me you are going for surgery. It's time for you to accept that and get ready. You've got two hours."

"I could barely stay inside my skin," Sara later wrote about her experience. "I felt like a caged animal, in agony on the hospital bed, hooked to two fetal monitors, an IV Pitocin drip, a head monitor through my vagina on my son's scalp. Tethered to a three-foot radius. I wasn't allowed to drink or eat, I was barely allowed to walk."

Sara had vocalized through every contraction to give sound to the pain. But when Bayer left the room, she let go a wail that came from a different place—a place of longing and heartbreak. She sobbed and shook so hard she vomited. "I closed my eyes, went inside, and asked my guides in that moment, is this what you really want? Surrender again here? And the answer came back clearly, 'No. Go for your dream.' It was time to pray."

So her husband and women friends held hands and knelt by her bed, shifting their focus away from the physical immediacy of the contractions to pray for a miracle. "Let the gates be open, let the miracle happen," they prayed. "Let these babies come through now."

Sara thought she was beyond help. Demoralized by the doctor's attitude, she had all but given up. But the experienced women supporting her, and her husband, retained their faith. Their confidence buoyed her. And so did their actions: Joe, who had been at the births of his three older children, helped Sara get off her back and out of bed so she could labor vertically. She decided to give it one last try, reminding herself that she was the daughter in an unbroken chain of women who had given birth to healthy babies, and that she could do it too.

Ninety minutes later Bayer returned—wearing blue scrubs and a facemask in anticipation of surgery. "Unbelievable!" she said when she checked Sara's cervix one last time. "I never would have predicted this. You're almost ready to push these babies out! We're back on track. Whatever you're doing keep it up." Then she left the room. Soon after, Sara was wheeled into the operating room where her twins, Sam and Maya, were born vaginally.

Through sheer willpower and with the unwavering support of her loved ones, Sara managed to overcome hospital protocol and birth her twins her way. But most American women, especially first-time moms, don't get that chance. On today's labor and delivery floors women are expected to dilate in the amount of time dictated by the hospital. At the same time their movements are often restricted by fetal monitoring equipment and IVs, and they are usually forbidden from eating. Many American hospitals will not allow women water during labor, only ice chips. (In contrast, in the Norwegian hospital I visited, laboring women are encouraged to buy food at the cafeteria downstairs. And snacks— including fresh fruit, fish spread, caviar, and crackers—are available to them in the common room on the labor ward.) The prohibition against food and drink during labor, in defiance of common sense (could you effectively climb a mountain while fasting?), began in the 1940s after one doctor hypothesized that a laboring woman who needed surgery and was put under general anesthesia might breathe in food particles from her stomach, which could cause pneumonia. Labor is hard physical work and laboring women are often hungry. Forbidding a laboring woman who is hungry from eating and drinking can cause dehydration, low blood sugar, and extreme stress. A 2010 review of the existing scientific literature concluded that there is no medical justification for the restriction of food and drink for women in labor.

In order to prevent dehydration caused by their outdated protocols, some hospitals hook laboring women up to IV fluids as a matter of course. Others dictate that every woman get a tube inserted into a vein in her hand (called a Hep lock) when she is first admitted "just in case" she will need fluids, pain medication, or emergency surgery later during the childbirth, even when she specifies she does not want a medicated birth. When women try to refuse, they find themselves at odds with their medical attendants, told they are jeopardizing the safety of their baby, and coerced into accepting unwanted and unnecessary procedures.

YOUR DOCTOR BELIEVES BIRTH IS AN ILLNESS

Stuart Fischbein, M.D., thinks the problem begins in American medical schools, where students are taught to fear birth.

"You're taught the model that birth is a disease or an illness. Your mentors in residency are maternal-fetal medicine specialists. Many have never done a normal delivery in their entire medical career," Fischbein explains. He's sitting on his living room couch in a South Park T-shirt, with five o'clock shadow and tousled hair. His fifteen-year-old daughter wanders in and out, wondering when her dad is going to get off Skype and make some dinner. "They look at every patient as a potential problem."

"Birth *is* an inherently dangerous process," insists Kurt Wiese, an obstetrician in private practice in Valparaiso, Indiana, a state where certified professional midwives may not legally attend homebirths. "And things happen in a heartbeat. Postpartum hemorrhage. Shoulder dystocia. The kid can be dead if you don't get him out in five minutes."

"Childbirth is not safe," agrees Mary Elizabeth Soper, sixty-two, a recently retired Indianapolis-based obstetrician who practiced obstetrics for thirty-five years. "If you really honestly want to be one hundred percent safe, you should have your pregnancy go to thirty-nine weeks and have a scheduled C-section. That's the best we can do. And even with that you haven't got any guarantees."

But Dr. Fischbein argues that obstetricians blame the birth process for being unsafe instead of scrutinizing their own dangerous practices. "No one's talking about how unsafe hospital birth has become because they are too busy trying to protect their livelihood," Fischbein says. "Normal pregnancy and labor is not a medical problem, so why are highly trained medical specialists dealing with normal birth? That's a big part of the reason it turns sour." Admittedly arrogant when he first started practicing, Fischbein learned over time that there was a gentler way to deliver babies. When no other doctor was willing, Fischbein agreed to be the backup physician for a group of midwives in Culver City, California. He would see the midwives' patients once during pregnancy, and he was surprised to discover that his stereotypes about homebirth moms weren't true: Most of the women choosing homebirth were not crazy or misinformed or idiotic, he tells me, but thoughtful and well educated. Fischbein was also impressed by the relationship the homebirth midwives developed with their clients and the way they took their time instead of rushing through appointments.

"If you look at other mammals in labor, they go off to a field, under the stairs, in the barn. They pace. They walk around. They eat if they're hungry. They're not bothered by other members of the herd. No one walks up to them and asks them to sign consent forms." Fischbein talks quickly, leaning forward to emphasize his point. "In fact, if they are disturbed by a predator, they come out of their primitive brain and go into their cognitive brain. They get up and run away. Their contractions stop for a time because they know giving birth at that moment just isn't safe. Better to stop now and try again later.

"But in humans, from the moment you get in your car and leave your home for the hospital, everything that's done is the antithesis to what nature has intended. In the hospital you're treated like a sick person, put in a wheelchair, changed into a hospital gown, stuck in a room where you're constantly interrupted . . . All the things that make birth normal in mammals we do the opposite of in humans." Fischbein tells me about Maria, who had one C-section in Mexico followed by three uncomplicated vaginal births and is pregnant again. Because she had one C-section more than ten years ago, Maria cannot find a hospital in Los Angeles that will allow her to birth her baby vaginally. "If she's forced to have a C-section and there are complications, whose fault is that?"

Before going into labor with twins, thirty-three-year-old Laura Swaminathan was told by her obstetrician that policy mandated that she give birth in the operating room and that she have an epidural. Laura informed the doctor that she did not want an epidural and didn't think she would need one—her first son was born at home and Laura felt confident about her ability to manage pain. When Laura and her husband arrived at the hospital they purposefully parked far away and took their time walking across the parking lot. In the triage room, the nurse asked her to rate the pain of her contractions on a scale from one to ten. Since Laura kept giving low numbers (she didn't find her first or second labors particularly painful), the nurse ignored the objective signs (timing, duration, and intervals between contractions) that Laura was dilating quickly. She and her husband stayed in triage for so long that Laura started pushing just after she waddled her way to an assigned labor room. The staff's inattention helped her avoid the operating room entirely. Laura's twins were born ninety minutes apart vaginally without anesthesia but with a team of medical professionals watching: a pediatrician for each twin, the attending obstetrician, the resident, and several labor and delivery nurses. Laura sends me a photo taken during labor. She's smiling and calm, and looks like she's having the time of her life.

The next day Laura found herself showered with gratitude. "The resident came back to see me and said, 'We hardly ever get an experience like that, it just doesn't happen at the hospital,' " Laura tells me. " 'That might be the only time in my experience that I get to see a natural delivery, so thank you.'"

THE ICELANDIC ALTERNATIVE

On a flight from Boston to Reykjavik, without any prompting a stewardess asks a solo traveler to change seats so a mom and her toddler can have the row to themselves. "You're a tired little angel, aren't you?" another stewardess coos to a five-month-old standing on his mother's lap. It's September 2011, and I'm flying to Iceland and then to Norway, to investigate how mothers and babies fare in childbirth overseas.

Norway has one of the lowest maternal and infant mortality rates in the world. Women in the United States are *more than three times* as likely to die in childbirth than women in Norway. At the same time, more than 70 percent of the births in Norway, as in all of Scandinavia, are attended by medically trained midwives. Anne Flem Jacobsen, M.D., the head of the obstetric section at Ullevål University Hospital, the largest hospital in Scandinavia, tells me one big difference between America and Norway is that there are *ten times more* midwives attending births at her hospital than obstetricians. Jacobsen and two of her colleagues explain that Norwegian obstetricians pride themselves on being highly skilled. They are trained to do vaginal breech births, vacuum extractions, forceps deliveries, and, of course, C-sections. But they do not see low-risk women in labor unless there are complications. In the Netherlands, where the infant and maternal mortality rates are higher than in Norway, but still lower than in the United States, one third of births take place at home. Indeed, in industrialized countries that enjoy better birth outcomes than the United States—including Japan, Denmark, Sweden, Finland, England, Ireland, and Switzerland—the majority of birthing women are attended by highly trained midwives, who work in collaboration with doctors and often outnumber them (in Europe, 75 percent of births are principally attended by midwives).

Even though I've read about it, it is still surprising that birth on a cold volcanic island in the middle of the Atlantic Ocean is safer than birth in the United States. A woman in America is *almost five times* more likely to die in childbirth than a woman in Iceland. Some argue that the good outcomes are because Iceland has a relatively homogeneous population and women are healthier. But obesity in this country is on the rise and immigrant women from diverse backgrounds enjoy outcomes that are just as good as native Icelanders. While in the United States doctors have financial incentives to deliver more women faster, in Iceland a doctor's salary is set by the state and in no way tied to how many women he delivers. Lawsuits, though they do happen, are not nearly as common as in the United States, since the universal health insurance system ensures that any baby who suffers birth-related injuries—for whatever reason—will get state-of-the-art medical care for the rest of his life. There is also a transparency around maternal and infant demise in Scandinavia that is lacking in the United States. While in the United States we have no centralized federal system of tracking maternal deaths, every poor outcome in Iceland is a matter of public record.

Iceland's C-section rate in 2010 was 14.6 percent, while in the United States it was 32.8 percent. Their maternal and infant mortality rates are also a fraction of ours. New parents get a total of nine months paid leave (three months for the mother, three for the father, and three for the parents to share as they wish).

In Iceland I interview two homebirth midwives, several doctors, including the equivalent of the American surgeon general (Geir Gunnlaugsson, a pediatrician), more than half a dozen hospital midwives at two different birth centers, and five new moms. The Icelanders I talk to say theirs is a baby-friendly culture, and my experience on the plane and traveling with an infant of my own seems to prove that generalization true. "GO AHEAD & BREASTFEED," reads a colorful sign in English and Icelandic outside a café on Austurstræti in downtown Reykjavik. "WE LIKE BOTH BABIES AND BOOBS!"

At the largest hospital in the country, Landspítali, where 70 percent of the births take place, I meet with Helga Sigurðardóttir, the head midwife of the post- and prepartum units. Helga has straight blond hair, dark gray eyes, and a serious affect. She tells me that in the last few years there have been some major improvements at the hospital. "We used to put every woman on a monitor when she came in for a birth. We took readings upon admission and then intermittently throughout her labor," Helga says. "But then research showed that in a normal delivery, this is not particularly beneficial. So we stopped."

It took several months of meetings and many long conversations for the staff to agree, but in May 2009, routine fetal monitoring for women in labor was discontinued. At first the older midwives and the doctors hesitated, fearing the change would make birth less safe despite the scientific evidence to the contrary. But since the hospital keeps meticulous records of every birth, they could easily see if the change had negative repercussions. It did not. Since doing away with routine electronic fetal monitoring, the C-section rate has gone down from 16.5 to 14.6. The infant mortality rate has remained virtually the same.

"I recently watched the film *The Business of Being Born,* and I was very surprised at the amount of Pitocin they used," Helga says. "It seemed like no women could give birth unless they got Pitocin. Here it is a last resort under normal circumstances. You try and get the woman moving and try to figure out what is slowing the contractions. We consider it high risk to give women Pitocin. Midwives and doctors first try to figure out why the woman is not progressing normally. We bear in mind that if you give her the Pitocin, it could overstretch the uterus, and it could burst, so it might not be the best solution."

I ask Helga what happens when a low-risk woman birthing in Iceland "fails to progress." She takes me upstairs to the labor ward and pulls out a light purple *rebozo,* a Mexican cloth the hospital midwives in the low-risk labor unit (called "The Nest") are using more and more during labor if they suspect the baby's not in an optimal position. A midwife puts the *rebozo* around the hips and buttocks of the laboring woman and shakes it in between contractions.

They've been having a lot of success with this technique, Helga says. She wraps the cloth around her own hips and wiggles it, a broad smile lighting up her face. They haven't been using the *rebozo* long enough to have anything more than anecdotal evidence. Maybe the shaking really does help, or maybe the sheer absurdity of having a cloth wrapped around your bum takes a woman's mind off her contractions. In Iceland the majority of midwives, who first study to be nurses and then spend two years training to become midwives, are also trained in acupuncture, which has been shown to be an effective pain management technique during labor. Laboring women are also encouraged to use the birthing tubs, take showers, move around as much as they want, and eat if they are hungry.

Dr. Hildur Harðardóttir, M.D., Landspítali's head of obstetrics, believes the reasons for the low maternal and infant mortality rates in Iceland, the high rates of vaginal birth, and the low Cesarean rates are because doctors, like her, work in close collaboration with midwives, value the patient-client relationship, have clear protocols based on scientific research, and champion vaginal birth. Icelandic doctors train for several years outside of Iceland (going to Sweden, Norway, Great Britain, Denmark, and the United States) and bring back techniques from various countries. Hildur tells me that the hospital's doctors and midwives regularly attend emergency medicine trainings together where they practice solving obstetric crises that might arise, such as postpartum hemorrhaging and shoulder dystocia. Doctors and midwives have mutual respect for each other, Hildur says, and these regular trainings keep the doctors and the midwives working as a team.

"We all promote normal vaginal delivery," insists Hildur, fifty-four, a practical woman with light blue eyes who has more than twenty-five years of experience delivering babies, including nine years training and working in the United States. Hildur, however, does not champion unmedicated birth ("Would you go to the dentist and have a tooth extracted without anesthesia?" she asks) or homebirth ("We think it's dangerous"), but she says every obstetrician in Iceland believes vaginal birth is the safest option. She points out that there are more than twice as many midwives delivering babies as obstetricians in Iceland, and that medical students start by shadowing midwives to observe normal births. "We believe every woman should have one-on-one care with a midwife," she also tells me. Later Hildur emails me a photograph of her new grandson, Gudmundur, her twenty-nine-year-old daughter's second child, born on November 13 after three hours of labor in the hospital with an epidural. A midwife delivered the baby.

It is this midwifery model of care, says Dagný Zoega, a midwife in Selfoss with fourteen years of experience caring for pregnant women, that is why Iceland enjoys such good outcomes. "Doctors-in-training train with midwives. They watch three midwife-delivered births and then do three deliveries themselves," Dagný explains as we sit in her red and pink living room surrounded by books and photographs of her five children and two grandchildren. "They get another perspective on birth because they see normal birth and ways to tackle the birth that enhances the experience."

No system is perfect: Dagný's daughter Helga says she felt bossed around during her first labor, perhaps because she was an eighteen-year-old unwed mother, so she chose to have her second baby at home. Another mom shares that after her birth ended disappointingly in a C-section, a hospital midwife quipped within earshot, "That's what happens when middle-aged women decide to have first babies." She was thirty-five. Emma Swift, an Icelandic-born midwife and mother of four, tells me she was given Pitocin at the hospital with her first baby even though the entire labor lasted only four and a half hours. Later she was told there was no medical indication in her chart for Pitocin, but it was administered shortly before the shift change. She too birthed her second baby at home.

But when a woman in Iceland has a hospital birth she does not feel good about, she has recourse. Six years ago, in an effort to make birth more woman-centered, the hospital began a program called *Ljáðu mér eyra,* "Lend an Ear." Any mom who needs to process her birth experience or better understand what happened can meet with a hospital midwife and a psychologist, as many times as she wishes. She can also meet face-to-face with the midwife or obstetrician who delivered her baby. Lend an Ear is available to pregnant women and women considering pregnancy if they want to talk about their fears before giving birth. Guðrún Eggertsdóttir, who is the head midwife of the labor ward at Landspítali, says the midwives and the doctors take the feedback very seriously. Instead of getting defensive, they try to improve their practices. "We are human. We make mistakes," Guðrún says. "We are always trying to do better."

Average charge for C-section birth in United States: $51,000

Average charge for vaginal birth in a hospital in the United States: $32,000

Average cost of homebirth in southern California: $4,500

Cost of an unassisted homebirth: $15.90

Average time to deliver a baby via C-section: 45–60 minutes

Average time to deliver a baby vaginally: 15–20 hours

Cost to a woman in Iceland for prenatal care and vaginal birth: $0

Cost to a woman in Iceland for a C-section birth: $0

Lauren Shaddox: The Birth I Wanted

Lauren Shaddox, who has worked as an advocate for immigrant victims of domestic violence, is studying to become a nurse practitioner. Lauren, twenty-seven, never thought about where she would give birth. When she got pregnant, she and her boyfriend lived in Ridgway, Colorado, about a forty-minute drive from the nearest hospital, so she started researching her options. Her daughter, Penelope, was born at home on July 9, 2009, in a 115-year-old farmhouse.

Early in my pregnancy I went to see an obstetrician and I didn't like the way I was treated. They acted like anything could go wrong at any moment; they gave me a "freebie" bag from a formula company; and I felt like they considered my pregnancy a problem waiting to happen. I started thinking about how women have been giving birth naturally without any intervention for eons. Then a friend introduced me to a homebirth midwife. I met with her—not because I was going to have a homebirth, but because I wanted to find out more about what she did. It just made so much sense. I don't really like hospitals. The more I found out about how births happen in the hospital, the more I felt like having a homebirth was a better option.

When I first told my mom, she was really upset. She said I was born with the cord wrapped around my neck and that they had to give me oxygen in the hospital or I would have died. But I've read that babies are often born with the cord wrapped around their necks, and that it usually isn't a medical emergency. Unless the cord's very tight, or wrapped several times, all you have to do is lift it off. I made my mom watch some documentaries about birth. We saw Ricki Lake's *The Business of Being Born,* and Debra Pascali-Bonaro's *Orgasmic Birth.* My mom didn't realize that the midwife carries oxygen and other medical supplies in case there's an emergency. Once she learned more about it, my mom really got behind it. My friends were all really supportive. In our area it's not that uncommon to have a homebirth—generally everyone knows someone who has had one.

Alejandro, Penelope's father, is Colombian. Doctors are very respected in Colombia, but in Latin America people are less judgmental about alternative choices. So he supported what I thought was best and decided the idea of having the baby at home was great! It's hard to explain because most women give birth in hospitals in Colombia, and usually the women who have homebirths are very poor. I think he just felt like giving birth is the woman's role, so he wasn't going to have a strong opinion about what was my choice.

During labor I became very introverted. I didn't really care about what anyone else was doing or the fact that anyone else was there. I just kind of did what my body told me to do—I spent a lot of time walking and a lot of time laboring on the toilet. I was in labor for twenty-five hours and it was painful, especially at the end, but it's not the pain I remember.

I started having contractions around midnight. It was very dark and quiet in our house. Though the contractions woke me up, I tried to sleep them off at first. I was lying down but whenever I had a contraction I would stand up until it was over.

Our house was very small. There was only one bedroom and I walked back and forth a lot between the bedroom and the bathroom. The bathroom had two big windows so I could look out and see the trees and horse pastures nearby. In the morning I walked around outside the house with Alejandro. We didn't spend too much time outside because it was a hot and muggy July day and the mosquitoes kept biting me.

The midwife came in the morning after that first night of contractions. She said she could do a pelvic exam to see how dilated I was. I was only at two centimeters. I was still thinking it would just happen all of a sudden, but the midwife said she thought I'd be in labor for a long time. She left to run errands and meet with other clients.

Right after she left, the contractions started to get really uncomfortable and I started to wish she was there. She came back that afternoon and stayed with us. My daughter was born at one a.m. The pushing really hurt and I screamed a lot! Penelope was born with her elbow bent and her arm up by her head so her elbow had been pressing on my cervix at the same time as her head, which was probably why I had felt so much pressure. I don't know if I pushed for a long time. I never looked at a clock. I never thought about how long it was taking. I knew that eventually my daughter was going to be born.

Penelope was born in an inflatable birthing tub we set up in the living room. She was the most beautiful thing I'd ever seen. It was a total rush. Alejandro cried when he saw her. Holding the baby, I got out of the tub, waddled fifteen steps to our bedroom, and lay down in our bed with the baby on my chest. The midwife and her assistant had put pads down to protect the bed. I held my daughter for a long time before delivering the placenta or cutting the cord. I had a tear from my daughter's elbow, which the midwife sewed up. Then I tried to eat something—I hadn't been able to eat during labor because I was too nauseous. I tried to nurse her that night but she wasn't interested. We were all so tired! So then we just went to sleep. Penelope slept right between my husband and me.

The midwife and her assistant put a mattress in the utility room and spent the night at our house. They were there in the morning. They checked my blood pressure, examined my daughter to make sure she was okay, and helped me change a big poopy diaper.

Being able to have my daughter without anyone messing with me was the most empowering experience of my life. I had the birth I wanted. I knew that being in the hospital would greatly increase the chance of someone taking the control away from me and my body. Had something gone wrong with Penelope, I don't think I would have attributed it to being at home, since the kinds of things that often go wrong can happen whether at home or at the hospital.

It's common to hear birth spoken of as something unnatural, difficult, excruciating. It was painful, but it was definitely not unnatural or excruciating, and not really difficult in the sense that it was only a day of my life, and the outcome was worth the challenge! I have a long history of not standing up for myself, and of placing my worth in external sources (like having a relationship with a man), so having the birth I wanted was a good reminder of how capable I am and how my worth is contained in myself, just as I am.

When Penelope was two years old, I witnessed a hospital birth. Until then, I would have told a pregnant friend looking for advice to "just do what you're comfortable doing." But after seeing how this laboring woman was treated in the hospital, now my advice is, "Give birth at home." It's safer for you and your baby, and you can do it. Our bodies are made for it. The pain is part of the process and embracing it makes you stronger. Giving birth is an amazing rite of passage for a woman.

I think my mom was really proud, because she still tells total strangers—sometimes at inappropriate times—that Penelope was born at home. Sometimes, when women hear that I gave birth at home, I get the sense they're thinking, "Wow, you must be something else." I'm not anything special. I'm not even very athletic. I just believed that it was possible.

CHAPTER 4

CUTTING COSTS:
The Business of Cesarean Birth

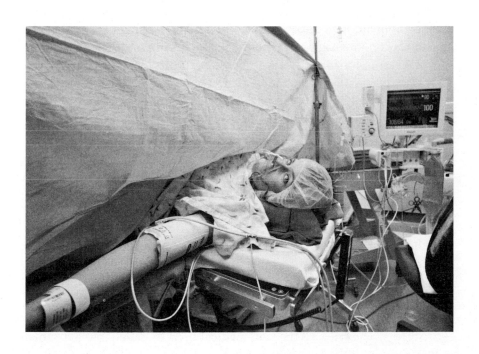

"Cut time," which is what these doctors call the time surgery will begin, is 7:15 a.m.

The young pregnant woman, wearing a light purple hospital gown, sits on the edge of the operating table. She has a tattoo right below her left shoulder: a baby's footprint just above the name of her firstborn. That baby was born via Cesarean section. Since this hospital's policy is not to allow women to attempt vaginal birth if they've had any previous Cesareans, this mom has no choice but to have surgery again.

There are six medical personnel in the room busy laying out metal instruments on trays, filling out paperwork, and gathering supplies. One is the obstetrician who will perform the surgery. The second is the doctor who will assist her. The third, a nurse anesthetist, will monitor the woman's vital signs. A scrub nurse stands at the ready to hand the doctors their equipment. A circulating nurse walks around bringing equipment to the doctors and managing the room. The last person in the room (besides me) is a respiratory therapist who will be responsible for evaluating the baby once he is pulled from the mother's womb.

Head down, the young woman leans her weight into her obstetrician as the nurse anesthetist inserts a small needle filled with numbing medication into the skin overlaying her spinal cord. She flinches and cries out when the needle pricks her back.

"You're doing great, honey," the doctor tells her soothingly. "You're doing great."

It's very bright in the operating room and the mother looks scared and exposed. She is alone. There is no one—not a friend or even a partner—to hold her hand.

Now that he's numbed the area, the nurse anesthetist pokes a much longer needle into the woman's spinal column.

"Ouch, ouch," she cries again. "It stings."

"Where does it hurt?" the nurse anesthetist asks.

"My legs," she cries. As the pain medication is threaded into her spinal column, some of the nerves in her back are irritated. She feels shooting pains down her legs.

They lay her down on the operating table. She's a big woman with a white mound of pregnant belly, making the table seem small. She has a blood pressure cuff on her arm, an oxygen tube running through her nose, and a monitor pinching her index finger that records her oxygen levels and her pulse.

As soon as she's flat on her back, everyone springs into action. Two doctors wrap cuffs around her legs to strap them down to the table. The nurse anesthetist keeps the mother calm, explaining that her legs and her bottom will start to feel numb. Two nurses strap the mother's midsection to the table with a black strap that has metal buckles like an airplane seat belt. The mom gags and chokes and the nurse anesthetist elevates her head for a minute, carefully observing the monitors to make sure her blood pressure isn't dropping too low.

"With spinal medication the patient can't move below the chest," the obstetrician explains to me. Like the doctors, I'm dressed in baggy cornflower blue scrubs with the requisite light blue shower cap hairnet and facemask. I scribble furiously in a black notebook, trying to take notes without looking down so I don't miss anything. "It sometimes makes them really nauseous. If you put it in too high it can paralyze the respiratory muscles. That's a bad outcome." She laughs. "We don't want that."

The obstetrician shaves the pubic hair that has grown up around the mom's previous C-section scar as a nurse inserts a catheter into her urethra attached to a yellow rubber tube, which immediately fills with urine and empties into an attached plastic bag. Iodine is painted around the scar and on her lower abdomen. This all happens in a blur of activity. A blanket is laid over the mom's legs, and then two nurses shake out a big blue disposable sheet that has a clear plastic center to go over her belly. The sheet is attached to two metal IV poles, hanging like a curtain so it breaks the line of vision: Now the mom or anyone near her head can no longer see what the doctors are doing to her body. The circulating nurse leaves the OR to "get Daddy," who comes into the room looking sheepish.

A light is focused on the mom's abdomen so the doctors can see their work. The lead obstetrician uses a small scalpel to make an incision following the line of the previous scar, cutting through layers of fascia— fibrous connective tissue—and red muscle. It is an unpleasant mix of whitish and blood red strands, like a layer cake made of plump mealworms and blood. As fresh blood pours out of the wound the scrub nurse and doctors rush to wipe it away with gauze and suction it through tubes. The two obstetricians work quickly and without hesitation, but they have to tug the flesh apart really hard to open up the young woman's belly. The plastic overlay that covers her torso above where they are working puckers onto her skin. They use retractors to pull the tissue of the sidewalls of the abdomen out of view and a larger retractor (called a "bladder blade") that looks like a cross between an enormous metal spatula and a crowbar to hold the bladder away from the uterus.

"We have some adhesions," the OB murmurs. These adhesions are scar tissue from her previous C-section that has grown together too tightly and attached to other organs. It's eerily quiet in the room. Neither the young mom nor her husband says a word and the nurses are silent, watching the doctors yank at the woman's flesh. The only sound is the beeping coming from the monitors.

Then lead OB turns on a small handheld device that makes a whirring noise. Along with the noise comes the smell of burning meat. She is cauterizing the mom's flesh.

Two nurses wince and step away, turning their heads in disgust. The smell is overwhelming.

"I'm taking out some of these adhesions," the doctor explains gently. She doesn't like to use this cauterizing device, called a Bovie, before the baby's out because there's a chance she could scald the baby, she tells me later, but she has to use it this time because there is so much scar tissue that she can't separate the bladder from the uterus without it.

The doctor and her partner continue to cut through the layers of flesh. They are working more slowly now. A moment later amniotic fluid starts gushing from the 10-centimeter incision that the assisting obstetrician is holding apart with his hands. The fluid gushes down the woman's belly and is caught by a plastic sleeve and then suctioned away through a tube.

The respiratory therapist stands at attention with a receiving blanket. I realize they have finally reached and cut through her uterus. The lead obstetrician puts her hands inside the incision and tugs on the baby's head while the assisting obstetrician pushes on the mother's stomach from the outside, to help move the baby out of the uterus. The baby's head emerges. His skin is gray, his eyes are closed, and there's a mass of black hair plastered to his head.

"We've got a head!" the doctor cries. Both doctors are working hard. They tug more on the baby's head in order to get the shoulders out. His body pops out of the hole in one swift motion. They clamp and cut the cord as the baby gives a piercing wail.

"Here you go. Happy Birthday!" the doctor quips, holding the baby up for the parents to see before handing him off to the respiratory therapist with the blanket.

The OB puts her hand almost up to her elbow into the incision in the woman's abdomen as she manually grabs and dislodges the placenta from the walls of the uterus. Once the placenta is out she takes a cloth and wipes out the inside of the uterus to make sure there are no leftover membranes. Then she pulls the young mom's uterus entirely out of her body and flips it onto the woman's plastic-wrapped stomach.

The uterus is round and purple. At first I think it resembles a trussed chicken, but as I get used to what I'm seeing I realize it is shaped more like a ball. "We exteriorize the uterus in order to better visualize the hole we've made in it," the doctor tells me later.

The doctor must now sew together the hole she has made, using a metal needle that she holds in a clamp in one hand and another metal clamp to catch the thread in the other. The baby, wrapped snugly in a blanket with a beige cap on his head, hasn't stopped crying and is taken away by the circulating nurse. The dad follows behind, leaving his wife alone with just the medical personnel, her guts inside out on the operating table.

The obstetrician closes the uterus in two layers of stitches. "There are different schools of thought on this one," she explains. "I'm not convinced having one layer is a bad idea. It takes more time to do two. But they tell us now we should do two, so I'm doing two."

The mom squirms.

"Everything okay?" she asks the doctor.

"You had a lot of scar tissue."

"That bad?"

"No, it just makes it more difficult for me. That was a challenge. I earned my money on that one."

In a quieter voice she explains to me that there were so many adhesions that the young woman's bladder was actually stuck to her uterus and she shows me the spot where it had adhered. The uterus has shrunk in size, contracting down. Now it is white and purple and looks like a bald head. The doctor also shows me the fallopian tubes and the ovaries. They don't look anything like any of the pictures I've seen. The tube looks like strands of blood-red hair, the ovaries a mass of something nondescript. The doctor handles them matter-of-factly. The young woman starts shivering.

"Irrigation," the doctor orders and a nurse pours water directly into the woman's abdominal cavity. A tube suctions the water out again and the doctor stuffs a cotton cloth inside the woman's body to dry it. She gets out her cauterizer, and the smell of burning flesh permeates the room once more. After several more suctionings, the doctor stuffs the woman's uterus back into her body, pushing it back into place. Then she begins sewing the muscle layers back together. What looked to my inexperienced eyes like a mass of blood and tissue starts to come together as the doctor sews. This part is so routine that she and the scrub nurse (the assisting obstetrician ducked out earlier) chat about upcoming vacations.

"Where are you going?"

"Mexico! We were thinking of renting a condo but we decided to stay in a resort. It's just so much easier, you know?"

Finally they rip open the plastic overlay and the obstetrician sews the skin shut, leaving the scrub nurse to apply sticky Steri-strips that will hold the wound closed.

The baby may as well be a hundred miles away. He's down the hall in the nursery lying on his back in a plastic bassinet, wailing loudly. Then he stops shrieking, as if in defeat, and starts making little peeping noises, a baby bird hoping to be fed. His father looks at him but does not pick him up. If all goes well, his mother will be able to see him in ten or fifteen minutes.

"The nurses like to observe them after a C-section, make sure everything's okay," the obstetrician explains. It will be about half an hour after giving birth before this mom will get to hold her baby, count his tiny fingers and toes, inhale his scent, and nurse him. Her legs will still be numb, she may feel nauseous, and she'll be woozy from all the medication.

When will she be recovered enough from this clinical, frightening, and almost completely disembodied experience, I find myself wondering, to have the image of this baby's foot tattooed next to his brother's?

THE RUNAWAY RISE IN C-SECTION DELIVERIES

The mother I watched that morning is one of more than 1.3 million women in America who gave birth via Cesarean section in 2010. She is part of a trend, one that is moving in the wrong direction.

As recently as the 1960s, the Cesarean rate in the United States was about 5 percent. By the mid-1980s, when the C-section rate in the United States was still less than 20 percent but had been starting to steadily rise (particularly among white women with health insurance), doctors, birth advocates, government watchdogs, and the media started criticizing the "Cesarean epidemic" and insisted it must be stopped. Yet Cesarean birth rates in this country have continued to climb. For twelve years in a row, from 1997 to 2009, the C-section rate in the United States rose. In 2009 it reached an all-time high of 32.9 percent. Some doctors predict that if this trend continues, soon *more than 50 percent* of births in America will be done via Cesarean section. Others are asking their colleagues in professional journals if vaginal birth will become a "relic of the past in bulldogs and women" (the C-section rate for purebred bulldogs is 86 percent). What was once an operation done to save a woman or her child from a life-threatening situation is quickly becoming the method of choice, preferred by many obstetricians, and even requested by pregnant women.

"I don't get what's wrong with it," a high-powered thirty-eight-year-old New York literary agent told me recently. "My friends are totally blasé about it. They all want one. They talk about C-sections like they're the easiest thing in the world."

A woman who undergoes a C-section in an American hospital is led to believe the operation saved her life or saved the baby's life. In a small percentage of cases this is true: When a baby is in severe distress, or when the placenta has grown in the lowest part of the womb and is covering the cervical opening and blocking the baby from descending (a condition called placenta previa), or when a baby is lying transversely (that is, horizontally across the cervix, instead of head down or butt down) and does not turn during labor, a Cesarean may be the only option.

A timely Cesarean section can also protect a fetus from suffering a life-threatening injury during childbirth. If the mother has an active first-time herpes infection, or some other severe vaginal infection, for instance, the baby may be at higher risk from a vaginal birth than a C-section. If a baby is too large to fit through the vagina because of gestational diabetes or because the mother has an unusually small pelvis, a Cesarean may also be necessary.

Most women and their partners don't realize that a Cesarean section is major abdominal surgery that involves risks to both the baby and the mother, even when performed under the best possible conditions. Because a C-section is such a drastic intervention, the World Health Organization recommends that the optimal C-section rate not be higher than 10 to 15 percent. Delivering a baby by C-section increases the likelihood of having both short- and long-term problems after the birth for both the mother and the baby.

Seventy-nine percent of the 121 women who died in childbirth in New York City from 2001 to 2005 died after a Cesarean. A woman is as much as four times more likely to die if she gives birth by C-section than if she gives birth vaginally. A 2003 study analyzing causes of maternal deaths, published in *Obstetrics & Gynecology,* showed that about 36 women in every 100,000 died giving birth by Cesarean, versus approximately nine women giving birth vaginally. Other studies corroborate this finding:

Data compiled from more than two million births in the United Kingdom showed that a woman was nine times more likely to die from an emergency C-section than a vaginal birth, and three times more likely to die from a planned C-section.

As opposed to the idea of them being the epitome of lifesaving, "At worst, C-sections can kill," says Patji Alnaes-Katjavivi, an obstetrician at Oslo University Hospital in Norway who did his medical training in England and who has delivered babies in both developing countries and in Europe. "They won't kill the baby right then, perhaps, or the mother. But they increase the risk of a mother losing a second baby through uterine rupture and other lethal complications. I've seen examples of that many times. C-section isn't obstetrics, it is the surgery that is required when obstetrics has failed."

As Dr. Alnaes-Katjavivi points out, the problems with Cesarean birth increase with each subsequent operation. C-section deliveries have myriad serious side effects, including accidental cuts to internal organs, emergency hysterectomy because of uncontrolled bleeding, complications from anesthesia, chronic pain, and endometriosis. A study of 94,307 women found that women who delivered via C-section had twice the risk of a prolonged hospital stay, four times the risk of hysterectomy, and were five times more likely to need postpartum antibiotic treatment compared with those delivering vaginally. This study also found that "compared with vaginal deliveries, the risk was three to five times higher for maternal death."

In the past ten years researchers have come to realize that the human body is a complex ecosystem that contains literally trillions of microorganisms; much of these are "good" bacteria that inhabit our skin, genitals, mouth, nose, and particularly our intestines. Beneficial bacteria not only help us have a healthy digestive tract, but they also help us have a healthy immune system. Good bacteria crowd out harmful bacteria, synthesize compounds that are not present in our bodies at birth (like vitamin K), aid in digestion, and train the body not to overreact to outside substances that may not pose a danger. There is new and rather disturbing research that shows that infants born via C-section have markedly different bacteria on their noses, mouths, and bottoms than infants born vaginally. In one study, a team of researchers found that while babies born vaginally are colonized by the mother's beneficial bacteria, babies born via C-section are colonized by sometimes lethal hospital bacteria, including staphylococcus, corynebacterium, and propionibacterium. Another study found that the digestive tracts of infants born via C-section were disturbed for up to six months after birth.

Other studies have found that babies delivered via C-section are at greater risk for having breathing problems (because the baby is not being squeezed through the birth canal, which reduces the fluid in a baby's lungs and because babies born via C-section are often premature), difficulty breastfeeding, infection from antibiotic-resistant hospital bacteria, and severe childhood asthma. Researchers at Children's Hospital Boston found that babies born via C-section are twice as likely to be obese by age three. There is also a risk of the baby being nicked or otherwise harmed by the doctor during the delivery, which happens to more than fifteen thousand babies a year. In one recent case report of an elective C-section birth, doctors mistakenly amputated the baby's finger.

Though she had an uncomplicated vaginal delivery with her firstborn, Karen Bridges, thirty-eight, of Union, South Carolina, was told she had to have a C-section at the end of her second pregnancy because the baby was breech and her doctor refused to deliver him vaginally. The operation at the hospital in Spartanburg, South Carolina, seemed routine. Then, on September 19, 2009, two days after Karen brought Bryce home from the hospital, she was in agony. She spiked a fever of 104, the C-section scar turned an angry raised red, and she started vomiting uncontrollably. Her husband called the doctor's office several times. Nurses dismissed his concerns: Swine flu was going around and she probably just had that virus. Finally her husband took her to the emergency room. By the time they got there, the incision site had filled with so much pus it was the size of a softball.

Karen had contracted a virulent antibiotic-resistant infection from the hospital. She spent sixteen days in the hospital without her baby. She needed two more abdominal surgeries to clean out the C-section wound. Each day a team of doctors, including a surgeon, Karen's obstetrician, and an infectious disease specialist would come in, stand by her bedside, read her chart, and look worried. Karen, who continued vomiting even in the hospital, was too weak to respond to their questions. She had a special device inserted into her body (called a Wound V.A.C.) for almost three months after she was discharged. She could not breastfeed and she was in too much pain to hold her baby. She was terrified that the infection necrotizing her skin (to this day she has no feeling in the skin on her abdomen) might spread to and harm her baby.

"I was totally robbed," Karen says. "Everyone will tell you there's nothing to a C-section. That's total bull. It's major surgery. I will be scarred for life."

A complication like Karen's is relatively rare. In one seven-year study of more than five thousand C-section deliveries at the University of Alabama Birmingham Hospital, severe wound infection—called necrotizing fasciitis—occurred in only nine women, two of whom died. Postoperative infections, however, are much more common. An average of 6 percent of women who have C-sections will have a postpartum infection, which causes a high fever, and often does not respond readily to antibiotics. Like Karen's, these infections are often caused by microorganisms found almost exclusively in the hospital.

Women who wanted a vaginal birth but ended up with a Cesarean report emotional scarring as well as physical discomfort. These feelings of failure can be devastating. And like other kinds of grief, the negative feelings, self-doubt, and sadness often last for a long time. Every year on her son's birthday, Denise Schipani feels depressed. Daniel recently turned nine. Though Denise wasn't expecting a natural childbirth—she liked the idea of an epidural—she wanted a vaginal birth. But things were so busy at New York–Presbyterian Hospital that she and her husband were left alone for hours, first in the hallway and then in a triage room.

"I never saw my doctor or any of her partners," Denise remembers. "It was just residents." In all that time no one said an encouraging word to the nervous couple. After finally being admitted, getting an epidural, and dilating to 10 centimeters, Denise was told to push. She wasn't really sure what the nurse meant. She had no feeling from the waist down, she'd had nothing to eat for two days (the nurse would not allow her even to have a sip of her husband's orange juice). She was exhausted, hungry, and uncomfortable. Though the baby showed no sign of distress, after one hour of unproductive pushing, doctors decided Denise needed a C-section.

When she became pregnant for the second time sixteen months later, Denise was determined to do things differently. She had felt so removed from the birth of her first child—lying motionless on her back on an operating table as the doctors chitchatted while slicing through her abdomen and then seeing her son's startled, bloody face for the first time over a blue drape. At her twenty-week appointment, when the doctor announced, "I see we have your C-section scheduled," Denise switched practices. The doctors at the new practice were willing to let her try to deliver vaginally. "I think you're crazy," one young doctor (who delivered both of her own children by Cesarean) chided, "but suit yourself." Just one hour after she arrived at Saint Catherine of Siena Medical Center in Smithtown, New York, the doctor on call insisted there was "no way" the baby could fit under her pelvic bone and wheeled her into the operating room.

Years later, the skin above her C-section incision has no feeling and her voice catches when she talks about it. She feels like she failed at childbirth and was robbed of one of the most important experiences in a woman's life. She felt unable to stand up to the doctors. Twice. When she shares these feelings with other moms and her family, they belittle her grief, insisting she should be happy because she has two healthy sons.

"After you've had a C-section you're not allowed to say anything but the party line: The outcome is a healthy mom and a healthy baby and that's all anyone wants," Denise confides. "Of course I love my baby and I'm glad he's healthy, but that doesn't mean I don't have unresolved feelings about his birth."

CONVENIENCE RULES

When Poppy Street-Heywood, a mother of four (with a fifth on the way), was pregnant for the first time, her doctor told her she needed a C-section if she didn't deliver by the end of the month, even though she was barely eight months along. Poppy was young, healthy, naïve, and eager to meet her baby.

"I wanted to have my baby, do it, and be done," Poppy remembers. "I figured if the doctor said it's okay, then it's definitely okay."

But when Baby Kailleah was sectioned she weighed only 4.8 pounds. Her lungs were so immature she could not breathe on her own. She had to be helicoptered to a bigger hospital, where it took three weeks for her to gain two ounces. Now she has cerebral palsy as a result of oxygen deprivation at birth. It was only years later, after tracking down the obstetrician (who had no memory of her), that Poppy realized the reason he scheduled a C-section when she was only at week thirty-five of her pregnancy: this doctor was only practicing one week a month in Illinois and spending the rest of his time in Arizona. That was the most convenient time for him to deliver her baby.

Though most American doctors aren't consciously trying to inflict Cesareans on their patients, the sad truth is that an untold number of Cesarean sections are performed mostly—if not entirely—for the convenience of the doctors. Normal vaginal childbirth is often a long, drawn-out affair. The average time for labor in a first-time mother is from less than an hour to significantly more than twenty. It's not uncommon (or unsafe if the baby and the mom are both tolerating labor) for a woman to be in labor for *several days*. As long as the baby is tolerating the labor, which can be monitored by checking the baby's heart rate once every fifteen minutes, and the mom is not exhausted, a protracted labor is not an indication for Cesarean birth. There is no way to predict how fast or how slow a woman will give birth or what time of day her child will be born. Gestating human infants are born when they are ready. They flout the convenience of the doctors, labor and delivery nurses, and their own parents. For an eager-eyed young couple the unpredictability of childbirth is a good initiation into parenting and an early lesson in how infants tend to do things on their own schedule. But for an overworked obstetrician who has young children at home to tuck into bed, meetings to attend, or weekend plans, the inability to control when a laboring woman will give birth becomes a source of stress. A planned Cesarean section or induction, on the other hand, is entirely controllable.

Rebecca Zavala's doctor pressured her to induce labor a week before her due date, telling her the baby's head seemed large. The doctor also told Rebecca she was leaving soon on vacation. The doctor didn't mention that induced labor is much more likely to end in Cesarean birth, or that once you have one Cesarean in much of southern California it is virtually impossible to have a vaginal birth, since the majority of hospitals in Ventura County and Los Angeles either have written or unwritten policies (called de facto bans, where no doctor on staff will agree to attend a VBAC) against allowing women to try for a VBAC. After just four hours of labor at the Santa Monica–U.C.L.A. Medical Center, the doctor broke Rebecca's waters and increased labor-stimulating hormones. Then she told Rebecca that the baby was "showing signs of distress." Rebecca agreed to the "emergency" C-section without question. The nurses congratulated her on being a compliant patient.

Laboring women report being rushed into "emergency" C-sections before shift change at the hospital, and researchers have long remarked that the number of surgical births rises sharply before long weekends, holidays, and school vacations. As much as obstetricians deny that C-sections are done for reasons of convenience, government statistics that analyze when babies are born tell the story in black-and-white: In the past twenty years the timing of births in America has shifted dramatically. Today, according to the Centers for Disease Control, the vast majority of American babies are born Monday through Friday, with a much smaller number born on the weekends.

Stuart Fischbein, the doctor I interviewed via Skype, who has been practicing obstetrics in the Los Angeles area for more than thirty years and is one of the rare doctors who will, under the right circumstances, deliver twins and breech babies vaginally, believes that doctor convenience is one of the main reasons for the runaway rise in C-section rates. "If I do a breech I have to sit with a patient for six to ten hours because hospital policy says I have to be there," Fischbein explains. "I can read a book and watch *Sports Center* in the lounge. But I can't be making any money or covering the overhead I need to run my office or spending time with my family. Who's going to do that when they can do a C-section and be out in forty-five minutes? It's more expedient for doctors to do C-sections. It's just so much easier for them."

FINANCIAL MOTIVATIONS

A recent analysis of the rise in Cesarean rates in southern California, Orange County Health Care Agency's *Complications of Pregnancy and Childbirth in Orange County,* examined 144,584 local births from 2006 to 2008. The county's C-section rate had risen to 33 percent in 2008 from 22.5 percent in 1999. While the report cites medical indications like a rise in maternal obesity as a factor in the rise in Cesarean sections, it also found that C-section rates varied considerably based on where a woman lives and where she delivers her baby. The communities where women had the highest numbers of Cesarean operations were places where women also had the lowest rates of prenatal medical conditions that would put them at greater risk, leading the investigators to observe that the lack of medical indicators identified in this study "suggests that other non-medical factors may be responsible for the higher incidence of Cesarean deliveries in these cities."

The more money and insurance coverage you have, the more likely you will undergo a C-section. Another report found that women at for-profit hospitals were 17 percent more likely to have a Cesarean, despite having fewer risk factors, than women at nonprofit hospitals. This dovetails with earlier data showing that women with private insurance are 20 percent more likely to have a Cesarean birth than women without private insurance. The Orange County Health Care Agency's report also found that women who had C-sections stayed in the hospital nearly four days—almost twice as long as women who gave birth vaginally. An average C-section delivery costs $20,228, while a vaginal birth costs $11,114. Healthy middle-class women in their thirties and forties in private hospitals have about a 50 percent C-section rate, whereas healthy women in public hospitals have only a 10 to 15 percent rate. As writer Naomi Wolf explains, "women whose health plans can afford to reimburse the hospital for a C-section are more likely to be told they must have one."

Those numbers are right in line with what Fischbein tells me. "Hospitals make twice as much money for a C-section as opposed to a vaginal birth. Why do they want to lower the C-section rate in their institution?" Fischbein says bluntly. "They don't."

Mark C. Hornbrook, Ph.D., a health economist who works for Kaiser Permanente Center for Health Research, agrees that money is one of several factors driving up Cesarean rates. "Reimbursement rates for C-sections are higher for hospitals than for vaginal births," Hornbrook writes in an email, "usually because a higher intensity of care may be involved."

But one former CEO of a major hospital in New England disagrees that the rise in C-sections in the United States is tied to hospital finances. He says the decision about whether a woman has a Cesarean—at his hospital, anyway—was only between the laboring woman and her doctor.

"It doesn't have anything to do with hospital finances," this former CEO insists when I reach him by phone at his home. "When I talk to OBs about this, putting aside cases that are medical requirements, the ones that are on the border are often cases of the parents expressing a preference for Cesarean and the doctor saying, 'Is it worth the fight? Even if I think the baby could be born naturally? Do I want to take that chance?' It's a kind of defensive medicine that's going on. Doctors are concerned that if they do it naturally and something goes wrong they'll be sued for malpractice."

Obstetricians worry for a reason. While pediatricians and psychiatrists have the lowest incidents of claims filed against them, *obstetricians and surgeons are sued more frequently than any other medical professional.* Before they turn forty, more than 50 percent of obstetricians have been sued. According to Fischbein, most obstetricians have "certified-letter phobia." "What doctor gets a good certified letter?" he laughs cynically.

But being sued isn't a laughing matter for obstetricians. The former hospital CEO I interviewed knows two doctors at another hospital, both young women, who decided to stop practicing obstetrics and only do gynecology because they were so emotionally distraught over being sued.

"To be sued for doing the thing you've chosen to devote your life to—to help patients when you've used your best judgment—is a searing experience, it's a terrible experience. It's awful and the cases last for months and months," he says, "and you have to go up in front of a jury and judge and you're accused of being a terrible doctor. It's considered a stigma among doctors, even if they think they're not guilty, that they take extremely personally. It's traumatic. It's exhausting. You have lawyers who make you feel incompetent."

Because they are sued so often, obstetricians are considered "high risk" by insurance providers. As a result they pay enormously high malpractice insurance premiums. While rates vary widely by provider and state, one Virginia-based obstetrician reports being quoted $84,000 a year for medical malpractice, despite never being sued in the twenty-one years he has practiced. A report to Congress in 2003 about the skyrocketing rates of medical malpractice insurance found that obstetricians in Texas were quoted $92,000 a year, where the same coverage in one county in Florida would cost $201,000 a year.

If a doctor performs a C-section, the perception is that the doctor did everything he could. If a baby suffers an injury in a vaginal birth, doctors believe that they will be found guilty for not anticipating the problem and performing a C-section. "Most of the large malpractice cases result from a poor fetal outcome, that is, an expected 'normal' baby is born with health problems or has a bad outcome for whatever reason," explains Dr. Jeffrey Spencer, a former fellow in maternal fetal medicine at the University of Connecticut's Health Center. "The M.D.s get sued because they didn't do all that was possible for the baby—meaning perform a Cesarean."

But doctors are also being increasingly sued for performing Cesareans, especially when there is a bad outcome for the mother. In July 2011, a suit was filed on behalf of Kelly Casstevens Jarvis, whose bowel was severed during a C-section. A first-time mom, Kelly died from a sepsis infection and kidney failure five days after her son Ethan was born. In November 2009, Jana Pokorny filed suit again Froedtert Hospital in Wauwatosa, Wisconsin, for causing her pain and suffering, disability, emotional distress, and loss of enjoyment of life after administering Pitocin that caused her baby's heartbeat to slow down and resulted in an "emergency" C-section. In 2001, Maria Guerin's family was awarded $7.62 million after Maria died from bleeding following a Cesarean section.

Though doctors still believe they are less likely to be sued for performing surgery than for a bad outcome during a vaginal delivery, as C-section rates continue to rise, these lawsuits may become more common. At the same time, doctors are seeing more unusual complications as a result of America's skyrocketing C-section rate. "[Y]ou are sowing the seed for more complications down the line," admits Arthur Fougner, a Queens, New York–based obstetrician and commissioner of communications and publications for the Medical Society of the State of New York. In 2006 New

York had the tenth-highest C-section rate in the country and a 45 percent higher rate of maternal death than the national average. Though in thirty years of practice Fougner had never seen a life-threatening form of ectopic pregnancy, he has seen three or four in the past few years. One thirty-two-year-old woman needed a hysterectomy after she began hemorrhaging when the fetus attached to her C-section scar.

Felicia Cohen, an obstetrician in private practice in Grants Pass, Oregon, and a mother of three, sympathizes with families who suffer bad pregnancy outcomes. "I feel really bad for them," she says. "I really want them to get the help they need, and hurt for the hardships they have to go through." Still, Cohen is concerned that attorney greed, not medical need, drives too many of these lawsuits. "I have real issues with the medical-legal system. It encourages people to think that if they have an undesirable outcome, someone should pay," she continues. "It's one thing to get help and make sure their baby and their family get the care they need, but lawyers sometimes encourage them to think of it as a bag of gold. People think we should always have perfect babies, and if it doesn't turn out that way, there are plenty of lawyers out there ready to help them sue you. Lawyers have their purpose, and there certainly is legitimate malpractice out there. But it shouldn't be a lottery ticket whenever a pregnancy has a less-than-perfect outcome."

The medical malpractice lawyers that I've interviewed do not sound like money-grubbing mercenaries, as they are often depicted by obstetricians. From what these lawyers tell me, they care about having a medical system that is safe for babies and mothers. While being sued is not proof of wrongdoing, it *is* a way to get information about what happened out of a medical system shrouded in secrecy, with no legal requirement for transparency. A public trial is a way to force doctors to explain their actions and to hold them accountable.

"Doctors don't admit that they make mistakes, and the whole point of litigation is to identify and prove that a doctor indeed made a mistake," explains Dr. Marsden Wagner, former director of Women's and Children's Health at the World Health Organization, in his book *Born in the USA*. "This is humiliating and shatters an obstetrician's inflated sense of security. In an obstetrician's daily world, everyone with whom he comes in contact looks up to him and follows his orders. In a courtroom, an obstetrician may even be looked down on."

But there's a bigger issue at work here as well: Because we don't have affordable health insurance for every American, if a family has a birth-injured baby, they will be spending the rest of their lives struggling to pay for medical care. Often the money obtained from a lawsuit is the only money they have to care for a damaged baby.

In the UK, where the overall number of maternal deaths has been steadily decreasing, a detailed report with an account of every maternal death is released to the public once every three years. In countries where there is more transparency in the medical system and babies with extensive medical needs will be cared for by the state, obstetricians are sued much less frequently. "If something goes wrong, you don't have to worry about medical bills, because the government will take care of them. There's not really a financial motivation to sue," points out Emma Swift, the Iceland-born midwife who birthed two of her children in Iceland and two in America.

That a health care provider might care more—whether unconsciously or openly—about being sued than about what is in the best interests of an individual birthing woman, infuriates some doctors. "A fundamental principle of medical practice is that whatever the doctor does must be done first and foremost for the benefit of the patient," argues Wagner. "If a doctor picks up a scalpel and cuts open a pregnant woman's belly because the doctor is afraid of being sued or afraid of rising insurance costs, that doctor is not practicing medicine. He is practicing fear and greed."

WADDLING AWAY FROM AN UNNECESSARY CESAREAN

In 2003, when Patricia Roe was thirty-three years old and pregnant with her fifth child, her midwife sent her to Evergreen Hospital Medical Center in Kirkland, Washington, for a sonogram. As with her previous pregnancies, Patricia had slightly elevated blood pressure. She was in her fortieth week of pregnancy and her hospital midwife wanted to make sure the baby was okay. When Patricia (Buffy to her friends) closed the passenger's side door, she felt her baby startle and kick. She put her hand on her belly for a moment, enjoying the movement of the tiny being inside her.

But when she lay on her back for the sonogram, the baby didn't move. The sonogram technician furrowed her brow nervously as she moved the goopy probe over Patricia's pregnant belly.

"Something's wrong," the technician said in a tense voice. "The baby's not moving."

"He's just asleep," Patricia, who was used to the baby's patterns of sleeping and wakefulness, told the technician. "I can get him to move. If you'll let me drink a cold glass of water, he'll start kicking for sure."

The technician ignored her. "You need to be seen by an obstetrician right now," she said. "This is a very bad situation. You may need an emergency C-section. Stay right here while I call for a wheelchair."

When the technician left, Patricia and her husband looked at each other. They had five pregnancies' worth of experience, they were sure nothing was wrong, and they had no intention of being admitted to the hospital. With their four other children, the babies had waited until forty-one or forty-two weeks to be born.

"Let's go," Patricia whispered. She grabbed her husband's hand and they left the hospital together through a side door. Her heart in her throat, she heard her name being paged as she waddled down the stairs and out to the car as quickly as she could.

A week later Patricia gave birth vaginally with no complications, at a different hospital, under the care of a different doctor.

Women's health advocates are fond of pointing out that having an obstetrician attend a low-risk pregnancy is like having a pediatric neuro-surgeon babysit your toddler. Patricia chose to birth her next three children at home. When a pregnant woman, giddy with anticipation, goes to the hospital to have a baby, she isn't thinking about medical malpractice premiums, lawsuits, or the convenience of the doctors. A woman in labor wants what all her health providers want as well: a healthy baby and a safe outcome. Too often that is not what she gets.

C-section rate in America: 32.8%

C-section rate in Norway: 16.6%

C-section rate in Iceland: 14.6%

Maternal mortality rate in America: 24 per 100,000

Maternal mortality rate in Norway: 7 per 100,000

Maternal mortality rate in Iceland: 5 per 100,000

Number of midwives to doctors attending births in Scandinavia: 10 to 1

Number of midwives to doctors attending births in America: 1 to 12

Number of European countries (England, France, Germany, Iceland, Holland, Belgium, Denmark, Sweden, Norway, Finland) with higher maternal mortality rates than the U.S.: 0

Number of European countries with higher C-section rates than the U.S.: 0

George Denniston, M.D.: The Case Against Obstetrics

George Denniston majored in biology at Princeton, earned an M.D. from the University of Pennsylvania Medical School in 1959, and a master's degree in public health from Harvard in 1961. Specializing in preventive medicine, Denniston started seven birth control clinics in Seattle, Washington, all of which are still running. He practiced medicine in the Seattle area for more than thirty years. His experience in American medicine has led him to believe that modern obstetrics needs a massive overhaul.

There is little recognition by obstetricians in America that it is the woman who is giving birth, not the doctor. She has been given the skills, evolved over millions of years, to do this job properly, and in all cases until the last two hundred years, without the assistance of a doctor. Today's doctors, despite the best good intentions, have a total lack of awareness of nature. Their behavior is often arrogant, disrespectful, and sometimes even abusive. Obstetricians should not be permitted to practice the way they do in America today.

I recognized early in my career that doctors delivering babies were doing many things wrong. I have followed these issues with dismay for some fifty years. As a coproducer of the film *Birth As We Know It,* I'm confident I know what I am talking about.

We credit doctors with saving lives, but the truth is that doctors in the recent past have also been directly responsible for untold numbers of fetal and maternal deaths. For almost two hundred years—from the mid-seventeenth century to the mid-nineteenth century—doctors transmitted puerperal fever to millions of women before it was discovered that the doctors were to blame. In the 1840s, a Hungarian doctor-in-training in Vienna, Ignaz Semmelweis, discovered that homebirth was safer than doctor-attended birth in maternity wards. He hypothesized doctors were spreading infection because they did not wash their hands between patients. His findings that 98 percent of childbed fever could be prevented by handwashing were largely ignored. Instead, Semmelweis was ridiculed by his colleagues. Even after the cause of the epidemic was proven scientifically, doctors continued to deny it and transmit the disease.

There is also the famous case of the Chamberlen brothers, who invented forceps in the seventeenth century but refused to tell anyone their secret. The family set up elaborate ruses to hide their knowledge, blindfolding laboring women to prevent them from seeing the forceps. Keeping their invention hidden for more than a hundred years allowed them to charge patients higher prices. They used the forceps under a drape for years, while other doctors had to let their patients die.

Today's obstetricians lack both patience and skill. Training programs no longer teach young doctors the skills necessary for vaginal breech births or twin deliveries. Many do not learn how to do forceps or vacuum deliveries either. But even if they did know how to help women deliver vaginally, obstetricians lack patience to wait while nature takes its course. They have other interests—vacations, sports, and family—that supersede the importance of bringing a newborn into the world. The training they do get in medical school takes away their ability to recognize the miraculous. As a result of impatience, new mothers are often deprived of the empowerment that comes with having a baby surrounded by sympathetic assistants. An empowering childbirth is often the most important and exciting time of a young woman's life. But for women interfered with by doctors, their baby's birthday is remembered as a disaster.

Michel Odent, arguably the world's most effective obstetrician, has told me he believes the maximum C-section rate should be 6 percent, which is the rate he achieved at the maternity unit at Pithiviers Hospital in France, where he collaborated closely with midwives. But many American obstetricians have C-section rates higher than 50 percent. The overall rate in America these days is 32 percent and rising.

Cesarean sections were originally used during the Roman Empire to remove a living baby from a dead mother. Now they are used for convenience and money. With a C-section, the infant fails to complete the journey down the birth canal, into the vagina, and out into the world, and is instead yanked out suddenly. Instead of coming out wide-eyed and peaceful, as he often does during a water birth, he comes out screaming, eyes tightly shut against the painfully bright lights in the operating room.

Sitting in bed is probably the least effective position to be in during labor. Yet that is the position that permits a doctor to most easily examine his patient. So for the doctor's convenience (often justified under the guise of "hospital policy") a laboring woman in the United States is forced to remain flat on her back.

After the cervix is dilated and a woman's contractions change from opening the cervix to pushing the baby through the birth canal, a laboring woman is attended by a labor nurse or two, a doctor, and a well-meaning albeit poorly informed spouse, all looking at the electronic screen that indicates when the next contraction is peaking and shouting at the woman to "PUSH!" But telling a woman to push can be counterproductive. The uterus is doing the pushing, contracting regularly every few minutes, and there is little in the world that can stop it—except possibly exhaustion and dehydration. When the laboring mother is told to push she contracts her voluntary muscles, including muscles that the uterus is trying to push the baby through! A woman knows instinctively what to do, given the chance.

Episiotomy became popular in the United States in 1920. I believe this procedure is akin to sexual assault. The doctors who promoted it claimed it would reduce the rate of third- and fourth-degree lacerations. They had no evidence. Now, ninety years later, studies have clearly shown that episiotomy *increases* the risk of third- and fourth-degree lacerations, the very serious complication it was supposed to prevent. It also can cause fecal incontinence. This is another example of a complete failure to have faith in nature. Nature designed the vaginal opening to be capable of letting an infant head through. All that is required is to allow the woman to be in a position that is comfortable for her, and to properly control the rate of speed of the head, and tearing will be rare. This monumental error has unnecessarily harmed millions of women, in many cases making lovemaking unbearably painful or even impossible.

As soon as the infant is out, the doctor cuts the umbilical cord. The cord, if left alone, will pulsate for some time, pumping blood from the placenta into the infant. We can be sure that it will deliver just the right amount that the infant needs. Cutting the cord prematurely deprives the infant of that blood. It means that he comes into the world in a state of anemia.

No one knows what all of the advantages of a vaginal birth are, but one can assume that there are many things that happen during travel down the birth canal that are designed to assist the infant in his new world. These advantages have been built in over millions of years of evolution.

Often it seems as though the obstetrician does something that gets a woman into difficulty, and then he takes over, often violating her express wishes, and appears to be the hero of the day. Sometimes he is but just as often it is his impatience or intervention that is the real problem.

The profession of obstetrics is bound by tight conventions. They all think they have to do things in a certain way so that, even if there is a bad outcome, they are protected from being successfully sued for malpractice because they did things by the current "standard of care," even if that standard of care has been shown to be harmful.

Statistics show that midwife-attended births have better outcomes. But when a midwife has one bad outcome, there is a huge outcry and her entire practice is shut down. In the hospital, it is much easier to hide or even deny mistakes.

Good obstetricians are awed by nature. They are experienced yet humble. They permit the woman to walk about during labor, eat and drink as her body wishes, make as much noise as she needs to, and get into the positions that are most comfortable for her. They never do an episiotomy. They do everything they can to reduce the risk of a Cesarean section. They patiently wait for the baby to decide when he will enter the world. And when he arrives, they welcome him with love.

CHAPTER 5

PERINATAL PRICES:
Profit-Mongering After the Baby Is Born

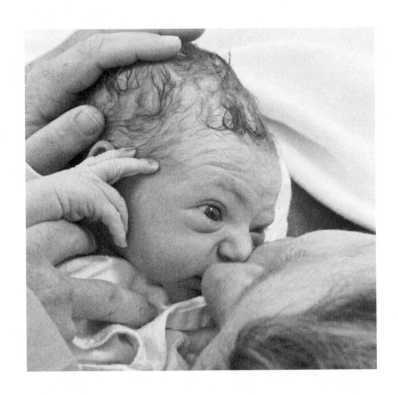

When a woman and her partner imagine the moments after their baby is born, they envision holding the baby in their arms, looking into his eyes, counting his fingers and toes, and spending as much time as they need quietly getting to know the small person who has just come into the world, a tiny creature they have never met before but who has somehow always been with them.

Yet this ideal is far from what actually happens during those precious first moments of life. In most American hospitals, after a baby is born there is a flurry of activity as both the mother and the baby are attended to by myriad medical personnel. The obstetrician or other attendant almost always inserts a rubber syringe into the baby's nose and mouth to suction secretions; the cord is quickly cut, the newborn is "cleaned up," antibiotic ointment is put into the baby's eyes, the baby is weighed and examined and assigned Apgar scores at one and five minutes after birth. A shot of vitamin K is injected into a baby's thigh. Eyes obscured with ointment, thigh pricked, nose and mouth irritated, the baby is then swaddled tightly in a hospital blanket and finally handed to the mother.

That was Cyndi Sellers's experience at Rogue Valley Medical Center four years ago. "It was hard because you just want your baby with you, but I didn't know any better. I didn't think I had a choice. I thought, that's just how they do it," Cyndi tells me over the phone. With her second child, Cyndi gave clear instructions to the nurses that she wanted to be the first person to hold him, so instead of taking him away first, the nurses handed the naked baby to his mom. Graydon's tiny face was swollen, his skin was purplish-pink, and Cyndi thought he was the cutest baby in the world. She breastfed him right away and marveled at the brown fuzz on his head. "It's just the best feeling. I don't even know what words describe it. You're so euphoric. There's not a better moment of seeing this baby that you've carried inside for so long."

Cyndi was allowed to hold her son for ten minutes until he was lifted out of her arms.

Separating a newborn from his mother at birth or shortly after birth is standard practice in most American hospitals. Even hospitals that allow moms and babies to have skin-to-skin contact in the first few minutes of life insist on taking the baby away within an hour of birth. Yet more than half a dozen scientific studies show that the outcome for infant health and for mother-baby bonding is better when a mother and a baby are not interrupted in this way. Dozens more nonhuman mammal studies show that early uninterrupted contact actually stimulates oxytocin and other hormones, which enhance the feeling that can only be described as being in love. Early uninterrupted skin-to-skin contact has been shown to reduce crying, improve bonding, keep the baby warm, and facilitate breastfeeding.

One of the most recent studies, published in *Biological Psychiatry* in November 2011, illustrates how much maternal-infant separation stresses the baby by measuring the physiological impact on newborns. When researchers compared stress levels in infants sleeping by themselves to infants sleeping with their mothers, they found something startling: newborns sleeping alone had a 176 percent increase in autonomic nervous system activity compared to newborns sleeping with their mothers. The almost threefold increase in nervous system activity is a stress response, showing that the separated newborns were anxious, even as they slept.

"Though they were sleeping, the quality of that sleep was poor," Nils Bergman, M.D., an independent researcher and one of the study's coauthors, explains to me. "Good-quality sleep consists of sleep cycling with approximately equal amounts of active sleep (when REM also happens) and quiet sleep, when the brain is wiring neural circuits. During separation, some of the infants had no sleep cycling at all, and of those that did had brief periods of cycling only."

Why were the newborn infants so anxious and sleeping so badly? Because humans, like other primates, need to be in constant contact with their mothers. "Maternal separation for a newborn primate is perceived as a very dire threat," continues Bergman, who worked as a mission doctor in a rural hospital in Zimbabwe, where the survival rate of premature infants who weighed between 2.2 and 3.3 pounds jumped from only 10 percent to 50 percent when mothers were instructed to carry them skin-to-skin in pouches on their fronts. (Survival rates for bigger babies jumped to 90 percent.) That experience changed Bergman's understanding of what human infants needed after they were born. "I realized that I was seeing the real biology of *Homo sapiens* at work, and that what I had learned at medical school was a result of modern culture equipped with amazing technology, but no understanding of human biology."

Bergman points out that animal research scientists separate newborn mammals from their mothers in order to study the *damage* on the developing newborn brain. So why do hospitals do this? One reason is for the convenience of the pediatrician, who is responsible for examining a newborn to make sure he is healthy. While it's entirely possible to examine a baby's reflexes, evaluate his condition, and even resuscitate a baby on the mother's chest, it's not as easy. The doctor has to get into an awkward position and is likely to get blood and other birthing fluids on his clothes.

"Inexperienced doctors, especially, may not feel comfortable," says Linda Hopkins, a high-risk obstetrician who works at Rogue Valley Medical Center and who believes newborn exams should be done without separating the mother and the baby. "It's harder to do. You have more trouble seeing the baby and hearing what you need to hear. If you're concerned at all about the baby, you want to assess it under lights with access to equipment."

Another reason is habit. As John H. Kennell, M.D., points out, the practice of separating mothers and babies started in the United States when birth was moved from the home to the hospital. Women doped with Twilight Sleep narcotics were unconscious, usually for several hours after their babies were born, so they were unable to hold or care for their newborns themselves.

"Based on no scientific evidence the solution chosen was to separate mothers from their babies, their partners and their families," writes Kennell, a professor emeritus of pediatrics at Case Western Reserve University School of Medicine, whose research on the importance of early baby bonding has helped change some hospital practices to allow infants to stay in the same room as their parents. "Evidence of the effectiveness of this drastic measure is lacking."

At York Hospital in York, Pennsylvania, separation of mom and baby is routine. "They move mom to the postpartum floor and the baby goes to the nursery. That's just how it happens. Period," a birth professional who worked at the hospital told me. "Once the mom is settled in and checked over and once they finish doing everything to the baby, then they are reunited. They are very 1950s. The postpartum unit is not even on the same floor as the nursery. I had a client who wanted her baby to stay with her and she was told, 'That's just not the way it's done around here.'"

Unfortunately, the change to an evidence-based way of treating babies right after birth and for the entire time they are in the hospital has been slow to be adopted in America. Power, ego, and fear factor into the reasons we do things contrary to what is in the best interest of the babies and their moms. For medical professionals to let newborns stay with their mothers—unless there is an emergency that indicates they need real help—is to acknowledge that the mother, not the health care provider, is the key to keeping the baby alive and healthy, the most indispensable part of the equation.

CORD CLAMPING

When a baby is born, he is still attached to the placenta by the thick ropy spiraling umbilical cord that is usually between about 20 and 24 inches long (50 to 60 centimeters) with a diameter of .3 to .8 inches (.8 to 2 centimeters), roughly between a dime and a nickel. The outside of the umbilical cord is a smooth white color. In the womb the umbilical cord supplies the growing fetus with oxygen-rich blood from the mother. After birth a newborn will stop getting oxygenated blood from the placenta and begins breathing through his lungs on his own.

Many first-time moms are surprised that labor isn't over when the baby is delivered. It's one thing to read about Stage 3: Delivery of the Placenta, and it's another to experience contractions *after* you've given birth. Sometimes these after-birth contractions—especially for second- and third-time moms, when the uterus has to work harder to shrink back to its original size—can be surprisingly intense. These powerful pushing contractions serve to expel the placenta and contract the uterus.

During the after-birth frenzy at the hospital the umbilical cord is immediately clamped and cut. But is this wise?

For the majority of his career, Nicholas Fogelson, M.D., clamped and cut the umbilical cord as soon as possible. When a family occasionally asked him to wait, he would humor them, letting the cord pulse for a few minutes in order to keep good patient-doctor relations. Fogelson, who works in the Department of Gynecology and Obstetrics at Emory University School of Medicine, had seen and been taught in medical school that the cord should be cut immediately. Without researching it or thinking about it, he assumed it was the right thing.

But then he started to change his mind. "[A]fter some research I found that there was some pretty compelling evidence that indeed, early clamping is harmful for the baby," Fogelson wrote in a blog post calling on obstetricians to stop the practice. "So much evidence, in fact, that I am a bit surprised that as a community, OBs in the U.S. have not developed a culture of delayed routine cord clamping for neonatal benefit."

The problem with cutting the cord while it is still warm and pulsing is that doing so actually deprives the newborn of its own blood, which is still in the cord and the placenta and has not yet finished circulating back into the baby. Up to 40 percent of his blood volume, including platelets and other clotting factors, is lost. Studies have shown that, among other benefits, infants whose cords are *not* clamped right away have higher iron stores in their blood and are less likely to hemorrhage. A systematic review of the scientific literature published in the *Journal of the American Medical Association* concluded that there are benefits to delaying cord clamping for at least two minutes that extend beyond the newborn period into infancy.

Waiting to cut the cord until after it stops pulsing means that the after-birth process has to be slowed down and individualized. This is inconvenient for medical professionals who are rushing to get to the next birth and for hospitals that want to move the birthing women along because they are being reimbursed at a fixed rate for a prescribed amount of time for delivery.

Moreover, the lucrative industry of banking cord blood, where parents are charged more than a thousand dollars to store their baby's cord blood in case it is needed later to generate stem cells to fight disease, would be impossible if we let the newborns keep the blood they need to live and thrive.

"Obtaining cord blood for future autologous transplantation of stem cells needs early clamping and seems to conflict with the infant's best interest," writes one researcher who reviewed the current data. "Although a tailored approach is required in the case of cord clamping, the balance of available data suggests that delayed cord clamping should be the method of choice."

Dr. Fogelson at Emory believes that antimidwife prejudice on the part of American obstetricians has played a part in continuing a practice that we now know is harmful:

"I wonder at times why delayed cord clamping has not become the standard already; why by and large we have not heeded the literature," he writes. It is sad to say that I believe it is because the champions of this practice have not been doctors, but midwives, and sometimes we are influenced by prejudice. Clearly, midwives and doctors tend to have some different ideas about how labor should be managed, but in the end data is data. We championed evidence-based medicine, but tend to ignore evidence when it comes from the wrong source . . ."

DON'T PUT THE BABY IN THE BATHWATER

A few hours after he was born, Cyndi's son Graydon had his first bath. A nurse and his dad washed his body and wispy brown hair thoroughly in a tub of warm water, using Head-to-Toe Baby Wash, made by Johnson & Johnson, from a small yellow bottle that the hospital provided.

Why give a baby a bath so soon after birth? "The message is that birth is dirty and the baby needs to be cleaned," says MaryBeth Foard-Nance, a doula and a mother of three who has attended more than 140 hospital births in Maryland and Pennsylvania. "I've had women tell me they think it's all gross and they don't want to see the baby until he's clean."

But not only is there no reason to wash a healthy newborn, the practice is actually harmful. It interferes with the natural bonding process between a mom and her baby, and depletes the baby of skin-protecting vernix (the sticky white substance that coats his skin). Washing the baby immediately after birth, which is done in many hospitals, also washes away the smell of amniotic fluid, which has been found to soothe the baby and help him connect with his mom.

In fact the smell of amniotic fluid on a baby is so individualized and distinct that human mothers (and even new fathers) can differentiate their amniotic fluid from that of another baby's. We also know, via research published in *Behavioral Neuroscience,* that newborns prefer the smell of their own amniotic fluid, which rubs off of them onto the mother's breasts when she holds her unwashed newborn. Furthermore, researchers at the Department of Medical Biochemistry and Biophysics at Karolinska Institutet in Stockholm, Sweden, found that the vernix on a baby's body at birth contains powerful antifungal and antimicrobial agents that protect the fetus and newborn against infection.

Pediatrician Susan Markel points out yet another downside to baby's first bath: Bathing a newborn causes him to lose body heat. To remedy this, he will then be placed in an artificially warmed box (a "warming bassinette") instead of in his mother's arms.

But there's an even more pressing reason to forgo that first bath. When a hospital bath takes place in front of the parents, it creates an excellent opportunity for companies that manufacture products to showcase their wares. But although it is advertised as "the #1 choice of hospitals" and "milder than baby soap," Johnson's Head-to-Toe Baby Wash contains a host of unpronounceable chemicals, some of which are known toxins. Michael Pollan, best-selling author and professor of science and environmental journalism at the University of California at Berkeley, suggests consumers read ingredient lists and not buy food made with unrecognizable or unpronounceable ingredients. The same advice holds for buying baby products. Yet the ingredient list of Head-to-Toe Baby Wash is barely recognizable to the average mom or dad: "Water, Cocamidopropyl Betaine, PEG-80 Sorbitan Laurate, Sodium Laureth Sulfate, PEG-150 Distearate, Tetrasodium EDTA, Sodium Chloride, Polyquaternium-10, Fragrance, Quaternium-15, Citric Acid, Sodium Hydroxide."

Unfortunately some of these ingredients are all too recognizable to environmental and child-welfare advocates. Quaternium-15 is a chemical preservative that kills bacteria by releasing formaldehyde, which is known to cause cancer in humans. Quaternium-15 is also a well-known allergen and the single most common cause of contact dermatitis to adult hands.

Another chemical with an unpronounceable name and a baffling numerical designation, 1,4-dioxane, has been found to cause cancerous tumors in dozens of animal studies and has been banned in Europe. This carcinogenic by-product, a side product of converting sodium *laurel* sulfate, a detergent and irritant, into the less irritating sodium *laureth* sulfate, was found in several Johnson's baby products, including Johnson's Baby Shampoo and Johnson's Aveeno Baby Soothing Relief Creamy Wash when tested in 2009 by the Safe Cosmetics Action Network. Yet parents have no way of knowing: Though present in baby products, 1,4-dioxane is not included on ingredient lists. Neil Carman, Ph.D., an environmental scientist based in Austin, Texas, thinks parents should be worried about 1,4-dioxane. "Babies are much more vulnerable to a chemical like this than a grown adult," Carman told the *Dallas News*. The safest way to avoid it: not to use on children's skin any products with PEG compounds listed in the ingredients. PEGs are made from polyethylene glycol, a petroleum-based compound that is often contaminated during the manufacturing process with 1,4-dioxine and ethylene oxide, another known carcinogen. Head-to-Toe Baby Wash contains PEG-80 and PEG-150.

Most people don't realize that your skin is your largest organ and that the lotions, body washes, and cosmetics you put on your skin—or your baby's skin—are partially or wholly absorbed into your body. This is the reason so many medications—from birth control to nicotine patches—are administered as skin patches.

"When you are slathering stuff on your face and skin and hair, it is going right into your bloodstream," contends Rex Rombach, who owned a $20 million hair care distribution company until his wife's father died of cancer and he began researching toxic chemicals. "Everyone thinks the FDA is all over that, and we're safe. It's not true."

Will Johnson's Head-to-Toe Body Wash harm your newborn? Toxic chemicals build up in a baby's body, which means the more chemicals a baby is exposed to the greater the risk of getting sick. Though we don't know why, the incidents of childhood cancers have gone up at least 26 percent since 1973. Between 1973 and 1999, incidents of acute lymphocytic leukemia have increased 62 percent; brain cancer, 50 percent; kidney cancer, 14 percent; and bone and joint cancer, 40 percent. In 2007 about 10,700 children under age fifteen were diagnosed with cancer. Childhood cancer is the second leading cause of death (after accidents), claiming more than 1,500 lives every year.

Samuel Epstein, M.D., professor emeritus of environmental and occupational medicine at the University of Illinois at Chicago School of Public Health, chairman of the Cancer Prevention Coalition, and former congressional consultant, claims there is substantial scientific evidence that fetuses, infants, and children are much more vulnerable to being exposed to carcinogens than adults. The author of 270 scientific articles and 18 books on the causes, prevention, and politics of cancer, Epstein has shown that carcinogens cause other toxic effects, including endocrine disruption and neurological and immunological damage. One of the biggest dangers for children: ingredients and contaminants in household lotions and shampoos.

When news of the toxic ingredients in many brand-name baby soaps became public in 2009 as a result of the Campaign for Safe Cosmetics report, there was consumer outcry in countries around the world. Responding to government pressure, Johnson & Johnson reformulated their products in Denmark, Finland, Japan, the Netherlands, Norway, South Africa, Sweden, and the UK. In those countries they are now selling baby products with nonformaldehyde preservatives. Although in the fall of 2011 the company announced they would phase out *some* of the carcinogens in their baby washes sold in America within two years, they continue to sell and promote products containing carcinogenic ingredients to use on sensitive newborn skin in the United States while using different formulas in Europe. At the same time, Johnson & Johnson's reputation as a baby-friendly company continues unabated. A Forbes online survey of 2,500 consumers conducted in August 2011 rated Johnson & Johnson the most trusted brand in America.

One of the reasons Johnson & Johnson is such a well-loved company is because of their effective product placement inside American hospitals. When a hospital washes a baby with Johnson's Head-to-Toe, that hospital is endorsing the product—sending a message to impressionable new parents that Johnson's is a brand that can be trusted. Johnson & Johnson donates the body wash free of charge or at deep discounts in exchange for what is nothing more than effective and lucrative free advertising to more than four million parents a year.

Our confidence in this company is misplaced. "You can't just trust brands and what they are telling us. We find so many toxins in so many baby products," says Stacy Malkan, communications director for the Campaign for Safe Cosmetics, who has been working on this issue for ten years. "We are so trained to be trusting of brands and to listen to their marketing and commercials. But we have to take the responsibility ourselves to find out what's really in these products and what these companies are really doing—to the planet and to our kids. Moms are realizing we have to take it upon ourselves to become educated . . . to do our own research and find what's safe."

VITAMIN K

Since 1961, it has been standard practice in American hospitals for newborns to be given vitamin K supplementation soon after birth. In a 2003 policy statement, reaffirmed in 2006, the American Academy of Pediatrics (AAP) asserted that synthetic vitamin K, which helps in blood clotting, is an effective way to avoid brain hemorrhage. It has been found that approximately 1 in 10,000 (.25 to 1.7 percent) babies will have what's called a vitamin K deficiency bleed.

Human babies get a bad rap, don't they? According to the medical establishment, they are dirty *and* deficient at birth, and it is only with our intelligent intervention that they can be protected. But the AAP recommendation overlooks the fact that human breast milk, especially the colostrum (first milk), is naturally rich in vitamin K. Vitamin K is also synthesized by healthy bacteria in the small intestine. By about a week after birth, as a baby continues to breastfeed, he builds up healthy bacteria in his digestive tract that synthesize vitamin K. Studies show that immediate cord clamping makes it more likely for newborns to hemorrhage. We also know that bottle-feeding—even just one bottle of formula—makes it harder for newborns to develop healthy intestinal bacteria and that infants born via C-section are colonized by bacteria found in the hospital instead of by a mother's beneficial vaginal and fecal bacteria. Several studies have shown that Cesarean birth can have a long-lasting negative impact on the human digestive tract. Blood draws and circumcision also cause unnatural blood loss in newborns. Perhaps, as many birth professionals and pediatricians believe, it is the unnecessary and even harmful medical interventions, not an evolutionary deficiency, that has brought on hemorrhagic disease that must then be cured by more intervention.

Do we really want to cause a newborn pain within an hour of birth? Do we really want to be jabbing him with a concoction of synthetic ingredients (including sugar) that may cause a metabolic imbalance? According to Merck & Co., which manufactures it, the shot contains a synthetic form of vitamin K (phytonadione), as well as polyoxyethylated fatty acid derivative, dextrose, water for injection, and benzyl alcohol (.9 percent) added as a preservative. The insert cautions that "Benzyl alcohol as a preservative in Bacteriostatic Sodium Chloride Injection has been associated with toxicity in newborns. Data are unavailable on the toxicity of other preservatives in this age group." The insert's list of warnings, precautions, and adverse reactions is more than a page long. Reported side effects include severe hypersensitivity reactions, death, anaphylactic shock (from an allergic reaction to the ingredients), "flushing sensations," "peculiar" sensations of taste, dizziness, rapid and weak pulse, profuse sweating, brief hypotension (lower than normal blood pressure), dyspnea (shortness of breath), and cyanosis (blue or purple coloration of the skin), as well as pain, swelling, and tenderness at the injection site. Another reported adverse reaction: death. Since colostrum is rich in vitamin K, as is breast milk, many would argue that those precious hours after birth are better spent establishing a breastfeeding connection without the disruption of a needle.

JAUNDICED PROTOCOLS

When Rachel Zaslow's daughter was six days old, her skin looked yellow, she seemed lethargic, and she wasn't pooping. Concerned that Amaya's bilirubin levels were too high, Rachel brought her to the pediatrician.

"Oh, I don't think she has jaundice," the pediatrician said dismissively, "and we can't get the results until tomorrow." Rachel, a twenty-eight-year-old health care professional, urged the doctor to do a blood-draw to test Amaya's bilirubin levels anyway. The next day the pediatrician telephoned. "Take your baby to the hospital," she cried into the phone, without saying hello. "Now!"

Virtually all newborns have some jaundice after birth. Their skin gets a yellow hue and the whites of their eyes may look yellowish also. This lemony skin tone comes from bilirubin, which is formed by the breakdown of hemoglobin in old red blood cells. A newborn's liver, which is the organ that processes bilirubin, is less mature and does not work as efficiently as an adult's at removing bilirubin from the bloodstream. Though the discoloration can be worrisome to parents, it appears that elevated bilirubin levels after birth may actually play a protective role in helping a newborn adjust to the higher oxygen levels outside the womb.

There is nothing to worry about when a baby is jaundiced unless, as was baby Amaya's case, the levels of bilirubin in the bloodstream are so high they could potentially move out of the bloodstream and into brain tissue. Amaya was six weeks premature and African American, which are both risk factors for more serious jaundice. In severe cases a buildup of bilirubin in the bloodstream can cause a form of brain damage called kernicterus. Amaya's bilirubin level was 23, more than twice as high as what would be considered normal. So Rachel, still recovering from the birth, took Amaya to the NICU at Long Island College Hospital in Brooklyn Heights.

The one-room NICU, filled with row after row of baby beds, had no place for Rachel to sit down. Standing awkwardly, she put her hands through the holes in the incubator so Amaya could feel her touch, and talked and sang softly to her. Soon after Amaya was admitted, the nurses asked Rachel to leave, informing her that visiting hours were over. Unwilling to be separated from her baby, Rachel insisted on staying. The supervising nurse was called to usher Rachel out, maintaining that having mothers in the NICU was against hospital policy. "What's the matter? Don't you trust us?" the supervising nurse said when Rachel again refused.

Rachel explained she wanted to be there to reassure and comfort her baby.

"You don't really think she knows you're here, do you?" the nurse scolded.

Though they did not call security to escort Rachel out, the nurses treated Rachel with disdain. They threatened to give Amaya formula if Rachel did not pump enough milk, but refused to help her fix the hospital breast pump (the pumping "room" is a closet with no door, just a curtain and a stool) when it malfunctioned.

"I was clearly an annoyance to the nursing staff. They were used to sitting around and gossiping and doing whatever they wanted with the babies," Rachel recalls. "They have a sense of ownership over the newborns that they couldn't have with a mom in the room."

When Amaya was first admitted, the NICU pediatrician recommended immediately operating to insert a central line into Amaya's belly button in case she needed a blood transfusion. When Rachel suggested they wait three hours to see if the bilirubin levels responded to phototherapy (light helps the body break down bilirubin, which is why doctors advise new parents to place a jaundiced baby in a sunny window for ten minutes twice a day or more), the doctor insisted. He told Rachel his shift was about to end, there was no one as skilled as he to do the operation, the surgery was difficult, and the central line would also be used for all subsequent blood draws.

Rachel agreed reluctantly, hoping this just-in-case operation would ultimately save her infant some unnecessary pain. But after the doctor left, the nursing staff in the NICU refused to take Amaya's blood from the central line, telling Rachel it was "not sanitary," and doing heel pricks instead.

Amaya's bilirubin levels responded immediately to the phototherapy and she did not need a blood transfusion. By the evening of her second day in the NICU Amaya's bilirubin level was down to 12, on the high range of normal. In all that time, Rachel had gone home once for three hours to take a shower, change her clothes, and sleep for an hour and a half. Amaya's numbers were low enough now that the neonatologist took her off the lights.

Since there was nowhere for Rachel to nurse or sit down, and since she was told they would not be testing Amaya again until the next morning, Rachel asked if she could take Amaya home and bring her back in the morning. The doctor refused. Since the NICU was an inpatient clinic they would be unable to test Amaya unless she remained in the hospital. Rachel suggested she take Amaya in the morning to the outpatient clinic and they could share the test results with the NICU. That's never been done before, she was told.

Rachel called a pediatrician friend in Oregon to get a second opinion. The pediatrician confirmed Amaya was well enough to go home. A quick Google search on her iPhone and a call to an attorney revealed that there was no New York State law forbidding Rachel from taking her baby home against medical advice. Keeping her overnight was just hospital protocol. But when Rachel insisted she and Amaya were going home because that would be the best place for her newborn, she was threatened with jail.

The pediatrician—whom Rachel had never met and who had never examined Amaya before—told Rachel that if she took Amaya home against medical advice, the police would be called and child protective services would take her baby away. "You could be separated for months," a nervous, well-meaning resident told her, pulling Rachel aside to urge her not to take Amaya. "I agree with you," the resident said, but did not publicly challenge the supervising doctor.

Exhausted but not defeated, Rachel called the managing editor of the *New York Daily News,* New York City's second-largest newspaper. She let the doctors and nurses know that her lawyer and the editor were on their way.

Within five minutes it was decided Rachel and Amaya could leave.

"There were something like thirty babies in the NICU, and in all the time I was there I only saw three moms," Rachel says, still upset by the ordeal two years later. "Almost every baby in that NICU is African American and the parents are completely disenfranchised. Nobody knows their rights in the hospital. People believe the doctors are the end all be all, they take their word as the truth. When your baby's in the NICU, you're so vulnerable, your baby's sick—of course you want to trust what the doctors say. But people aren't questioning why there aren't even chairs for mothers to sit down. I really thought the doctor and I could talk together and make a plan. But he was so dead set on asserting his authority that he refused to listen. It felt shocking and violating to me."

Rachel's lawyer suggested pressing kidnapping charges. Amaya was completely out of danger. There was no valid medical reason to keep her in the NICU. But to discharge her at eight o'clock at night would have been inconvenient for the staff and a break from protocol. It also meant a significant loss of revenue for the hospital. If Amaya had stayed overnight the hospital would have been able to bill the insurance company between $3,000 and $10,000 for another day in the NICU.

THE BUSINESS OF THE NEONATAL ICU

About 15 percent of infants are diagnosed with jaundice, but only 1 to 2 percent of newborns will have extremely high bilirubin. Most babies do not get jaundice as extreme as Amaya's and most parents will not have to navigate an ordeal of that magnitude just days after their baby is born. The majority of American parents will never step foot in a NICU, where tiny babies hooked up to beeping machines fight to stay alive.

Still, according to a three-year investigation by the WHO, the March of Dimes, and Save the Children, the United States is among the top ten countries, along with Nigeria and India, with the highest number of premature births in the world, and the number of babies in the United States born premature has risen 30 percent since 1981. Today, some half a million babies a year are being born prematurely in the United States, a higher percentage than in any Western European country or Japan, putting the United States on par with countries like Kenya and Honduras.

While in developing countries, the high number of premature births often stems from little access to prenatal care, poor nutrition, and very young mothers; in the United States our premature birth rates are fueled by the use of fertility drugs (resulting in twins and triplets), high C-section rates, and labor induction as well as by noniatrogenic causes. In 2000 nearly 40 percent of premature births in the United States were found to be provider-initiated (that is, caused by the doctors). With the continual rise in C-section rates, there is reason to believe that number, if measured today, would be much higher. Though the technological advancements in hospital care mean that the United States excels at keeping very premature babies alive, premature birth is one reason America has a higher infant mortality rate than forty-eight other countries.

Since the 1980s there has been an enormous rise in the number of neonatal intensive care units in American hospitals. These NICUs offer lifesaving interventions, especially to those babies born so severely premature that they cannot breathe on their own. As the rates of twin births, labor induction before forty weeks gestation, and early Cesarean sections have increased, there has been an increased need for NICUs. Still, evidence suggests that the steady and continuing rise of NICUs in America is not rooted only in need. Rates of reimbursement for stays in the NICU— especially for infants who need to be there over a long period of time—are significantly higher than reimbursement for routine postpartum care. This makes sense: Babies in the NICU usually need a high level of attention and often must have costly interventions to stay alive. But these high reimbursement rates give American hospitals an unfortunate incentive to overuse their NICUs in order to increase profit streams.

When the Texas Health and Human Services Commission investigated how NICUs were being used in their state, they found that the state could save $36.5 million by better managing which babies ended up in the NICU, prohibiting elective C-sections, and refusing to finance elective preterm inductions before thirty-nine weeks. The average NICU stay in Texas costs Medicaid $45,000, *eighteen times* more than a hospital birth with no need for the NICU.

"The NICU is a moneymaker for most hospitals," says Stuart Fischbein, the obstetrician based in Los Angeles. "Hospitals are not unhappy when babies have to go to the NICU. One hospital I worked for in Ventura County had a policy that every baby born by C-section had to go to the NICU for four hours."

Thomas M. Suehs, the Texas commissioner of health and human services, has two grandchildren who were both put into Texas hospital

NICUs in the last two years, despite being born healthy after uncomplicated deliveries.

In 2005, when the Seton Family of Hospitals in Texas decided to prohibit elective inductions before thirty-nine weeks to make birth safer, the number of babies admitted to their NICUs dropped significantly. And so did NICU-generated hospital profits; they dropped by 96 percent. The chain went from generating about $4.5 million a year from NICU stays to making just $186,000 annually.

"When we look at the data, it indicates that, yes, there is overutilization of NICUs," Suehs told a reporter for the *New York Times*. "More babies are being put in NICUs than need to be in NICUs."

The World Health Organization report reveals that more than 75 percent of deaths of infants born preterm can be avoided without the need for NICUs. The report's recommendations for optimal survival for premature infants includes early and exclusive breastfeeding, kangaroo care (meaning the mother or another adult carries the infant close on her chest), and skin-to-skin contact, three interventions that are difficult to carry out—if not impossible—in most American hospitals.

RECOMMENDATIONS VERSUS REALITIES

The AAP's first recommendation for treating severe jaundice is that clinicians "promote and support successful breastfeeding," yet Long Island College Hospital's NICU, as do many others in the United States, virtually ensures breastfeeding failure.

The AAP guidelines also caution doctors against making new moms overly anxious or generating unnecessary costs, advising they "minimize the risk of unintended harm such as increased anxiety, decreased breastfeeding, or unnecessary treatment for the general population and excessive cost and waste." In the case of Rachel and Amaya, it was only the threat of a negative news story and a lawsuit—not what was in the best interests of the health of the eight-day-old baby in their care—that changed the pediatrician's orders.

Most parents don't realize that much of what is routinely done to American infants in the hospital just after birth is not based on best evidence or best practices. "Parents are often not aware that many aspects of 'routine' postnatal care are not grounded in evidence-based practice," points out Susan Markel, a Connecticut-based pediatrician who has been practicing for more than thirty years. Sadly, parents who do request that things be done in a gentler, less invasive, more evidence-based fashion are often met with astonishment, hostility, and even, like Rachel, the threat of a call to the Department of Social Services.

FOLLOW THE MONEY

While normal, healthy babies after birth usually need nothing more than uninterrupted skin-to-skin contact with their parents, a gentle touch, and a quiet environment, they are instead examined, washed, injected, and subjected to antibiotics and bright lights.

"Babies are manhandled even in the best hospitals," says MaryBeth. "They are treated like they're footballs, not little humans who are already soaking up what's going on around them, and already internalizing how they're being touched and talked to. There's a gross lack of soothing touch and words. Everything is bright lights and big city. Everything is an accident waiting to happen."

MaryBeth has seen this both in her doula practice and in her personal experience as a mom (her children are fourteen, eleven, and nine months old). After two hospital births, MaryBeth and her husband decided to have their third baby at a birthing center in Gordonville, Pennsylvania. They were both delighted by how peaceful and quiet the birth was and surprised by how few medical attendants there were. While there had been at least half a dozen medical professionals in the room after each of her first two births—a hospital midwife, a labor nurse, a nurse for the baby, a scrub tech to clean the room, and a supervising pediatrician who came to check on the baby—at Birth Care, MaryBeth was attended by only a nurse midwife and a labor nurse. The price tag surprised them as well. Her insurance covered the entire cost of her water birth at Birth Care, which cost less than half as much as a hospital birth.

"When the medical system sells parents on the idea that their birth and their newborn baby are an accident waiting to happen, then it seems natural for the family to have all these people there," MaryBeth points out when we talk on the phone. "Are all those people needed at a normal birth? Absolutely not. All it does is drive up the cost of everything—after all, all these people have to have salaries. If we make it normal to have your baby at a birth center, too many people will be out of a job."

Even a normal, uncomplicated birth carries a high after-birth price tag in the hospital. When her son Jackson was born in July 2011, Sarah Vaile's itemized hospital bill contained a charge of $41.85 for eight 325 mg tablets of acetaminophen, the generic name for Tylenol ($5.23 per tablet). At CVS a box of one hundred 325 mg acetaminophen tablets retails for $8.49, about $.08 a tablet. The hospital charges *6,000 percent* more for the drug. At the hospitals in California and Oregon where Linda Hopkins, a high-risk obstetrician, has worked, protocol dictates that hospitalized patients must use the (usually much more expensive) hospital pharmacy and may not bring their own medication.

Though it is standard practice for insurance companies to negotiate with hospitals and pay a smaller markup than what is originally billed, these high medical costs don't come back to the profit-driven insurance companies but to American parents in the form of higher premiums. Since most insurance companies make a percentage off the total amount they pay out, they have no reason to keep costs down. When health costs are high, insurance companies simply raise monthly premiums.

The infant mortality rate in the United States and the cost of our health care system are both shamefully high. If we stopped unnecessary after-birth interventions for healthy infants would we be doing a disservice to newborns? The evidence suggests the opposite: If American infants were treated more gently, spent more uninterrupted skin-to-skin time with their mothers after birth, and were exclusively breastfed, we would improve infant outcomes and lower the cost of health care.

A GENTLE AFTER-BIRTH ALTERNATIVE

"You wanted to catch the baby!" Angelina cried.

"I know, but I can't leave you!" Angelina's husband, Simon, cried back.

It was a rainy day in late January 2012, and Angelina and Simon Mendenhall had traveled two hours from Fort Jones, California, to birth their baby on a queen-size bed at the Trillium Water Birth Center in Medford, Oregon. Simon was holding Angelina's hand, looking into her eyes, breathing with her, and smoothing back her curly brown hair. Two midwives, a midwife apprentice, a doula, a photographer, Angelina's older sister, and Simon's mom were all in the room for the birth. Amy Winehouse's sultry contralto played in the background. The mood was festive.

The head midwife and owner of the center, Augustine Colebrook, had briefed family members beforehand that after Angelina pushed the baby's head out, there would be a pause. The midwives would be there in case they were needed but, if everything went normally, the baby would rotate his body on his own time so each of his shoulders could slide out of the birth canal and he could come into the world. No one would announce the gender of the baby. Everyone would be there "holding the space," taking their cues from Angelina, encouraging her and experiencing the joy of her baby's birth with her without interrupting or distracting her.

"Something's wrong!" Angelina said to Augustine when she felt the pressure of the baby's head emerging from the birth canal. "That's just your baby's head," Augustine replied calmly, supporting Angelina's perineum so it wouldn't tear. "Now something's definitely wrong," Angelina said again. "That's just your baby's shoulders," Augustine replied kindly.

Augustine suggested Angelina touch the baby. Reaching down to feel the head, Angelina got very excited. The baby was out in one more push and as it emerged Angelina put her hand under his shoulder with Augustine's help and lifted him onto her chest. She hadn't planned to catch her own baby and no one guided her: She just followed her instincts. Her mouth open wide with awe and joy, she held the wet baby to her body. The midwives dried the baby off and pulled the soiled linens from under Angelina without disturbing the new family. They put a green blanket over the baby, listened to the lungs and heart, and took the temperature with a forehead swipe while Angelina hugged the tiny naked body to hers, admiring the mop of dark hair, the curled-up fingers, and impossibly big eyes.

That's when she and her husband realized they had forgotten to see if the baby was a boy or a girl. They peeled back the blanket and peeked. "It's a boy," Simon said to Angelina. The new parents had uninterrupted skin-to-skin time with their newborn, inhaling his scent, smiling into his wide-open eyes, and finally passing him to proud family members who were dying for a turn to hold him. They named him Becker.

When the midwives finally did a thorough newborn exam, three hours later, they did it with Baby Becker in his mother's arms. No one took the baby away. No one put antibiotic cream in his eyes. No one gave him a vitamin K shot. No one scrubbed his soft new skin with brand-name chemical-laden soaps. They didn't cut the cord for twenty-one minutes. (At Trillium Water Birth Center the cord isn't cut until it has entirely stopped pulsing, from five to fifty minutes after birth.)

"You have so much oxytocin in your system, you have all the endorphins that create bonding. We have an alive, alert, undrugged baby and a conscious, alert, blissed-out mom. Their whole purpose in life is to meet each other right then," says Augustine, who has delivered more than four hundred babies in her twelve years of experience as a midwife (as well as having three homebirths of her own).

"Moms have nothing but eyes for their baby at that moment, and that's the way it's supposed to be. Anything we do would interrupt them. The longer we can wait before we ask the mom to raise back into her thinking, functioning, conscious brain, the better the bonding is for a lifetime."

Countries with highest premature birth rates: India, Nigeria, United States

Babies born prematurely in United States: 500,000

Number of European countries with a lower percentage of premature birth rates: 100 percent

Average cost of one-day stay in the hospital: $3,949

Average cost of stay in the NICU: $45,000

Average charges for first three months of newborn care after C-section birth (with NICU stay): $83,000

Average charges for first three months of newborn care after vaginal birth (with NICU stay): $55,000

Revenue lost in 2005 by Seton Hospitals when they *prohibited* inductions before 39 weeks in 2005: more than $4.4 million

Hospital mark-up on pharmaceutical products like headache medicine: 6,000 percent

Postbirth homebirth facility charge: $0

Cost to forgo neonatal interventions: $0

James di Properzio: As Surreal as a Science Fiction Movie

When our first daughter was born in Atlanta, Georgia, after twenty-two hours of labor, I was only allowed to hold her for a few minutes. Then the nurse took the baby out of my arms. Here's my husband's story of what happened after the nurses whisked our daughter away.

I was thirty years old when our first child was born. Going to the hospital seemed the safest thing to do. After arguing with OBs throughout twenty-two hours of labor, my wife delivered a healthy 7-pound, 1-ounce daughter. The nurses got us ready to leave the birthing room and move to a private room for the night, and my wife, still wobbly from the spinal, was helped into a wheelchair.

The baby, as a matter of course, was put into a clear plastic box on a trolley and whisked off to the nursery for "a few things before joining us in the room." The nurse was clearly brushing me off, but in the first few minutes of my child's life I was too interested in her not to follow. I pursued the hurrying nurse down a labyrinth of long corridors with the strange feeling that she was trying to ditch me, and eventually my daughter and I were braceleted and scanned into a secure nursery area. We had to wait outside the metal door—no telling why, since the room was windowless, with a red light over the door like a photography darkroom. The two desk attendants told the nurse to wait with the trolley parked there, and told me it would be a while, in case I wanted to check on my wife.

Check on my wife? Torn between two people in my new family, I scanned out and jogged down the corridors to the family overnight rooms.

When I came panting into the room, my wife, Jennifer, looked at me in alarm. "Where's the baby?" she asked.

"I just left her in the nursery to come check on you," I replied.

"Don't leave her alone! I'm fine; go back and get her!"

I sprinted back along my route, alarming the attendants guarding the nursery as I rushed up. I showed them my armband. They were reluctant to let me in, first assuring me that the baby would be fine and they would bring her "to see us" later.

"No, she'll be staying in the room with us; we have a room for families," I explained.

"Well, they're giving her a bath right now," they told me, as if I ought not to disturb her privacy.

If they had told me up front that she was just in the middle of being bathed and that if I would wait here a minute they would bring her right out for me, I would have gone for it; but the way the nurses acted showed me that I was being stalled, and that they didn't want me to go in there for some reason. That set off my bullshit detector, so I called their bluff and walked directly to the door, which they were obligated to buzz open for me.

The room beyond was nothing like I had expected. The public rooms in that wing all had wood moldings and pastel walls, while this nursery was undecorated and utilitarian, functional and hygienic. There were rows of newborns in a variety of complicated containers that reminded me of the biology laboratory in high school, while around the walls were large steel washbasins like the dishwashing area of an industrial kitchen. Several newborns in the bassinets were crying continuously, but the two nurses in the room ignored them. They each had a baby at a workstation by the wall, and both acknowledged my entrance only by looking askance at me from the corners of their eyes, scowling.

One of them held a screaming newborn over a washbasin. She was holding her with one hand while vigorously scrubbing her with the other, like she was washing a chicken in the kitchen sink.

I stood in shock at how ungentle she was, totally uncaring for the terrified baby only minutes old in her hand. The nurse herself was not much over five feet tall but well over two hundred pounds, scowling at the baby as if in resentment of her chore. Why were the newborns entrusted to these women? Was this a despised, low-totem job? Was Crawford Long Hospital so busy that they had industrialized the newborn procedures to such an inhuman extent?

Realizing that it could be my baby, I moved toward the nurse and asked if she was mine. The nurse shrugged, grunted, and pointed an elbow at the label, which showed another surname. When I asked where my daughter was, she didn't reply, so I moved off among the rows of babies. I realized that since I had been gone scarcely five minutes, the nurse who sent me off had been trying to get rid of me before whisking the baby in and getting her washed, out of my sight.

Under an unusually large infrared heater, I found my daughter naked but for a diaper, her skin clean and slightly chapped, crying, albeit less loudly than the baby being washed. I put my finger in her grip and talked to her, and she quieted when she heard my voice, familiar after my talking to her in the womb every day.

I was reluctant to interrupt the warming action of the machine, but eager to get her out of there, so I asked the nurses when she would be done, and when I could take her. They refused to speak to me. Obviously they were uncomfortable with parents invading this room, and the gatekeepers must normally have kept parents out altogether. This was not how a slick hospital presented a room that the customers would see—I was behind the scenes. It was as surreal as a science fiction movie, where you get a peek at what's going on behind the scenes of life in a high-tech society and it's horrifying. Was this how I wanted my child treated at the beginning of her life?

I picked up my baby, announced that I was taking her back to the room, and strode out the door, feeling like I was saving her from the fate of those other babies. I held her close to my chest and neck, touching her with as much warm skin as I could and covering her with my forearms. Her cool, dry skin felt papery, like it needed moisturizer. She had seemed fine right after they had toweled her off at birth, her skin moist, warm, and glowing. Did she really need to be sterilized?

Our son was born at home four years later. When the midwife caught him, she gently wiped his face, put a towel around him, and brought him straight to his mother's breast. My wife lay back in her own bed while the midwife and a friend tidied up, and I sat next to her and wrapped my arms around them both. We got to bond with our son immediately, while he had the comfort of being held and nursed after the shock of emerging into the chilly air. That was how I wanted to experience my child's birth, and how I wanted my child to come into the world.

CHAPTER 6

FORESKINS FOR SALE:
The Business of Circumcision

D r. Beth Hardiman, an obstetrician at Mount Auburn Hospital in Cambridge, Massachusetts, sets the timer on her iPhone for five minutes.

"This makes a huge difference," Hardiman tells me, taking a moment to stretch after a long day of delivering patients and making rounds. At five feet, with a stylish short haircut, Hardiman manages to look professional-chic even though she's wearing blue scrubs. Though her legs are shorter than mine, I practically have to jog down the hospital corridors to keep up. She tells me most doctors are too impatient to wait the five to five and a half minutes it takes for the anesthesia to numb the penis, and that circumcisions are excruciatingly painful without it.

The dad of the one-day-old baby, who lies quietly on a pink-and-green-striped blanket, unaware that he is about to lose part of his penis, wants his son to look like him. That's just what people in his family do, he tells the doctor. Hardiman challenges them, "We're not going to modify his appearance in other ways," she says. "We're not going to modify his hair or change the shape of his nose. You *are* clear that this is a completely elective procedure?"

Hardiman tells the new mom and dad in cringeworthy detail about possible side effects: bleeding, infection, injury to the penis, and a 1 percent rate of unsatisfactory outcome, which could lead to the need for another operation. But the father and mother remain resolute, even though she gives them ample time to change their minds ("Informed consent is a very important part of this process," she tells me after we leave their room). Grandma, who is lying on the extra bed in the room, urges her daughter and son-in-law to reconsider. All three adults shake their heads when Hardiman invites them to come along, preferring to entrust their newborn to strangers rather than squirm through watching the surgery.

The newborn has red puffy cheeks and a bit of fuzz on his head. Calm and alert, he starts to cry until the nurse gives him a glucose solution on her gloved finger to suck on. Hardiman waits until he's sucking hard. Then she swabs his penis with alcohol and uses a 29-gauge syringe to inject a 1 percent lidocaine solution at ten o'clock and two o'clock near the base of his penis, pinching the tip of the foreskin at the same time both to loosen it and to distract him from the pain of the shot. The injection makes a tiny bubble of blood, but the baby, still sucking, hardly flinches.

The nurse lifts him out of the bassinet onto a white Olympic Circumstraint, a plastic body-shaped contraption to keep his legs spread apart. She immobilizes his legs with light blue Velcro straps. His arms, which are swaddled in the blanket, are too short to reach his genitals so they are not strapped down. When there are twenty-eight seconds left on the timer, Hardiman begins. But the baby's penis isn't numb yet. He flinches, tenses his legs, and curls his hands into fists. She decides to wait a little longer.

When she's sure his penis is numb, Hardiman uses a blunt-edged metal probe to separate the skin that sheaths and protects the penis like a sleeve, called the foreskin or prepuce, from the head of the penis itself. She uses a mosquito clamp (which looks like a small scissors) to pull this foreskin tissue upward and out, lifting it past the head of the boy's tiny penis, like pulling a sleeve over a fist. Then Hardiman slides a clamp, called a Mogen shield, over the foreskin, making sure before she closes its jaws that she will not clamp the head of the penis by mistake.

"This gets very tight," she tells me. "This is a big pinch, lovey," she tells the baby. She shuts the jaws of the clamp and then uses a retractable disposable scalpel to slice off the compressed skin. After she cuts it but before she unclamps the Mogen shield, Hardiman checks to make sure she has removed both the outer layer of the foreskin, which looks like skin, and the inner layer of the foreskin, which is a slippery mucuslike layer that looks like the inside of an eyelid. Then she takes off the clamp, pulls the pinched skin apart, and pushes what is left down under the mushroomlike head of the penis. The exposed head is the bright purple color of raw liver. The stretchy tissue she has just removed is wrinkled. It reminds me absurdly of *ravioli nudi*. I'm surprised by how much tissue there is—this little boy has just lost skin that amounts to about a quarter of his penis. Hardiman puts pressure on the wound with a piece of gauze to keep it from bleeding.

"This tissue is very stretchy, and the amount really varies," Hardiman says. "Some babies have quite a bit. Some have very little. It's not erectile tissue but it is erogenous tissue. It has a lot of nerve endings."

The newborn, who has been calm, gives one high-pitched squeal once the operation is over. "If we're going to do it, let's do it respectfully to the baby and the parents," Hardiman says. "That's the way I make peace with this." Once his diaper is back on, the assisting nurse scoops up the newborn and rocks him against her chest.

A nursing student from SUNY Rockland Community College at Nyack

Hospital in Nyack, New York, had a very different experience watching a circumcision in March 2012. "There was no anesthesia," she tells me. "They just gave the baby glucose water." The procedure took about fifteen minutes to complete (the doctor went slowly for the purposes of demonstration) and the infant was in visible pain. "The poor baby was screaming and having all these facial expressions. Oh, my goodness, I would never have this happen to my kid! The ten of us all had the same reaction; we turned our heads away. We were like, 'that was like too much.' It was really bad."

Hardiman's circumcision in the hospital is also very different from two Jewish circumcisions I've attended as a guest, not a journalist. During one, done in a general practitioner's office with anesthesia, the baby screamed so loudly his cries could be heard through the closed doors. During the other, done with no anesthesia besides a vodka-soaked cloth for the baby to suck, the eight-day-old screamed for twenty-five minutes, crying the entire time his male relatives were dancing with him, writhing in so much pain when the rabbi finally handed him back that his mother had trouble settling him enough to nurse.

A GLOBAL PERSPECTIVE ON CIRCUMCISION

Globally, the vast majority of men are not circumcised. In Europe, South America, Central America, Mexico, Southeast Asia, India, and China fewer than 20 percent of men are circumcised. In some countries, like Spain, Denmark, and Finland, fewer than 2 percent of men have been circumcised.

But circumcision has long been the norm in the United States. Between 1979 and 1999, according to CDC estimates, about 65 percent of all American boys born in hospitals were circumcised. Circumcision rates vary by ethnicity (Hispanics have low numbers of circumcision, African Americans have higher numbers), geographical location (boys born on the West Coast tend to have lower rates of circumcision than babies born in the Midwest), and religion (Jewish Americans and Muslim Americans have higher rates of circumcision than others).

Medically sanctioned circumcision became prevalent throughout the British Commonwealth, including in Great Britain and the United States, in the nineteenth century primarily to control masturbation in adolescents and young boys. Though the procedure became widespread, when England privatized their health care system in 1948, circumcision rates fell dramatically. Now, approximately 5 percent of men in England are circumcised. The National Health Service does not routinely pay for the operation, since it is not considered medically necessary. According to the NHS, "[M]ost healthcare professionals now agree that the risks associated with routine circumcision, such as infection and excessive bleeding, outweigh any potential benefits."

Other English-speaking countries also have much lower circumcision rates than in the United States. In Canada, 31.9 percent of infants are circumcised, in New Zealand and Australia it is estimated to be between 10 and 20 percent. There is evidence that more American parents are choosing to follow the global trend to forgo circumcision for their infants. In the past ten years, circumcision rates in the United States have been falling. Today the CDC estimates that only a slight majority, 54.7 percent, of American newborn boys are circumcised in the hospital.

WHY PARENTS CIRCUMCISE

When Aseem Shukla, director of pediatric urology at the University of Minnesota Amplatz Children's Hospital, was in medical school, circumcision was the norm. "We believed it was what everyone should do," Shukla remembers. "There was a strong dogmatic belief that it was necessary to prevent penile cancer." Today Shukla has changed his thinking. He now believes it's ultimately a family decision based on what the parents want and he no longer pushes parents to circumcise. Shukla tells parents who ask him that there are both benefits and risks to circumcision. Circumcision has been found to protect against urinary tract infections, especially in boys who have renal tract abnormalities. It is also associated with a lower risk of penile cancer, cervical cancer for women whose partners are circumcised, and the spread of some sexually transmitted diseases.

But, Shukla is quick to add, urinary tract infections among infants in the first year of life are rare. "The risk of infection is very low: only one percent," Shukla informs parents. "That means we would have to do more than ninety-nine circumcisions to prevent one infection. Is that really worth it?" Penile cancer does not occur in infancy, usually striking men over age sixty, and the overall lifetime risk of penile cancer is rare. Furthermore, there is disagreement in the medical community about whether circumcision actually prevents penile carcinoma. Case studies indicate that circumcised men can get penile cancer, and epidemiological data shows that incidents of penile cancer vary greatly in populations where most men are not circumcised. One large Danish study showed that penile cancer decreased in prevalence from 1.15 per 100,000, which is what it was between 1943 and 1947, to 0.8 per 100,000 in 1988–1990, while circumcision rates remained constant in Denmark (1.6 percent), leading researchers to attribute the decline in cancer to improved sanitation.

The final medical justification is that infant circumcision offers protection against sexually transmitted diseases, both in the men who are circumcised and their partners. According to a newly revised statement on circumcision from the American Academy of Pediatrics, a review of the scientific literature reveals that circumcision reduces the risk of men getting HIV from between 40 to 60 percent in Africa, where transmission is predominantly among heterosexuals. Three randomized control trials, which were all halted before completion, suggested that circumcising adult men in African countries where AIDS is widespread can lead to less transmission of HIV/AIDS. Yet other researchers have found several flaws in these African studies, and a meta-analysis of the circumcision status of men who have sex with men concluded there is "insufficient evidence" that circumcision prevents the transmission of HIV or other STDs. One study in Zambia found that a quarter of the men who have undergone circumcision resume sexual relations without protection before their incisions have properly healed, often with multiple partners.

Critics have also pointed out that the zealous circumcision campaigns currently under way in parts of Africa undermine the use of condoms because men believe they have received the equivalent of a surgical vaccine and are therefore immune to HIV. For many health professionals, the idea that you would use penile surgery as a prophylactic measure instead of education and condom distribution is absurd. "The proponents of this speculation choose to ignore the obvious fact that AIDS infections were first recognized in American homosexual men who were overwhelmingly Caucasian, middle class, and circumcised," points out James L. Snyder, M.D., a urologist with more than forty years of experience. "[This is a] testimony to the failure of circumcision to offer any degree of immunity to AIDS infections."

Although the AAP now contends that the medical benefits to circumcision outweigh the harms, they do not recommend routine male circumcision. Instead the AAP asserts that the decision should be left up to parents and that health insurance companies should pay for the procedure if parents decide to have it done. None of the equivalent medical associations in Europe recommends routine male circumcision, and some actively oppose it. The Royal Dutch Medical Association has urged a ban on the procedure in Holland, both because it endangers an infant's health and because it is ethically questionable. In July 2012, the German Medical Association told doctors not to perform medically unnecessary circumcisions. In the words of one law professor writing in the *New York Times,* "All human beings should be able to make their own decisions about whether their genitals are to be injured. All the more so if such a procedure is irreversible and not medically necessary." Hospitals in other European countries, including Austria and Switzerland, have followed Germany's lead to stop performing medically unnecessary circumcisions. Norway, too, has been considering banning the operation in neonates.

When American parents choose to circumcise their sons, health is usually a secondary consideration. Religious circumcision is only a fraction of the circumcisions being done in America today, as only 2 percent of the American population is Jewish and less than 1 percent is Muslim. Instead, circumcision in America is a cultural practice, chosen by parents for social conformity or aesthetics.

ANTEATER AESTHETICS

When Deston Nokes was a senior at Willamette University, a freshman pledge taped sticky notes to the doors of each guy in the Delta Tau Delta frat house. "We all awoke to find that he had labeled us: helmets or anteaters," Deston remembers. "There were a couple of dudes who weren't circumcised. We all laughed like hell. My girlfriend was staying over, so I had to explain it to her. 'Gross,' she said. 'They look like anteaters?' "

Deston chose to have his son circumcised, because he wanted him to look like everyone else. "I wanted him to look like the other boys in the gym," Deston tells me. "I know how awful it is to be different when you're a teenager."

Robyn Eagles also circumcised her son for similar reasons. But when she got pregnant again eight years later, in 2011, her new husband, Jade, felt strongly that circumcision was unnecessary and cruel. "Statistically, if you look at the world population, the vast majority of men are uncircumcised," Jade points out. "Everyone has different body shapes. Your son's penis is not going to look exactly like yours anyway."

Yet Robyn fretted about having two boys in the house with different penises. I talked to Jade and Robyn on a three-way conference call when they were three months pregnant and still arguing about what to do. "It's a form of mutilation, isn't it?" said Jade. "I can't see that there's any benefit to it. If my child wants to decide to be circumcised, he can do so when he's a grown-up."

"No grown-up will ever decide that!" Robyn interrupted.

"Exactly!" Jade said. "So how can we do it to him as a baby?"

Robyn grew quiet. "I took care of my older son's pee pee for years, and it's what I'm used to," she confessed. "I'm having a mental issue thinking of having to take care of an uncircumcised penis. The penis is actually inside the foreskin and then the glans drops out. It just looks so weird."

A PAINFUL PROCEDURE

Adam Deutsch, an internist and cardiologist in New York City, chose to have his infant son circumcised with no anesthesia in a traditional Jewish ceremony outside the hospital. He dismisses the idea that circumcision causes pain.

"You put one drop of wine in the baby's mouth and they go to sleep. They're fine," Deutsch tells me. "He went right to sleep within a few moments after the procedure was done."

But other doctors, psychologists, and health professionals argue that falling into a deep sleep is one coping mechanism human babies use for dealing with extreme pain. Sylvia Fine, who practiced obstetrics for almost thirty years (and who is also Jewish), decided to stop performing circumcisions after she witnessed a baby writhing in pain from the procedure in a hospital in Honolulu.

"Doing things in sensitive parts of the body without anesthesia is cruel," Fine tells me over a cup of peppermint tea in her Cambridge apartment. She points out that infant pain reactions are similar to adult pain reactions: a newborn in pain will scream, have an elevated heart rate, and elevated adrenaline levels. There is also research that shows that a newborn's brain chemistry is altered in response to a painful event in the same way an adult's is. One study comparing pain responses of infant boys found that those who had been circumcised at birth were much less able to cope with the pain of immunization when they were a few months old. "They exhibit all the physiologic changes that grown-ups exhibit when they feel pain, and they scream," says Fine, pouring me another cup. "So it's going really far to say they are not mentally developed enough to feel pain just because they can't talk.

While debate often centers around the pain of the procedure as it is taking place, circumcision opponents point out that there is also the pain a baby feels in the days afterward. In the hours after the circumcision a baby's penis becomes swollen and angry-looking, so much so that Hardiman says many parents think, "Oh, God, what have we done?" But, she counsels parents beforehand, this kind of swelling is normal, and the angry purple scab will quickly fade to pale pink. In the meantime, a soft yellowy scab forms on the head of the penis, and should be left alone. She instructs parents to put Vaseline on the scab, since the wound on the penis may adhere to the diaper, which can also be painful.

"The baby goes home from the hospital with a wound where perfectly healthy skin used to be," says Georganne Chapin, the executive director of Intact America, a nonprofit health advocacy organization that opposes circumcision. "It will take one week to ten days to heal. Even if you kill the pain of the surgery with an anesthetic, it still stings when he pees, and interrupts breastfeeding."

Another consequence of neonatal circumcision is when the surgery itself is botched and the head of the penis, or even the entire penis itself, is mistakenly removed. In January 2003, a one-week-old baby lost the head of his penis when he was circumcised at the Maternity Center of Vermont in Los Angeles. In 2010 a Florida couple was awarded $10.8 million after their son lost the head of his penis during a Jewish circumcision using a Mogen clamp. Severely botched circumcisions are considered rare but between 2000 and 2011, the FDA received 139 reports of problems from circumcision clamps, including fifty-one injuries.

Another serious side effect of circumcision is blood loss. "We have about two babies come into the ER a month with penile bleeding," an ER doctor, who chose to circumcise her son, tells me. When she sees the surprise on my face—two babies a month at a small emergency room at a regional hospital sounds unusually high to me—she shrugs. "It's usually no big deal," she dismisses. "The parents freak out but, honestly, all you do is apply pressure and it stops. Sometimes I have to put in a stitch or two, but rarely."

But in a tiny baby whose entire blood volume is only a little more than an eight-ounce glass of water, even small blood loss can lead to anemia, dehydration, vitamin deficiency, and other complications, including death. Babies, especially fragile newborns, need the blood they were born with to stay in their bodies. Although small amounts of blood loss usually do not lead to severe complications, in 2008 Eric Keefe, a six-week-old born at Rosebud Hospital, an Indian Health Service Hospital in Rosebud, South Dakota, died from unnoticed blood loss following a circumcision. Another baby, Ryleigh, died in August 2002. Though his father, Brent

McWillis, was a medical professional who worked for Penticton Regional Hospital in Canada where Ryleigh was circumcised, he did not recognize the signs that Ryleigh was bleeding to death. "I didn't realize a baby can die from losing as little as one ounce of blood," his dad told a journalist.

Most bleeding after a circumcision resolves itself without the tragedy suffered by these families. But some doctors believe that the blood loss from circumcision, along with premature cord clamping, may actually be *causing* the vitamin K deficiency observed in newborns. Dr. Susan Markel, a Connecticut-based pediatrician, notes that ritual Jewish circumcision is performed when a baby has had eight days to begin building up natural vitamin K levels.

It is very difficult to tally the number of infant deaths directly related to circumcisions. It is not standard in American hospitals to perform autopsies (even when requested by parents, who must then shoulder the expense). And deaths following routine neonatal circumcision are often blamed on other causes. In England in 2007, when a baby stopped breathing just moments after being circumcised and later died in the hospital, Jewish leaders insisted that his heart failure was unconnected to the surgery. Another baby, Joshua, died in October 2010 at St. Vincent Women's Hospital in Indianapolis, unable to recover from being circumcised because of a congenital heart condition. In 2009, seven-month-old Bradley Dorcius died in the recovery room of SUNY Downstate Medical Center in Brooklyn after being circumcised as well as having corrective surgery on his penis. When David Reimer was an infant his circumcision at St. Boniface Hospital in Winnipeg, Canada, was so badly botched his penis had to be amputated. His parents named him Brenda and tried to raise him as a girl, giving him feminizing hormones to promote breast development at puberty. Confused and depressed about his identity, David (who had never felt like a girl and went back to being identified as a man) committed suicide in May 2004 when he was thirty-eight years old. Circumcision opponents, including the outspoken pediatrician Paul M. Fleiss, argue that David's death should be attributed to circumcision.

It's easy to dismiss what happened to David Reimer as extreme. But one recent estimate concludes that as many as 117 deaths from complications related to neonatal circumcision occur annually in the United States. Is any risk of death for a cosmetic elective surgery on a newborn really worth taking?

ARE BOTCHED CIRCUMCISIONS MORE COMMON THAN WE THINK?

David Llewellyn, a medical malpractice attorney based in Atlanta, Georgia, is currently litigating a second case against a doctor who cut off a portion of the head of a boy's penis several years ago but did not change his method of circumcising and, it is alleged, committed the same error again. Llewellyn has worked on more than sixteen botched circumcision cases and is currently preparing to file five more. He believes, mostly from what doctors themselves admit, that botched circumcisions are much more common than is reported.

In a Texas case he litigated, a boy had too much shaft skin removed during his circumcision. The child's own pediatric urologist offered video testimony for the defense stating that such a result was not malpractice because he repairs at least one case like the boy's a week.

"Now that's illogical," Llewellyn points out. "If he's repairing a circumcision a week then that doesn't mean it's not malpractice, it just means there's a lot of malpractice. That should indicate how risky it is." The daily experience of Dr. Andrew Freedman, a pediatric urologist based in southern California, supports Llewellyn's assertion: Freedman recently told a reporter that 20 percent of the patients he sees come to him because of issues related to their circumcisions.

The consent form that every parent must sign before a circumcision at Brigham and Women's Hospital in Boston states: "I/we understand that . . . significant complication occurs about once in every 500 circumcisions." Since approximately 1,250,000 boys are circumcised each year in the hospital, some 2,500 newborn boys suffer significant complications annually.

Even the most skilled doctors can easily botch a circumcision. Here's why: Though the word *foreskin* suggests that this skin is only at the tip of the penis, foreskin is actually also skin along the shaft that is loose and slides up and down. At birth, there is enough extra skin there that it is doubled over (think of the folds of neck skin on a pug). It stretches past the end of the penis and doubles back to rejoin the shaft skin. A circumcision is a circular cut that takes out the middle length of shaft skin that is between the base of the penis and the head. There is no line, however, that clearly delineates the foreskin from the shaft skin, so it is easy for a doctor to fail to estimate correctly the amount of tissue to remove. Though many circumcised men don't realize it, doctors often cut off too much shaft skin. If you cut off too little it looks like the penis has a short collar. If you cut off too much, the remaining skin on the penis has to stretch to make up for it, pulling hair-bearing skin from the base of the penis and the scrotum up onto the penis. "Any man who has pubic hair on the shaft of his erect penis has had a botched circumcision," Llewellyn contends.

In his practice in Minnesota, pediatric urologist Aseem Shukla finds he does about a hundred circumcisions and related procedures, including fifty-five to sixty do-overs, per year. Sometimes this surgery is cosmetic: Parents unhappy with how their son's penis looks, because the pediatrician was worried about removing too much foreskin, come to Shukla to get more foreskin removed. But just as often the problems are more serious. Circumcision can result in meatal stenosis, an abnormal narrowing of the pee hole at the end of the penis that can lead to blocked urine and other complications. Meatal stenosis is caused by injury as the newly exposed tip of the penis rubs against a diaper or a baby's own skin. One common symptom is a drop of blood at the completion of urinating. One study found that 24 out of 329 circumcised boys assessed developed meatal stenosis, numbers that suggest it may be the most common complication of circumcision. The procedure to fix meatal stenosis is called a "meatotomy" and is done by using a sharp instrument like a hemostat (which looks like skinny scissors) or a scalpel to split open the underside of the glans in order to enlarge the infant's pee hole. Though the procedure is fairly routine, it is not always successful.

Vicki Usagi, a mother of four from Charlotte, South Carolina, had her oldest son circumcised in 2000 because her mother said it was cleaner and that "everyone likes them to have little helmets." She started noticing something was wrong with his penis when he was about six months old and cried in pain during an erection. A few months later her son had blood in his diaper—extra skin on the shaft of his penis where he was circumcised was adhering to the head of his penis so tightly that it caused bleeding during an erection. After two years of pain, he was diagnosed with a buried penis, which is considered a circumcision complication. When her second son was born, Vicki did not circumcise him.

Skin adhesions on the head of the penis are a common problem that Shukla sees; a result, he says, of poor postoperative care. He also had a patient from Minnesota's Somali community whose doctor, a medicine man, inadvertently made a second hole in the urethra during circumcision, leaving the boy peeing out of two places. Though less experienced doctors working with him are aghast when they see how a boy's penis needs to be fixed after a pediatrician botched the circumcision, Shukla says the problems are mostly easy to repair: "Ninety-nine percent of the time it's just a nip and tuck," he tells me. Parents like Vicki (who now considers herself an anticircumcision activist) are less cavalier. They think it's time to put an end to routine infant circumcision in America so that we are no longer creating a problem with an unnecessary operation that then requires more surgery and more money to fix.

FORESKINS FOR SALE

What American parents do not realize, when they decide whether or not to remove a part of their son's penis at birth, is that the foreskin is tremendously valuable. Hospitals provide foreskins donated by parents at a cost (masking the fact that they are selling them by only charging shipping and handling fees) to the biotechnical industry, which uses them to create medical products that are then resold to the consumer. Infant foreskins are used to make high-end beauty products like injectable wrinkle treatments; artificial skin for burn victims, diabetics, and plastic surgery; wound dressings; hair regrowth products; and spa products. Danik MedSpa in Pembroke Pines, Florida, a luxury doctor-run resort for the affluent, boasts in rather scientifically inaccurate prose that its foreskin-derived products will help clients have healthier skin: "Medical skin care products contain a great concentration of active ingredients such as human dermal fibroblast, a naturally secreted and balanced combination of growth hormones. The molecules of these hormones strengthen the skin's natural ability to repair itself."

Though Kris Ghosh, a gynecologist-oncologist in private practice in San Diego, has successfully used Apligraf, a living skin substitute derived from infant foreskin, to reconstruct a woman's vagina after she had cancer, he is bothered by the price. The artificial skin, ordered through the hospital, costs about $2,000 for each six-inch circle. "Hospital systems make money on every end of it," Ghosh tells me. "They get money from cutting it in the first place, some profit from harvesting and selling the foreskin, and others for supplying the product made from it back to you."

Plug the words *neonatal human dermal fibroblast,* which is what cell lines made from infant foreskins are called by those who profit from them, into an Internet search engine and you will come up with more than a hundred and sixty thousand hits. On the first page of Google the majority of these search results are companies offering their foreskin-derived products for sale. At Genlantis, a San Diego–based for-profit company that sells human and animal tissues to scientists around the world, one petri dish of neonatal human dermal fibroblast (HDF)—fewer than 500,000 cells—costs $511. Cells are delivered on dry ice. The same product from ATCC (American Type Culture Collection, Inc.), a nonprofit clearinghouse of human and animal tissue that has a secure temperature-controlled warehouse with more square footage than two football fields in Manassas, Virginia, costs $399. In 2009, ATCC reported income of over $85 million.

Human dermal fibroblasts and human keratinocytes—both derived from foreskins—are listed among the top ten best-selling products for Cell Applications, Inc., a privately owned San Diego–based company. Like the other companies, Cell Applications, Inc., tests each of their lots of human skin cells cultured from infant foreskin for HIV, hepatitis B, and hepatitis C and guarantees that they contain no mycoplasma, bacteria, yeast, or fungi. According to their promotional materials, their customers include scientific researchers and biotech, pharmaceutical, and consumer product companies. For a first passage of these cells on a twelve-by-eight-inch tray, the cost is $730.

Other companies that sell skin cells cultured from infant foreskins include Lonza; Cell N Tech Advanced Cell Systems (Swiss companies that have branches or distributors in North America); Lifeline Cell Technology (a subsidiary of International Stem Cell Corporation), located in Frederick, Maryland; ZenBio, Inc., located in Research Triangle Park, North Carolina; Life Technologies Corporation, a global biotechnical company based in Carlsbad, California; PromoCell, based in Heidelberg, Germany; System Biosciences (SBI) in Mountain View, California; Applied Biological Materials (ABM), based in Richmond, British Columbia; AllCells, based in Emeryville, California; Millipore, in Billerica, Massachusetts; Allele Biotechnology, in San Diego; and ScienCell Research Laboratories, in Carlsbad, California—to name just a few.

As much as foreskins are used to make expensive dermatological consumer products, they are also used to conduct biomedical research. "If you want to study normal cell biology, you have to study normal cells," says David Bermudes, a microbiologist at California State University at Northridge. "So if you want to study people, where are you going to get the stuff? From placentas, umbilical cords, foreskins, surgeries, and dead bodies. Foreskins, placenta, and umbilical cords are very much sought after." Considered a waste product by some, foreskins are actually of enormous commercial and scientific value.

SHOW ME THE MONEY

After she finishes the circumcision, Hardiman completes the paperwork to bill the insurance company. She charges $300, though she says she expects to be paid between $100 and $300 depending on the patient's insurance. The doctor's fee is in addition to what the hospital will bill for the circumcision. These charges are sometimes ridiculously inflated and health care advocates advise parents to request an itemized bill of all charges, and to read those bills carefully. In the summer of 2011, when Stephanie Bottner, a mom from Somerville, Massachusetts, began researching how much a circumcision for her fourteen-month-old son would cost, she was given the runaround for months. When a ballpark figure finally came through from Massachusetts General Hospital in Boston, Stephanie was shocked. The young mom was told the total charge could be as much as $23,000: $18,000 to $20,000 for the surgery procedure, OR outpatient fee, and anesthesia. She would have to pay an additional charge of $762 to the doctor's office for the procedure, which takes less than thirty minutes. When a seasoned health journalist, Rachel Zimmerman, did a subsequent follow-up investigation of circumcision fees in Boston-area hospitals, she was told by Mass General that the total cost of a circumcision would range between $9,000 and $17,000. But "[a] complex case with an overnight stay would drive a higher charge," Mass General said in a statement. At Mount Auburn Hospital a billing specialist said a recent circumcision on a three-day-old baby cost $2,700. An area rabbi told the journalist he charges between $900 and $1,600 for an eight-day-old baby.

Overcharging for circumcisions (and other procedures) is one way hospitals generate more revenue. "We see mistakes on almost every bill we look at," says Pat Palmer, founder of Medical Billing Advocates of America, a consumer health care advocacy group that helps individuals and companies dispute inaccurate hospital bills. Pat remembers one bill for a circumcision sent to parents with a newborn girl. Dr. Ghosh says this "mistake" is a common one: "I see that happen all the time, that people get billed for things that are not services rendered. No one looks at their insurance bill as closely as they should. There are so many hidden costs."

But Dr. Shukla insists he has no financial motivation to do circumcisions, specifying that he bills between $75 and $468 for the procedure, depending on the patient's insurance.

"As a pediatric urologist, it's the least-paying procedure that we do, but we do them as a service," Shukla says. "I laugh when people think we do it for the money. If every one of my circumcisions went away tomorrow, maybe I'd lose $10,000 to $15,000. . . . The only way you break even is if you have to do a bunch of circumcisions, you schedule them together. You have to do three to four in an outpatient surgery center, and have to do them efficiently. I'm at a medical school and I train residents. The residents grumble and say, 'Oh, another circumcision.' I tell them, 'Look, you do the circumcision and the pediatrician will send you the undescended testicle or the obstructed kidney. You do the little cases so the referring doctor will send you the big cases."

When I toured Beth Israel Deaconess Medical Center and Mount Auburn Hospital, I was told at both that they used to provide foreskins to the area's biomedical companies. Neither hospital does so anymore. Hardiman tells me Mount Auburn stopped several years ago after their internal review board decided it was unethical. Kelly Lawman, senior media relations specialist at Beth Israel, steers the conversation in a different direction the minute a nurse brings it up. When I ask her about it later she declines to comment.

Most people outside the biomedical industry are unaware of the robust need for foreskins. Were these hospital review boards right? Is it an ethical violation for hospitals to sell—or even donate—a baby's foreskin to biomedical companies so they can make tremendous profits without *explicit* parental knowledge and consent?

Robyn and Jade's son, Merrick Matthew Eagles, was born on December 30, 2011, via a scheduled C-section at Mission Hospital in Mission Viejo, California. Every day a different nurse would come into the room and ask Robyn if Merrick was going to be circumcised. One nurse they talked to said that even parents without insurance will scrape together the $350 in cash to have their sons circumcised. But though Robyn wanted to say yes one day while her husband was home showering, she refrained. She and Jade decided to leave their son's penis intact and they both feel happy with their decision (it's Robyn's father who has had the hardest time with it).

When my research assistant at Brandeis phoned more than forty American hospitals to find out which anesthesia is used during a neonatal circumcision, she was told by staff at hospitals in upstate New York, Georgia, and Tennessee that only sugar water or no anesthesia at all is used during the operation, depending on the doctor's preference. Often hospitals contacted in Pennsylvania refused to disclose (their wording) information about pain medication during infant circumcision. A labor and delivery nurse at Henry County Medical Center in Paris, Tennessee, said they use an injectable lidocaine block and Emla (a topical cream), admitting that, "a lot of places don't use anything, but we do."

Circumcision is an often painful and sometimes dangerous operation, performed without consent, which is medically unnecessary for a newborn. In our for-profit medical system both the doctors who perform it and the hospitals where it is performed generate income from the procedure. It is not surprising, then, that the AAP's new position, written by a task force of stakeholders, all of whom, given their ages and nationality, are presumably circumcised or have circumcized partners, that circumcision's health benefits outweigh the harms, calls upon insurance companies to pay for it. "All you have to do is follow the money trail," suggests Tora Spigner, a doula and registered nurse with eighteen years of experience who works at Alta Bates Summit Medical Center in Berkeley, California, and describes herself as mother to an intact daughter and son. "If the intent is to circumcise, you can always find a reason to do it, no matter how trivial or unfounded. If the intent is to keep normal, healthy sexually sensitive tissue, there are more factual reasons to do so. But the AAP was not looking for facts, they needed support for their view and found it. If people look at the information objectively, they can see that."

After five years of examining the new scientific literature, this AAP Task Force could not find enough scientific evidence to justify recommending routine neonatal male circumcision. Dr. Beth Hardiman herself is outspoken in her dislike of the operation she was kind enough to let me watch: "It's a completely cosmetic procedure," she says. "I'm Jewish, and that's why we did it. Had I married a non-Jewish man, I would have thrown my body in front of the doctor before I would have let him circumcise my son."

Cost of circumcision: $105–1,420

Number of circumcisions performed yearly in United States: 1 million

Money made by hospitals performing circumcisions: more than $100 million

Circumstraint Newborn Immobilizer: $298.70

Surgical Gamco Circumcision Clamp: $532.40

Mogen Clamp (probe and scissors not included): $251.13

Disposable PlastiBell Circumcision Device (eliminates need for repetitive and costly sterilization): $183.38 (box of 25)

Twelve-by-eight-inch tray of foreskin-derived neonatal fibroblast: $730

Award by Fulton County jury to parents of a boy whose penis was severed during a botched circumcision in 2004: $2.3 million

Number of medical associations that recommend routine neonatal circumcision in America and around the world: 0

Cost to leave a boy's penis intact: $0

Skadi Hatfield: A Bloody Mess

Skadi Hatfield, a forty-year-old mother of two, works for a software development company. Her husband owns his own contracting business, making customized decks and bathrooms for clients in the greater Washington, D.C., area. They have a six-year-old daughter and a one-year-old son. Skadi's husband felt strongly their son should be circumcised.

I tried to convince my husband that we shouldn't circumcise our son. I was born in Germany and we don't do that there. But my husband is American and he said that he was circumcised, he was happy with what he had, and he wanted our son's penis to look like his. I tried to convince him and get support from my OB, but she told me I should leave the decision up to my husband because he is the one who has a penis.

While I was in labor, I had an epidural. I put my arms on my husband's shoulders. He was sitting in front of me on a rolling chair and fainted and hit his face on the foot of my bed. He had to go to the ER for a broken nose, so he was not there for Nicholas's birth. I told the OB that I didn't want to circumcise our son, but my husband did, and I asked her what happens if we don't agree. She said they would not do it if I said no. But since my husband and I had decided we would do it, I felt like I had to go through with it. I told the doctor to go ahead, even though I didn't want to. They had me put an anesthetic cream on his penis. I put it on but I didn't know how to do it, I'm not a doctor. I kept thinking that I shouldn't be the one putting on the cream. What if I didn't put it on in the right places? What if I didn't use enough? What if I used too much? A doctor or a nurse should take care of that, it's their job. But I guess they couldn't take two minutes out of their busy day to do it.

My husband had promised to be there when they did the circumcision. But he had been put on heavy pain medication and was at home. I went into the shower, which was across from the nursery, so I would have something to do after they wheeled him away. But even in the shower I could hear him. He screamed like crazy. It was a high-pitched scream that went through the walls. It was horrid. I've never heard a scream like that since. It just tears your heart apart. They brought him back to me and I couldn't stop crying. He had a perfect little penis beforehand and it was a bloody mess afterward.

Before the operation he was a pretty quiet baby. He would sleep a little and wake up and mutter for me, not really cry. But after he came back he cried so much. I tried to convince myself that it was because he was just two days old and he was hungry and my milk hadn't come in. But now I realize there had been more crying due to the circumcision. It's an open, raw wound. It must have hurt him, especially when he peed on it.

Now we are having a problem with what is left of my son's foreskin reattaching to the shaft of his penis. His penis is irritated. It looks like he has little pus pockets where smegma, the white stuff that naturally occurs under the foreskin, is getting trapped and accumulating. At his six-month checkup the pediatrician told me he had adhesions, which is a common complication from circumcisions. She said I should use a steroid cream to see if I can loosen up the skin. So once a day for the past six months I've been giving my baby steroids. That can't be good for him. The cream hasn't been working. But they told me that, as he gets older, the re-adhered skin will loosen up. If it doesn't, he might need another operation.

One of the reasons we did it in the hospital, even though I wanted to wait, is because the insurance would only pay for it if it were done there. I found out later that is because they consider circumcision a cosmetic surgery. But if it's cosmetic, why do we do it at all? The foreskin is there for a reason. It protects their little penises. And when you are older it helps keep your partner from drying out when you have sex. My other partners, not that I've had very many, were in Germany and were not circumcised. I never had a problem with vaginal dryness until now. My husband does not know how he would experience his sexuality with an intact penis because he is not intact. But sex is more pleasurable for both the man and the woman.

You get only half the information you need when you are deciding to do it. I know one woman who plans to have her breasts cut off because she has the gene for breast cancer. But that is her decision. We are cutting off a piece of a penis from a baby who doesn't make the decision for himself.

It was such a mistake to have done this to my son. I am so sorry. I didn't want to put him through that pain. I would tell other women to read as much as they can about it, educate themselves, and find someone who supports their decision even if their husbands disagree. You have to be strong for your baby. I hope other moms can be stronger than I was. If we were to have another child, that boy would not be circumcised.

CHAPTER 7

BOTTLED PROFITS:
How Formula Manufacturers Manipulate Moms

A nyone who has inhaled the scent of a newborn knows that the smell of a tiny baby is intoxicating; the sweetness of holding a baby skin-to-skin more addictive than fine chocolate. That almost primal feeling is intensified by breastfeeding. For Claudine Jalajas, a mom of three in Rocky Point, New York, breastfeeding was "the most amazing snuggle to have with your baby." For Jennifer Fink, a mother of four in Mayville, Wisconsin, breastfeeding gave parenting a sense of urgency: "If I'd been apart from my baby or if it had been a while since the last feeding, my breasts felt full, and I was as eager to return to my baby as my baby was eager to be close to me," Jennifer remembers. Jennifer's connection to each of her infants was both physical and emotional: "When I breastfed my babies, I felt relaxed and content. Breastfeeding gave me a perfectly reasonable 'excuse' to sit down, put my feet up, and cuddle my baby many times a day. The sight of a content, blissed-out baby at my breast brought me a deep sense of satisfaction, contentment, and blessing."

That deep sense of satisfaction was not just in Jennifer's imagination. Studies on both animals and humans show that breastfeeding provides a physiological mechanism to help a mother feel connected to her baby. As the milk flows from the mother's breast to the baby, oxytocin (also called the love hormone), which has been found in both human and animal studies to be beneficial in increasing feelings of trust, peace, well-being, and bonding, is released into a new mom's bloodstream. Nursing a baby actually makes women feel relaxed and drowsy. Since there are no bottles to warm or formula to mix in the middle of the night, breastfeeding helps new mothers who sleep next to their infants get more rest.

The benefits of breastfeeding for the mother continue long past the baby's infancy. Women who breastfeed are less likely to develop breast cancer, ovarian cancer, or endometrial cancer as well as rheumatoid arthritis, heart disease, and adult-onset diabetes. Exclusive breastfeeding keeps women from ovulating, providing a natural method for family planning. Since a pregnant woman's body stores fat during pregnancy in anticipation of breastfeeding, it has long been thought that exclusive breastfeeding also helps postpartum women lose weight. A Danish epidemiological study of more than 36,000 women confirmed this hypothesis, finding that moms who breastfeed more easily lose their pregnancy weight by the time their babies are six months old. While mothers who bottle-feed also bond with their babies, of course, scientists point out that women who do not breastfeed at all undergo the same sort of physiological changes as mothers who give birth to stillborn babies.

Virtually all new parents have heard the slogan "Breast is best," and most are aware that exhaustive scientific research has repeatedly shown tremendous benefits to breastfeeding for babies as well as their mothers. Some of the benefits to babies don't seem all that striking: Breastfed infants are less likely to have ear infections, flatulence, halitosis, digestive problems (like microbleeding of the intestines due to irritation or an allergic reaction to formula), gastrointestinal illnesses, respiratory tract infections, and even bouts of bad behavior in older infants. Others are a matter of life and death: An infant who has received some breast milk is 60 percent less likely to die of sudden infant death syndrome (SIDS, also known as crib death) than a bottle-fed infant; an exclusively breastfed baby is more than 70 percent less likely to die of SIDS. Breastfed babies are also at lower risk for developing a condition called necrotizing enterocolitis, where the lining of the intestinal wall actually dies and the tissue falls off. Less common in full-term babies, necrotizing enterocolitis affects 7 percent of infants who weigh between a little over a pound (500 grams) and 3.3 pounds (1,500 grams) and is "among the most common and devastating diseases in neonates," according to pediatric medical researchers from the Department of Pediatrics at the University of Florida and the Harvard Medical School. It is a condition so serious that 15 to 25 percent of the infants who get it will die. Very early feeding of breast milk has also been found to reduce significantly the risk of late-onset septicemia in premature babies, another life-threatening infection.

The health benefits of breastfeeding continue long past infancy: A breastfed child is less likely to become obese, develop diabetes, have asthma and other allergies, or get childhood leukemia. Breastfed babies have been found to have lower harmful cholesterol levels later in life, suggesting to researchers that "breastfeeding may have long-term benefits for cardiovascular health." An Australian study of more than seven thousand mother-infant pairs found that breastfed babies are less likely to be abused or neglected by their mothers. Another recent study, of fifty adolescents, showed that breastfeeding prompts the development of white matter in the brain (the tissue that contains nerve fibers and is responsible for transmitting messages throughout the body), particularly among boys. This study is consistent with myriad others that show that breastfed babies are more intelligent, scoring higher on tests of mental development between one and two years old, on intelligence tests during the preschool years, and even on intelligence tests at age ten.

Because of the overwhelming evidence that exclusive breastfeeding gives newborns the best start in life and is the healthiest option for mothers, both national and international health organizations urge moms to breastfeed exclusively. In 2005 the AAP issued a policy statement recommending babies be breastfed—*and only breastfed*—for the first six months, and continue to breastfeed for at least a year or "as long as mutually desired by mother and child." In its statement the AAP asserts that pacifier use may interfere with establishment of good breastfeeding practices and that formula or supplementation with other fluids should not be given to newborns unless medically indicated. The World Health Organization (WHO) concurs: "Breastfeeding is an unequalled way of providing ideal food for the healthy growth and development of infants . . . [I]nfants should be exclusively breastfed for the first six months of life to achieve optimal growth, development and health."

Yet despite these unequivocal recommendations, our breastfeeding rates are among the lowest in the industrialized world. In the United States only 77 percent of American women even initiate breastfeeding and of those only 36 percent are exclusively nursing when a baby is three months old. That means that of the 4.3 million babies born in the United States each year, only 1.5 million are nursing at three months. That leaves nearly 3 million babies still too young to sit up by themselves, who are no longer being nursed.

What's going on? Why is there such a discrepancy between what we know is best for babies and what American moms are actually doing? Why do so many American women stop nursing so soon while women in other countries breastfeed without difficulty?

IT ISN'T ALWAYS EASY TO BREASTFEED

As blood rushes to her chest after birth (providing the perfect temperature to warm a baby), a postpartum mom's breasts enlarge, first with colostrum, the precursor to milk that is rich in antibodies and nutrients for the baby, and then—anywhere from a few days to more than a week after giving birth—with milk, which will be mixed with colostrum at first. Some women have no real difficulties establishing nursing: Their baby seems to know what to do, and these first-time moms bumble along as best they can, responding to their infant's cues and figuring it out as they go. But most new moms and their infants need time, patience, and support to learn to breastfeed. Sore and tired from giving birth, in a state of heightened awareness and excitement from the flood of postpartum hormones and the fact of having just had a baby, and surrounded by people (some of whom are strangers if the birth has taken place in the hospital), a new mom has to get the hang of doing something with two intimate, private body parts that she has never done before. And she's not the only one learning. Her newborn too has to learn to extract sustenance from a breast that is probably much larger than his small, wrinkled head.

In contemporary American culture a bottle is a symbol of infancy, not a baby at the breast. Many, if not most, new moms were not breastfed themselves and most are unfamiliar with the process. I remember being astonished that there were several holes in each of my nipples. (I thought my breast would be like a bottle. Instead, breast milk is released through "milk duct orifices," which are the name for the many holes out of which human milk flows. There can be as many as eight or nine small holes in each nipple.) Many women may never have seen a newborn being breastfed when they were growing up. They may not know other moms who have chosen to breastfeed.

When I first tried breastfeeding, I was taken aback by how hard my baby sucked when she first latched on, how sore my breasts became, and how strange and new and almost surreal the experience was. In the first few days after she was born my nipples blistered and bled. I was frantic with worry: Would the blood harm the baby? A kind lactation consultant checked the latch and said it looked fine. She advised me to get plenty of air and sunlight on my breasts to help them toughen up and reassured me that what I was experiencing, albeit painful, was normal and would not cause the baby any harm. A few (very long and difficult) days later the blisters and cracks were gone.

When Annie Urban tried to nurse her son, who was born in Ottawa in 2005, nothing went right. Her son weighed nine pounds and Annie, twenty-nine, had a vaginal birth with no complications and an abundant milk supply. Still, baby Julian had trouble latching on.

Annie had not anticipated difficulties. "I just assumed that breastfeeding was really natural," she explains to me during a phone interview. "My mum had breastfed all four of us. She hadn't had any lactation consultant or anything like that. I always thought, 'Oh, those lactation consultants, that's kind of over the top.' I thought I'd listen to the nurses and what they tell me in the hospital and everything would be fine. But it wasn't. No matter what I tried I couldn't get him to latch on at all."

The hospital doctors suspected Julian had "tongue-tie," a condition that probably affects 3 to 4 percent of newborns where the membrane under the tongue that attaches it to the floor of the mouth (called the frenulum) is so tight that babies have trouble moving their tongue well enough to nurse. They clipped Julian's frenulum to allow him more freedom of movement. But the procedure didn't alleviate the problem. It wasn't until Julian was examined by an ear, nose, and throat specialist, when he was seven weeks old, and the physician reclipped his frenulum that the problem ameliorated. It still took another seven and a half weeks for Annie to be exclusively nursing.

NOT ENOUGH SUPPORT FOR BREASTFEEDING MOMS

If Annie had been living in the United States, there's a good chance her son would have been bottle-fed. A working mom, Annie was allotted a year of partially paid maternity leave from the Canadian government. Even though she credits her own stubbornness, Annie also believes Canada's generous leave helped: "If I had to go back to work right away, I don't know if I would have been able to work through all those latch and tongue-tie issues with my son," she admits.

The United States is the only country in the industrialized world that does not mandate paid leave for new mothers. In 2009–2010 alone, the AAP accepted more than $1,530,000 from the formula industry, including money from the International Formula Council, an industry-sponsored lobbying and public relations group that has "Infant Formula: A Safe, Nutritious Feeding Choice" as its Internet motto. The combination of no paid maternity leave for new moms, no regulation on formula-sponsored advertising campaigns, and corporate-sponsored medical societies helps explain why America ranked *last* (thirty-sixth out of thirty-six) on Save the Children's 2012 assessment of how well wealthy countries support breastfeeding moms.

Though we tend to blame new moms for not nursing, in our country, it is often doctors or other medical personnel themselves who undermine a woman's ability to breastfeed. Leslie Ott always knew she wanted to breastfeed. After her daughter Ella was born at a hospital in Glendale, Arizona, in 2008, a nurse gave Leslie a log to record every feeding. Breastfeeding was much more difficult than Leslie expected: Her nipples felt like they were being painfully pinched every time Ella latched on, and her daughter was waking at least once every thirty minutes to nurse. The morning after Ella's birth the pediatrician walked into the room as Leslie was attempting to breastfeed. "Oh, you must be hungry. You look like you're starving," Leslie remembers the doctor saying to her baby. The twenty-one-year-old new mom was devastated. "To hear the pediatrician tell me I was starving my baby and had no milk was just such a blow. I figured, I'm so young. I had no issues with my pregnancy. Why would my body stop now? Why wouldn't it do what it was supposed to do to keep this baby alive?"

Ella's birth had been difficult. Leslie's husband was out of town and the labor nurse had become angered when Leslie told her she did not want an epidural, informing Leslie that she had just had a baby and couldn't do it without one and Leslie wouldn't be able to either. By that point Leslie's contractions were so intense she couldn't answer. Dumbfounded by the nurse's hostility and mistakenly believing she was only three centimeters dilated despite the intensity of the contractions (she found out later she was nine), Leslie didn't know how to respond. The nurse left the room, coming back a few minutes later: "Are you done feeling what natural labor feels like?" she scoffed. "Do women who have natural labor walk around the mall with gold stars on their foreheads, do they get a trophy?" After the epidural, Leslie's labor stalled, her legs and pelvis went completely numb, the doctor noticed meconium in the amniotic fluid, and a team of neonatologists rushed into the room. "I couldn't feel anything," Leslie remembers. "The nurse is trying to tell me when to push. I'm pushing as hard as I can. I have no idea if I'm pushing, or what's really happening."

So when the pediatrician said Leslie was starving her baby, Leslie was already feeling like a failure. The next time she made rounds, the pediatrician told Leslie that Ella's jaundice was worsening and threatened to admit Ella to the NICU and discharge Leslie if she didn't agree to give the baby a bottle. She also told her that if she nursed for more than fifteen minutes on a side breastfeeding would not work. Feeling bullied but wanting to do what was best for her baby, Leslie gave Ella a bottle of hospital-provided formula. When she and Ella were ready to leave, the hospital loaded her up with artificial milk. "They sent me home with a huge clear garbage bag full of formula samples. I had enough to last me a whole month," she tells me when we talk on the phone.

If you ask Ruby Wentz, twenty-six, she will tell you she had trouble breastfeeding because her body did not produce enough milk. When Ruby gave birth to her daughter via C-section at a hospital in Casper, Wyoming, in 2010, Baby Jaelynn weighed almost nine pounds. Less than an hour later a nurse told Ruby the baby needed formula because she was so big. Though the nurse's instructions confused her, Ruby did not question them.

"She said she was really hungry and starving," Ruby explains. "I didn't want my baby to starve. They pushed for the baby to have formula, I don't know exactly why." Though she pumped her breasts for the first three months and fed the milk to her baby in a bottle, Ruby was only getting between two and four ounces at a time and the pediatrician advised her to add formula to Jaelynn's bottles, even though it made the baby spit up more. Ruby switched exclusively to formula when Jaelynn was four months old.

When she gave birth at a hospital in Downey, California, Lana Wahlquist, twenty-eight, was told by a nurse that her hours-old baby was choking on breast milk.

"I don't hear anything," Lana protested.

"It's silent," the nurse answered back. "Like drowning."

The health care professionals advising Leslie, Ruby, and Lana all, we have to believe, had the best intentions. Their goal was to assure the newborns in their care had the best possible outcomes, not to criticize or frighten these new moms. Yet their advice was not based on medical evidence but instead on personal prejudice and misinformation. Instead of doing everything they could to support the breastfeeding relationship, these providers did everything they could to undermine it.

Two days after Melissa Bartick's second son was born via scheduled Cesarean section at a major Boston teaching hospital, the hospital pediatrician was concerned that her son had lost 14 percent of his body weight. He told her that the baby's sodium levels could get high, he could have seizures, and that it was time to start supplementing because her son was not getting enough milk from her breasts. Melissa remembers the doctor saying, you know, you need to give him formula.

A medical doctor herself, Melissa had nursed her firstborn but only after great difficulty. Even once breastfeeding was finally going smoothly, she almost gave up. Her son was two and a half months old, she had gone back to work full-time, and she was exhausted. She used an electric pump to extract milk from her breasts but was discouraged by how little milk dripped into the bottles. The medical assistant at her pediatrician's office who answered her desperate after-hours call suggested Melissa keep the baby in bed and nurse lying down. Melissa hesitated because most of her friends had kept their babies in separate rooms, but the medical assistant (whom Melissa describes now as an angel) explained that with the baby in the same room, she wouldn't need to turn on a light or get out of bed in the middle of the night. Breastfeeding would be easier, and she could get more sleep. It worked: Melissa nursed her first son until he was nineteen months old.

In the intervening two years she educated herself about breastfeeding: She knew the importance of avoiding pacifiers and bottles in the first days of life because both can undermine an infant's ability to latch to the breast. She learned that breastfeeding in the first hours of life when a newborn has a period of alertness was essential for optimal infant-mother bonding. But the medication from the C-section made her so nauseous Melissa had thrown up twice while trying to nurse her new baby in those early hours. Now a medical colleague was instructing her to go against everything she read about the importance of exclusive breastfeeding. As both a doctor and a vulnerable woman who just had a baby, her instinct was to listen to the doctor's advice. But Melissa suspected the reason for her son's weight loss had more to do with him being excessively fluid overloaded than with being dehydrated.

"The weight was all the extra fluid that he'd gotten from the labor intervention I'd had," she explains. "When he was born he was so swollen with extra fluid, he looked like a round butterball. A fourteen percent weight loss is normally very concerning. But I could tell he was not dehydrated: his mouth was moist, he was making saliva, his latch was good, there was no pain at all, and I could hear him swallowing my colostrum. He seemed happy and healthy."

When the hospital pediatrician warned her that the baby could have convulsions without formula supplementation and insinuated Melissa was being a bad mother for disregarding his advice, Melissa requested he test her son's blood to see if his sodium levels were actually dangerously high. "You would never treat an adult for hypernatremia [high sodium levels] without actually checking it," she tells me. "Or at the very least you would draw the lab to confirm the diagnosis while treating the patient at the same time. Why would a baby be held to a different standard than anyone else? Formula is not a benign intervention. If you don't really need it, you shouldn't have it. If he really needed it, then he should have it, but I didn't want to give in without a diagnosis."

The pediatrician never ordered the blood test. The next day Melissa could feel her milk coming in and her son was guzzling like a champ. "He gained two ounces in one day," Melissa says. "There was no problem."

Unfortunately, the way Leslie, Ruby, Lana, and Melissa were treated is the norm rather than the exception in many American hospitals. Mothers are wrongly told by misinformed doctors or other medical staff that they must give their babies formula or their babies will die. "I work at Cedars-Sinai Medical Center, Saint John's Health Center in Santa Monica, and Santa Monica–UCLA," says Jay Gordon, a pediatrician with thirty-three years of experience in private practice in southern California and a noted breastfeeding advocate. "At every single one of those hospitals *in the last two weeks* I've had mothers who were told their babies were losing too much weight and they should supplement with formula." Dr. Gordon talks quickly as he pulls his car into the parking lot of Cedars-Sinai Medical Center, where he is going to make rounds. "It's 2012," he laments. "These are three of the best hospitals I know of, and still this is what women are being told every day, even though every piece of research says it isn't true."

When a woman is told before she even leaves the hospital that her breasts are not adequately producing milk, it makes it very difficult for her to successfully breastfeed. But what is really going on? "Every mother is told that it is good to breastfeed, but we make it as hard as possible for them to do so," declares Marsha Walker, an articulate, compassionate, no-nonsense registered nurse and an International Board Certified Lactation Consultant (IBCLC), who has been helping women breastfeed for more than thirty years and is the author of *Selling Out Mothers and Babies: Marketing of Breast Milk Substitutes in the USA*. Walker points out that trouble initiating breastfeeding is often medically induced in the first place. Medicated hospital birth—especially one that includes other interventions like vacuum extraction, epidural analgesia, separation of the mom and baby, delayed nursing, and nonmedical formula supplementation—often leads to women having breastfeeding difficulties. Though it sounds counterintuitive, Walker points out that breastfeeding difficulties actually start *during labor*. Epidurals can cause fever in the mothers, which necessitates the infant being taken to the special care nursery for a septic workup, separating the mother and baby and delaying breastfeeding. Then, if the newborn doesn't nurse within a prescribed period of time, the staff may insist on formula supplementation. "The epidural prolongs the second stage of labor. You wind up with a nice little baby who sits there and does nothing," Walker explains. "He's got a headache and all these extra fluids in him. But if he doesn't feed right away, someone is going to say, 'We need to feed him in order to prevent hypoglycemia.' Out comes a bottle of formula. Now you've intervened again.

"This baby already wasn't latching on because of all the labor and delivery interventions, or maybe he just needs a little more time to acclimate to his new environment," Walker continues. "Now you put an artificial nipple in his mouth and the chances of him actually nursing go down even further. Or maybe he does nurse right away but the mother gets sore nipples because the baby isn't positioned well and there's no one there helping her. She has minimal support in the hospital, gets discharged soon after the birth, and has no support when she leaves. It's like setting up a row of dominoes, the minute she walks into the hospital all the dominoes get pushed down. Then we wonder why she can't breastfeed."

PEDIATRICIANS PUSHING FORMULA

When Sandra Musial, a pediatrician based in Saunderstown, Rhode Island, who teaches pediatric residents at Hasbro Children's Hospital, first took a job at Narragansett Bay Pediatrics, she was thirty-one years old and had a feisty (later she would realize he was difficult) infant named Max. Musial liked the practice because all the doctors had young children, valued family life, and seemed really up to date.

But as she started practicing, Musial realized the doctors weren't as progressive as she had believed. Two had adopted their children, so they didn't breastfeed, and the male doctor had no personal experience to draw on. Though the practice *ostensibly* recommended breastfeeding, the pediatricians had little idea how to support moms having difficulties. "The breastfeeding education that pediatricians receive has historically not been good," Dr. Musial explains. "They weren't educated about the nuances of breastfeeding. They feared that solely breastfeeding would lead to dehydration; they were quick to recommend supplementation."

The least-senior doctor in the practice, Musial tried to explain diplomatically to her colleagues options they could offer new moms to ensure breastfeeding success—referral to a knowledgeable lactation consultant, pumping to increase milk supply—instead of the quick but ultimately damaging fix of formula. They reacted defensively. "When I gave them feedback they didn't say, 'Oh, thanks so much.' Instead they would say, 'Well, I was worried about the baby. His urine output was dropping and I was afraid his sodium would get too high.' "

The lack of breastfeeding knowledge is partially a result of inadequate medical training. "Most pediatricians don't receive adequate breastfeeding training in school," says Nancy Mohrbacher, an IBCLC with more than twenty-nine years of experience providing breastfeeding support in the Chicago area. "They don't necessarily know any more about breastfeeding than your next-door neighbor."

But it wasn't just the pediatricians' inadequate training and lack of personal experience with breastfeeding that was leading them to recommend formula. Salesmen visited the offices every week, leaving free samples, coupons, and posters. There was formula in every room. "We had samples everywhere and we had formula bags," Musial says. "We had it sitting out on the counter. We had coupons at checkout; and ROSS formula posters up on the walls. It was totally pervasive."

Musial's senior colleague, Dr. Maggie Kozel, had breastfed her own daughters. "Compared to other practices, we were so much better," argues Kozel. "We wanted the babies in our practice to be breastfed. We thought we were supporting breastfeeding. But then we'd fall short because we lacked the confidence to keep supporting a breastfeeding mom when she ran into difficulties."

It took Kozel years to realize how their practice was undermining nursing women. "The formula reps would hand-deliver it on huge dollies. We were always well stocked. We'd get these adorable teddy bears and insulated ice packs from them as well. We were constantly supplied with products and free lunches. They'd bring food or they'd order pizzas for us. It was hardly high graft," Kozel laughs. "If I were going to sell my soul, I'd do it for a lot more money. But it was very much a commercial intrusion into our practice."

HARDLY HIGH GRAFT?

Abbott Laboratories, headquartered in north suburban Chicago, employs 91,000 people worldwide. Abbott makes Similac brand formula for infants, Gain "growing-up" milk for toddlers, and PediaSure products, sugar-laden plastic packaged drinks that come in vanilla, chocolate, strawberry, banana, and "berry" for children over two. Abbott reported $38.9 billion in sales in 2011. In addition to robust pediatric beverage sales, Abbott made record earnings selling diabetes management products and pharmaceuticals. It's a good business strategy: manufacture sugar-laden pediatric "growth" products as well as medicine to treat the diseases they cause.

Mead Johnson Nutrition, the Glenville, Illinois–based company that makes Enfamil and other artificial milk products, reported earnings of $3.1 billion in 2010. Its annual report boasts that sales of "Enfa" brand products grew 17 percent worldwide. In 2011–2012, Mead Johnson and

Abbott Nutrition were two of the five companies in the highest donation bracket to the American Academy of Pediatrics, together donating $1.5 million to the AAP.

The third-leading formula manufacturer in the United States, Gerber, is now owned by a multibillion-dollar Swiss food company. On May 21, 1994, Gerber's board of directors voted to sell the company to the Swiss drug manufacturer Sandoz, Ltd., for $3.7 billion. By then the company had strayed so far from its original mom-and-pop all-American profile that no one in the business world was surprised to see it bought by a European pharmaceutical company. Two years later, Sandoz merged with another Swiss drug company, Ciba-Geigy, and became Novartis. Eleven years after the merger, in 2007, Nestlé bought Gerber from Novartis for $5.5 billion. Nestlé, the largest food company in the world, is based in Vevey, Switzerland. It has factories or operations in almost every country in the world, employs some 330,000 people, and did almost 83.7 million CHF (Swiss francs) in sales in 2011 (about US$88 million). But the multimillion-dollar Swiss food giant has the smallest share of the formula market in the United States (approximately 15 percent). Its donations to the AAP were similarly modest, between $25,000 and $49,999 for the fiscal year 2009–2010.

In the five years following the AAP's unequivocal endorsement of exclusive breastfeeding for the first six months, the formula manufacturers alone have donated more than $6.7 million to the organization. Formula manufacturers also contributed at least $3 million to build the AAP's headquarters in Elk Grove Village, Illinois.

Robert Block, M.D., the head of the AAP, agreed to speak to me by phone about the ethics of corporate sponsorship, but canceled a few hours before the interview because "something came up" in his schedule. Instead, the AAP sent me a written statement that corporate sponsorship does not compromise but rather enhances the organization's ability to meet the needs of America's children. "The AAP accepts corporate funding to allow it to carry out many of the programs that help pediatricians provide care for children in the U.S. and around the world," the statement reads. "Funds received from corporations for sponsorship represent a small portion of the overall budget. On average for the past four years, 8 percent of the AAP's income comes from corporate sponsorships and grants (federal and non-federal government grants account for 10 percent). When corporate contributions are accepted, the AAP ensures that safeguards are in place so that these donations do not influence AAP policies or recommendations.

By accepting outside funding with the right safeguards in place, the AAP is able to reach a much larger audience of parents and pediatricians with its messages. This amplifies the positive impact we are able to have on children's health and fosters the mission of the AAP, which is to attain optimal physical, mental, and social health and well-being for all infants, children, adolescents and young adults."

Can accepting millions of dollars from the formula industry really have no impact on the AAP? Despite what Dr. Block describes as a mutually beneficial relationship between the AAP and America's children, the trade organization for pediatricians refuses to be transparent about the actual dollar amount obtained from corporate donors. Abbott is listed as giving "$750,000 and above" in AAP's Honor Roll of Donors for the fiscal year 2010–2011. When asked for the specific amount, an AAP spokesperson responded: "I'm sorry, but that is proprietary information that is not given out to the public." If there is no conflict of interest, why the secrecy?

But it's not just that professional medical organizations like ACOG and AAP accept money from corporations that sell products that undermine breastfeeding (and children's health in general—Coca-Cola, PepsiCo, and McDonald's are also regular contributors to the AAP). Some doctors from these organizations are closely aligned with corporate America. Steven M. Altschuler, M.D., the current chief executive officer of the Children's Hospital of Philadelphia, is a member of Mead Johnson's board of directors, alongside former pharmaceutical executives, food industry analysts, business consultants, a private investor, and one former oil company president.

"The vast majority of doctors in my department believe it's wrong to take handouts from formula companies and corporations like Coca-Cola," says Stefan Topolski, M.D., assistant professor of family medicine and community health at the University of Massachusetts Medical School, when we meet at the Northampton VA Medical Center in Leeds, Massachusetts, where he has just finished an all-night shift. "It's wrong and it's embarrassing," Topolski continues. "The biggest danger is loss of professional standing with the public and our patients who see us as a profession that cannot act independently or stand on our own." In addition to his academic appointment and moonlighting at the VA, Topolski runs a rural nonprofit practice in western Massachusetts and is himself married to a doctor. "Exposure to free samples and financing biases every aspect of what we do."

MISLEADING FORMULA ADVERTISING

Artificial infant milk manufacturers have only one goal: to sell formula. In order to do so, they need to convince both health care professionals and parents that the products they sell are better than the cheaper natural human alternative, breast milk. They also have to convince parents that their brand of formula is superior to their competitors'. Corporate decision-making is targeted toward increasing formula sales and by extension decreasing breastfeeding. Aggressive advertising in the United States and other countries that do not regulate advertising toward pregnant women and young children (places like China and Niger, and elsewhere in the developing world) is one of their most effective ways of increasing sales. According to a 2010 financial report, Mead Johnson Nutrition says that their profits were "benefited significantly from [their] increased investments behind advertising and promotion. . . ." Mead Johnson tells its shareholders its mission is to "nourish the world's children for the best start in life." But its advertising and promotional campaigns have been so duplicitous that the company has lost three federal lawsuits over false advertising, been found guilty five times of running deliberately misleading ads, and been called the "poster child for corporate wrongdoing" by CBS News.

One Mead Johnson ad depicts an infant's vision at twelve months with a worryingly out-of-focus picture of a yellow rubber duck (infants who use other formula brands) juxtaposed with a sharp image of the same duck (to depict the vision of infants who drink Enfamil PREMIUM with Triple Health Guard). In another, a television actor playing a pediatrician tells moms to be picky about the formula brand they use, and then marks off a checklist: "Growth," "Brain and Eye," and "Immune System," suggesting only Enfamil provides those benefits.

No scientific studies have shown that any brand of formula is superior to any other. (The science is clear and overwhelming, with more than a thousand studies confirming that breastfeeding is the only way to ensure optimal infant growth, brain and eye development, and immune function.) A quick look at the ingredient list of branded infant formula reveals that it is virtually identical to generic formula. According to CBS, manufacturers even get their ingredients from the same suppliers. The only difference is cost.

How does Mead Johnson explain this deception? I called Christopher Perille, the company's longtime spokesperson, and arranged to do an in-person interview at their Glenville, Illinois, headquarters so I could hear Mead Johnson's side of the story. When Perille had not confirmed the time of our appointment the day before I was flying to Chicago, I called again. "You caught me off guard," Perille mumbled, talking to me from his cell phone in the car and claiming to have sent me an email (I did not receive an email and when I asked him to resend it he did not). Perille then said he'd "done some research" on me and was no longer available. I went to the Mead Johnson headquarters, in an unassuming building on Patriot Boulevard, the day we were slated to meet anyway, hoping to convince him to change his mind. Three larger-than-life photographs of a baby greet the visitor as you exit the elevator and turn right to where a heavyset security officer dressed in a coat and tie sits behind a desk. On both sides of the small lobby are displays of Mead Johnson products in English and other languages and a framed *Wall Street Journal* article from 2009 announcing that the company had a higher-than-expected IPO.

Wearing a pinstriped button-down shirt with short-cropped gray hair, Chris stepped into the lobby. He was much too busy to meet, flying to Atlanta next week and then Brazil and Mexico. No, I couldn't have a copy of the limited edition *A Century of Caring: Celebrating the First 100 Years of Mead Johnson & Company,* the corporate-sponsored book about the company's history, because they had none to spare, but if I left my contact information (which he already had), he'd see what he could do (he never sent me one). I had more than an hour before the next train to Chicago and it was more than ninety degrees outside so the security officer let me wait inside. Before I left Chris strolled out again, a coffee cup in his hand. He was visibly startled when he realized I was still there. "I'll catch up with you later," he called to the security guard, turning on his heel and rushing back through the door.

THE LACTATION CONSULTANT SCAM

When Margaret Pemberton was having trouble getting her newborn to latch, a lactation consultant sent by the hospital or her pediatrician (she can't remember which) came to her home in Wilton, Connecticut. The lactation consultant looked concerned as Margaret struggled to nurse, shook her head, and told Margaret she didn't know why it wasn't working.

"She wasn't helpful at all," Margaret recalls. "She would bring me huge samples of formula every time she came. She also mentioned to me that she never had a child, nor had she breastfed, and after seeing how hard it is, she would not breastfeed herself."

Margaret pumped for a month until her son finally managed to latch on. But even then breastfeeding was really hard—her son was allergic to the cow's milk in her diet so she went on a restricted diet. She produced so much milk that it was difficult for her son to nurse and uncomfortable for her. "It was so painful and so miserable and I hated every second of it," the mom of three admits.

Anyone can identify herself as a lactation consultant, even if she lacks the qualifications to help a new mother learn to breastfeed. Formula companies make "experts" available to advise women on how to breastfeed. In June 2010, Similac, owned by Abbott Nutrition, began advertising a service to new moms offering support from their staff of "Feeding Experts." At http://similac.com/feeding-nutrition/ is a toll-free number where new moms can get live one-on-one help. On the same page is a button to click to sign up for up to $329 in free stuff. Below the phone number are links to articles. On the day I looked at the site these articles were about how to get started with formula, what to do about baby spit-up, and how to sign up for Similac's free Baby Journal App. If you don't want to call the hotline right away, you can schedule a call for later.

According to the Similac website, "Someone from the Feeding Expert team will answer your call. This initial contact will be able to answer the majority of your questions. Nurses and dietitians also will be available to talk with you. If they all are busy when you call, a qualified expert will get back with you as soon as possible. . . . If you have questions related to breastfeeding, a lactation consultant will be available to help you. Lactation consultants are provided by a third party." When an Internationally Board Certified Lactation Consultant (IBCLC) from Florida called and asked the "Feeding Expert" she reached about her qualifications, she was told the "expert" was a Certified Lactation Educator (CLE) and had obtained that certification by taking an eighty-hour online course.

IBCLCs are required to complete ninety hours of education about breastfeeding and a thousand hours of clinical practice working directly with moms and babies and breastfeeding. Most candidates complete their clinical hours by finding an Internationally Board Certified Lactation Consultant in a hospital or private practice, shadowing them as they help new moms and babies, and then themselves help mothers under the supervision of a mentor. They must then pass a rigorous exam that tests their knowledge of breastfeeding practices.

Like American nurses, pediatricians, obstetricians, and others in the health care profession who have daily contact with women and children, Internationally Board Certified Lactation Consultants are held to a code of conduct: They agree to "disclose any actual or apparent conflict of interest, including a financial interest in relevant goods or services, or in organizations which provide relevant goods or services"; "ensure that commercial considerations do not influence professional judgment"; and adhere to "the principles and aim of the *International Code of Marketing of Breast-milk Substitutes* and subsequent relevant World Health Assembly's resolutions. Similac "feeding experts" do not.

On every page of Similac's advice site are advertisements for Similac. While some of the information is useful and accurate, most of it is written to make breastfeeding seem complicated, difficult, and even scary: New moms are advised to "position [a baby's mouth] over the pockets of milk located 1 to 1½ inches behind the nipple," warned against blocking his nostrils with the breast (in which case you should "pull his bottom upward and closer to you so his head will move back slightly"), and told to chart the baby's urine output, stool patterns, weight gain, and breastfeeding patterns. Just reading Similac's advice on how to get started breastfeeding is enough to intimidate a frazzled new mom and make her want to stop. This, of course, is Similac's intention: to appear to support breastfeeding while simultaneously undermining it in any way they can.

Another effective strategy used by artificial milk manufacturers to sell more formula is to mail free samples directly to new mothers. Erin Kotecki Vest, thirty-six, a mom of two in Valencia, California, was shocked when she received a package with two large canisters of Similac in the mail three weeks after she had been at the Henry Mayo Newhall Memorial. Erin has lupus. Inflammation in her pelvis destroyed her uterus, ovaries, and parts of her colon. Erin had been in the hospital for an unwanted hysterectomy.

"I have no doubt the company, given my recent stay in the 'Women's Ward,' assumed I had given birth. So they sent out their usual package attempting to deter women from breastfeeding and doing what is best for their newborn in order to make a profit. I found it disgusting not only that they would solicit new mothers this way, but also not take the time to see I was not a new mother, but a woman grieving," Erin, who was too sick to talk by phone, wrote me in an email.

Studies have shown that more than 95 percent of mothers given free formula samples in the hospital will purchase that brand of formula if they buy formula for their infant. So formula companies actively court American nurses who have daily contact with new parents in the hospital. In almost every hospital in America, with the exception of the some 143 hospitals and birth centers that have adopted the UNICEF/WHO Baby-Friendly Hospital Initiative policies (about 2 percent of all American hospitals and birth centers), formula sales representatives have often unlimited access to nurses, showing up at the maternity ward to ply doctors and nurses with branded calendars, crib cards, tape measures, buttons, notepads, pens, coffee mugs, name tags for stethoscopes, water bottles, and more. On one unit where nurses who collected the most formula bottle caps won a gift certificate to Victoria's Secret, there were reports of nurses walking into patients' rooms and opening a formula bottle, setting it next to the mother, and pocketing the cap.

The *International Code of Marketing of Breast-Milk Substitutes* is an international document written by the World Health Organization to protect new moms from unethical and potentially predatory advertising that might discourage breastfeeding. Since suboptimal breastfeeding has a global negative impact on health, the *Code* (as it is often called) was conceived as a public health document that could have global impact. In 1981, it was overwhelmingly approved by 118 countries at the World Health Assembly. Three countries abstained. Only one country—the United States—succumbing to pressure from corporate interests, actually voted against it. The *Code's* instructions are clear: "Health workers should not give samples of infant formula to pregnant women, mothers of infants and young children, or members of their families."

Sylvia Fine, an obstetrician who practiced medicine for almost thirty years, argues that any interaction between formula representatives and hospital nurses is detrimental to breastfeeding. "The people who work in postpartum and the nursery would never tell you they feel beholden on account of having received a cup of coffee or lunch from a formula representative," Fine argues as we drink spearmint tea in her Cambridge apartment. "People don't admit that, but it's been proven with studies on medical residents and interns—that one simple cup of coffee affects prescribing patterns."

The nurses become friends with formula salesmen and enjoy doughnuts or pizza delivered to the postpartum floor, often in the middle of the night when they are working long, lonely, exhausting shifts. These close ties, coupled with formula-sponsored continuing education courses, have helped turn American nurses—often despite their best intentions and without their conscious complicity—into formula couriers, ignoring the fact that *exclusive* breastfeeding is one of the most important factors in a newborn's health outcome, and that formula, especially within the first days of life, has a negative impact on a baby's health that can take months to undo.

Formula companies actually pay medical professionals to teach continuing medical education classes. On June 16, 2010, Mead Johnson Nutrition, which makes Enfamil, sponsored a nursing conference called

"48 Hours or Less: Identifying the Newborn at Risk." The keynote speaker, who was paid by Mead Johnson, was Madge Buus-Frank, a registered nurse who was touted as having a "passion for excellence in caring for infants and families." She exercises that passion as a neonatal nurse practitioner at the Children's Hospital at Dartmouth and also as an instructor at Dartmouth's medical school. Nurses who attended the event, held conveniently on Disneyland Drive in Anaheim, California, received continuing education credit hours required for their licenses.

I tour John H. Stroger Jr. Hospital on the South Side of Chicago, which has a state-of-the-art maternity ward and Level 3 NICU that serves many of Chicago's poorest families. Moms on Public Aid in Illinois can go to any hospital, Rosemarie Mamei Tamba, R.N., the head of the Maternal Child Nursing Division, tells me, but many suburban hospitals have quotas and refer Public Aid moms to John Stroger. About one third of the some one thousand moms who deliver at the hospital are high-risk—they have diabetes, heart problems, and even cancer or renal failure. We know that breastfeeding is especially beneficial for premature babies, but the NICU babies I see have ENFAMIL-branded pacifiers in their mouths or at the foot of their plastic bassinets. Stacks of name-brand formula line the shelves just under eye level.

"There's a bag we get from Enfamil," says Rosemarie, who's bustling, good-natured, affable, and obviously well intentioned, "it has some goodies in it." Rosemarie feels she is doing a valuable service for these less affluent moms, overlooking the fact that their health and budgets would be better off if they exclusively breastfed. According to the procurement office, there is no formal agreement between the hospital and the formula manufacturers, but Mead Johnson is the hospital's sole artificial milk supplier. They donate their products for free.

Nurses are under an ethical obligation to put their patients' interests first. The largest nursing association in the United States, American Nurses Association (ANA), has 200,000 members and a duty to represent "the interests of the nation's 3.1 million registered nurses." It has adopted a fourteen-page code of ethics that states: "The nurse's primary commitment is to the patient, whether an individual, family, group, or community," and continues, "The nurse promotes, advocates for, and strives to protect the health, safety, and rights of the patient."

Katie Brewer, senior policy analyst at the ANA, believes a commitment to supporting women breastfeeding is "one of the key things that nurses promote to give babies a healthy start." "Nurses are often the front line breastfeeding educators and advocates," she writes me in an email before we talk. "ANA has nurse representatives to the US Breastfeeding Committee, and we support a number of federal initiatives around breastfeeding, including First Lady Michelle Obama's campaign."

But when I interview her by phone, Katie admits that misguided nonevidence-based pressure from medical professionals is often the reason breastfeeding fails: "There's sometimes a rush to try to get the infant some nutritional supplement before the mom can breastfeed. That's not isolated to nursing. Unfortunately we're having this issue, we're trying to promote evidence-based strategies but there are a lot of bad practices that go on in certain hospitals, especially around the lack of support for moms trying to breastfeed."

"The nurses come to think of formula as equivalent to human milk because they are being educated by formula companies," explains Marsha Walker. "Formula companies do everything they can to generate goodwill and whitewash their corporate image. Some providers forget that and they are being seduced. The seduction is very subtle, but it's there. Ply people with enough food, friendship, praise, free education, and free stuff, and they start to believe whatever you tell them."

The campaign against breastfeeding continues after a family is discharged from the hospital. Margaret Cividino, who lives in Vancouver, B.C., received a phone call from Similac when her daughter was just two weeks old. The young woman on the phone congratulated Margaret on the birth of her baby and asked her if she was breastfeeding. When Margaret said yes, the telemarketer then asked how breastfeeding was going. Then the telemarketer asked her if she wanted more information about Similac because "there are some advantages in the long term, like convenience for you!"

Two months later another telemarketer called. The telemarketer asked Margaret if she was still breastfeeding. Margaret tells me that when she said yes the telemarketer congratulated her, " 'Well, that's great!' It sounded so condescending coming out of her mouth. Then she said, 'We just want to tell you that our formula is the closest to breast milk. For convenience's sake if you're ever traveling, you might want to use it.' "

This time Margaret got mad: "It's going to be more convenient for me to carry bottles, carry formula, have to find clean water, and somewhere to reheat it when all I have to literally do is undo my shirt and pull my boob out of my bra?" she cried before hanging up the phone. "Two steps instead of fifteen?!"

BREASTFED BABIES ARE SMARTER

Even though she's wearing a white lab coat and showing me her replica of a human brain, Lise Eliot doesn't look like a stereotypically "nerdy" neuroscientist. She has long blond hair and blue eyes, and is a tall and slender woman. Eliot is so down-to-earth it's easy to forget that she has a Ph.D. in physiology and cellular biophysics from Columbia University and is the author of the most extensive book about baby brain development in print today, *What's Going on in There? How the Brain and Mind Develop in the First Five Years of Life*. Until, that is, she starts tossing around words like "hippocampus" and "basal ganglia."

"No one disputes the association between breast-feeding and intelligence," Eliot wrote in her book. "The problem, however, is in figuring out the reason for it. . . ." Whether the increase in intelligence is caused by breastfeeding or only correlated with breastfeeding is still a matter of debate, but Eliot has found in her research that taken together the data persuasively demonstrates that when you control for all the other "confounding factors" (mothers who breastfeed are less likely to smoke, they tend to be older, more educated, and more affluent) there is something in the milk itself that aids a baby's brain development.

On a bright sunny day, I visit Eliot in her office in the Department of Neuroscience at the Chicago Medical School, where she is an associate professor, to discuss the neurological advantages to breastfeeding. "There's the social sensory interaction side," Eliot begins, running her hands over the large pink plastic replica of the human brain that has a furrowed cauliflowerlike surface, parts of which are color-coded. "A baby who's being breastfed has more contact with her mother, there's warm touch, the baby's positioned close to the mom's heart and her voice. Certainly there are benefits of massage, so contact in general is a good thing for all young mammals." Eliot breaks the plastic brain in half, revealing strangely shaped structures with unfamiliar names.

In addition to the benefits of skin-to-skin contact, Eliot theorizes there are benefits to breastfeeding that happen on a molecular level: "Mothers who are breastfeeding have higher levels of prolactin and oxytocin, to produce and let down milk," Eliot says. "Those hormones are known to have neurobehavioral effects, promoting bonding and affiliation." She pulls off the brain stem, which is surprisingly large. "So another theory is that mothers who are breastfeeding are biologically primed to be a little more sensitive to their infants' cues."

Then there is the milk itself, a substance so rich and complex that scientists are still trying to understand and analyze all its ingredients. "There are the components in the milk—the fats, proteins, lipids—and then also the non-nutritional components—hormones, growth factors, enzymes. . . . Oops, I dropped my lenticular nucleus," Eliot interrupts her explanation to retrieve the fallen brain part, a cone-shaped mass of gray matter that helps the brain coordinate muscle movements, before continuing: "You know, the mammary glands actually deliberately put into milk antibodies, growth factors, and other molecules that are there to continue the protective and nutrient function of the placenta. So all of that means breastfed babies are healthier, have fewer respiratory infections, GI infections, ear infections," Eliot says. "So it may be that because breastfeeding is healthier for that first year of life, babies learn more effectively, and lay down a stronger foundation for later cognitive abilities."

Formula is a highly processed synthetic product usually made by dehydrating cow's milk, rehydrating it in stainless-steel vats, to which synthetic vitamins, minerals, emulsifiers, corn syrup solids, and a variety of other ingredients, including synthesized omega-3 fatty acid (DHA) and omega-6 fatty acid (RHA) extracted from laboratory-grown fungus and algae with a toxic chemical, hexane. Breast milk, by contrast, is a living fluid. When you look at it under a microscope you see that it is teeming with life, full of shape-shifting white blood cells, fat globules, which look like shiny spheres of different sizes, and balls of protein. At much higher magnification the Y-shaped antibodies that are passed to the infant to boost his immune system can be seen.

Breast milk, which contains more than a hundred different known compounds and others that are still being discovered, has sugars called oligosaccharides that adhere to a baby's intestinal lining to allow beneficial bacteria in but repel harmful bugs. The fatty acids in the milk are used in the infant's brain, stimulating its development. J. Bruce German, Ph.D., a professor of food chemistry at the University of California at Davis, has been studying breast milk for twenty years, trying to imitate its effects. "The features that make human milk so unusual are that it's personalized and it's active," German told a writer for *Wired* magazine. "So almost by definition there's nothing on the horizon that would satisfy those criteria."

It takes only one look at breast milk under a microscope to see what a dynamically unique and interesting substance it is. While his Airedale terrier chews a hole in my toddler's hat downstairs, Carl Morten Laane,

Ph.D., a professor of molecular bioscience at the University of Oslo and a celebrated Norwegian microscopist, peers at breast milk under the microscope he keeps in his home office upstairs. A seventy-one-year-old man with no children of his own, Laane has never seen the molecules of breast milk before. With the glee of a kid in a candy store, he points out how the fat globules are different sizes, shows me an epithelial cell in the milk, and exalts at how if he takes this slide to his laboratory where he has a high-res camera he can get some "Buutiful images." Laane also takes a look at the part-skim cow's milk he has in the fridge and Nestlé Nan liquid formula, which costs 20 Norwegian kroner (about $3.50) for 200 ml (less than one cup), that I've purchased for the occasion.

While the breast milk is complicated and alive, the fat globules in the artificial infant formula are so small Laane has to use a higher-powered microscope at his laboratory at the university to see them clearly. "The cow milk has some resemblance to human milk, the Nestlé milk has not any, whatsoever, regarding content of physical structures," Laane reports to me later. "The Nestlé milk has only tiny structures in the liquid. It looks like a very unnatural kind of milk. Guess it is junk food for babies."

THE HEALTHIEST BABIES IN THE WORLD

The sky is an angry gray and gusts of wind scatter fallen leaves. It's raining on this September afternoon in Oslo, the capital of Norway. On the Metro I count two Somalis, nine Norwegians, and one dark-skinned Middle Eastern man. Though you'll hear American obstetricians argue Norway's outstanding maternal and infant outcomes are because they have good genes, in the past twenty years an influx of immigrants from Poland, Sweden, Germany, Lithuania, Somalia, Iraq, and dozens of other countries have come to live here, making up more than 13 percent of the population. The train zips along efficiently, whirring to a seconds-long stop to discharge passengers. I'm making the trek to the home of Gro Nylander, M.D., Ph.D., who lives in Høvik, a suburb south of Oslo.

Nylander, perhaps the best-known obstetrician in the country, is outspoken about the importance of breastfeeding. She teaches medical school, practices obstetrics at Oslo University Hospital, and also oversees the University of Norway's National Resource Center for Breastfeeding, an academic center that uses scientific research to disseminate information about breastfeeding to academics, health professionals, parents, government, and media. She tells me that today breastfeeding rates in Norway are among the highest in the industrialized world. Nearly 100 percent of moms initiate breastfeeding, and the majority of these women (about 80 percent) are still nursing when the baby is six months old. At one year of age, almost 50 percent of Norwegian babies continue to be nursed.

It didn't used to be that way. Married to a young doctor, Nylander had her first child in 1966. When her son was six weeks old she was producing so little milk he was barely at his birth weight. It took her years to understand what happened. "Now I know that everything that was done in the hospital at that time, with the best intentions, is what we do today to wean a baby," she explains. It was forbidden for moms to breastfeed more than once every four hours, they were instructed to nurse only on one side. At night the nurses made rounds asking, "What kind of sleeping pill would you like?" Babies were supplemented with formula. By 1968 nursing hit an all-time low: Only one out of five Norwegian babies was breastfeeding at three months.

Nylander links the decline of breastfeeding in Norway to the concomitant medicalization of childbirth. A laboring woman in the 1960s had her pubis shaved, was given an enormous enema, left alone for most of the labor, delivered flat on her back, and separated immediately from her infant. Newborns were kept in the hospital nursery except for twenty minutes every four hours during the daytime when they were brought to their moms.

But, Nylander tells me as I admire the view of the ocean from her home, Norway has always had a strong tradition of breastfeeding. Mother-to-mother support groups were started, and by the 1980s breastfeeding rates were again on the rise. Norwegian hospitals allowed babies to nurse on demand and stay with their mothers. By then Nylander had gone to medical school and was herself a young doctor. She realized moms still reported having trouble and tried to figure out why. No longer using formula, now the hospital protocol was to give babies sucrose water. In the first three days of life, babies received 600 mls of sugar water, the equivalent to a grown-up man drinking forty bottles of soda. "They had tummy aches," Nylander explains. "They were not hungry because they were receiving a lot of calories. Not interested in sucking because they were getting a bottle." Comparing a group of more than two hundred babies getting supplemental feedings of formula and sugar water to more than two hundred babies exclusively nursing, Nylander found that babies exclusively nursed, though they lost more weight initially, did not have higher bilirubin or blood sugar levels (hypoglycemia), and they quickly caught up to and surpassed their bottle-fed counterparts. Today in Norwegian hospitals, if a baby needs formula for a medical reason, she is fed from a syringe or a spoon, not a bottle, so as not to upset future breastfeeding.

"Formula is killing babies in the United States," Nylander says bluntly. A baby in the United States is almost twice as likely to die in infancy than a baby in Norway. One study showed that more than seven hundred American babies a year would escape or delay death if they were breastfed. Another demonstrated that exclusive breastfeeding would result in more than nine hundred lives and more than $13 billion in medical costs saved. Other studies have shown that breastfeeding, more than vaccinations or even clean water, is the single most important factor in saving children's lives. In Norway, following the WHO *Code*, manufacturers are forbidden from advertising formula. No free samples are given to new moms in the hospital, no telemarketers call them to peddle formula, and no pediatricians or obstetricians give them goodie bags with formula coupons and free samples. Manufacturers may not put pictures of babies on their packaging, and formula salesmen have no access to medical personnel, either at the hospital or doctor's office.

But restricting formula manufacturers is just one step. "Motivating moms is a big part of the work," Nylander says. Poised and articulate, Nylander smiles mischievously. "It's not enough if the nurses, doctors, and hospitals say so. You want to feel it's not only the right thing to do but it's a fun thing to do. It's a liberating thing to do. It's modern. It's smart. This is what the movie stars do and the sports idols!"

Nylander drops me at the bus stop and I begin the long trek back. As I watch the massive windshield wipers clear the rain from the bus windows I think about how the real problem for Mead Johnson and the other formula companies that "decline to comment" when reports reveal that Similac has been contaminated by warehouse insects (both larvae and adults), or a batch of Enfamil has been recalled for its unpleasant smell, or an entire lot of Good Start has been taken off the market for being off odor and causing gastrointestinal complaints, or when they are simply asked to share their marketing strategies and company values, is not the investigative reporter from Oregon who comes knocking at their door. It's the fact that no matter how aggressively you market it, no artificially manufactured product in the world will ever be as perfect for a baby as a mother's milk.

The only country to vote against a WHO document to protect women from unethical formula advertising: The United States

Amount of health care savings if American women followed AAP breastfeeding guidelines: $13 billion

Number of infant deaths that would be avoided if American women breastfed: 900 per year

Amount donated by Abbott to the AAP in the last eight years: at least $5.26 million

Formula industry donations to the AAP: at least $11,130,500

Net profits of Abbott (maker of Similac) in 2011: $38.9 billion

Net profits of Mead Johnson (maker of Enfamil) in 2011: $3.7 billion

Net profits of Nestlé in 2011: $10.1 billion

Cost of formula for an infant for 12 months: $2,366

Cost of breast milk for 12 months: $0

Maria J.: One Moment at a Time

The daughter of a military officer, Maria was born in El Paso, Texas, but grew up all over the United States. Maria had no intention of breastfeeding. The thought of nursing a baby made her uncomfortable. She was not aware if any of the women in her family, except her father's sister, breastfed their babies, and Maria, who is thirty-three, was not breastfed as a child. But after her son was born, everything changed.

I'm very conservative and I didn't feel like I could breastfeed. I felt like breastfeeding would be too sexual for me—I didn't want to expose myself to my son or to anybody who might see me if I needed to nurse in public. I was uncomfortable around nursing moms, and I felt that I didn't want to put other people in that position. There was another reason—I was sexually assaulted when I was seventeen. I spent most of my twenties trying to feel okay about it. I still go through moments where I don't want anyone touching me, when I just need my space. I didn't think I could have that space and still be a nursing mom.

My mom is Korean and she believed that breastfeeding was only beneficial if you ate nutritious food as a child growing up, and I knew I wasn't eating well in my twenties. My midwife and my doula both told me that's not true, that it's about what you eat now, not in the past. I went to prenatal classes and everyone told me how good breastfeeding is for the baby. My midwife and doula encouraged me to try. My midwife asked me, "How long do you think you could try it for? Three days?" I agreed reluctantly. "Okay," I said. "I think I can do three days, we'll try that."

Sammy came on his due date. Everybody told me he was going to be late, so for me he came early. Everybody told me I'd be in labor for two or three days but he was born in twelve hours. He was born at home, at 9:25 a.m., in the rec room downstairs. I was wearing pink pajamas and had towels and blankets around me. I could hear the sound of the neighbors installing their chain-link fence. The curtains were drawn but light was coming in around the shades and I could tell it was going to be a beautiful, sunny day. I remember thinking, "I wonder if anyone knows there's a new life in this world?" My husband sat beside me on our futon couch and the midwife and doula both helped me guide the baby's head to my breast. When Sammy latched on I was like, "Whoa, I have a baby, and he's nursing. This is weird! This is kind of cool! Oh my goodness! I kind of like this!" There was so much emotion . . . the floodgates just seemed to open—of love, of peace, of calm.

After the midwives left, my doula said, "You should learn how to nurse lying down, because you're really tired and you'll want to rest but the baby's hungry. Just try it."

Once we could nurse lying down, I was hooked. But five or six days after Sammy was born we hit a rough spot. My milk had not come in. Sammy was screeching, he wanted to nurse all the time, and my parents were in the house, which was stressing me out. My breasts were in pain. It felt like someone had taken all the skin off my nipples and dipped them in hot water. Everyone told me that when my milk came in it would be easier. I remember crying, rocking back and forth, holding the baby really tight against my chest, and saying, "I'm sorry, I'm sorry, I'm in a lot of pain," with tears pouring down my face.

My husband didn't like to see me like that. He said, "Can we just give him a bottle?" I said, "No! I said I'd make it a week and it hasn't been a week yet." The next day the milk came in. I remember feeling so much relief—a tingling feeling, a mixture of relief and a little pain, like a pinching sensation.

Honestly, if the milk hadn't come in within twenty-four hours, I might have given up. My doula told me not to worry, that the milk would come. She encouraged me to keep breastfeeding, even just for one more day, or one more feeding. After I got through that first week I made a goal to nurse for a month. The next thing I knew three months had come and gone and I thought we might as well go six months. Then it was a year.

Sammy's teeth came in in groups of two and four, and when he was teething he was always in pain and had a fever. Nursing was a way for me to comfort and reassure him. It was a way for him to focus and center himself. Sammy has dark brown eyes like I do. He would look up at me when he was nursing, and talk to me, making little groans and mumbles, and I'd talk back.

My family still to this day does not know about my sexual abuse. Neither does my husband's. It's one of those things where it happens to everybody else but it doesn't happen to us. But I talked it over with my husband and he encouraged me to tell you. It's been more than fifteen years since it happened. He said, "By sharing this you might help someone else either have a baby or be able to nurse that child."

Being a source of food, warmth, emotional support, and comfort helped me get over the assault and understand emotionally that my body is not just the sexual object that society makes it. Breastfeeding worked for us because I took it one day at a time, sometimes one hour at a time, or even one moment at a time. I gave myself small goals to accomplish. I would tell myself, "I have to do this right now but I don't have to do it tomorrow, whatever drop of breast milk you can get is worth it. Just take it one drop at a time."

CHAPTER 8

DIAPER DEALS:
How Corporate Profits Shape the Way We Potty

When Baby Kyra learned to walk—at just ten months—she would toddle everywhere after her mom, following her into the bathroom. "What doing? What doing?" Kyra would ask, cocking her head to the side and looking quizzical.

"I'd tell her, 'Mommy's going potty,' " her mom, Angela Akins, remembers. But despite Kyra's interest, Angela, who was twenty-two, was convinced it was too early to start potty training.

"I didn't want her to feel a sense of failure if she couldn't do it," Angela explains. "What do I say if she has an accident? She's so young. I didn't know if physiologically she could do it."

When Angela finally brought home a potty four months later, Kyra loved it so much she would talk to it, sit her dollies on it, and sit on it herself. Still worried Kyra was too young and even though the plastic diapers Kyra wore gave her a terrible rash, Angela did very little to potty train her daughter besides letting her be naked around the house.

But by eighteen months Kyra was out of diapers, dry at night, and going potty by herself.

"She potty trained herself, she wanted it that badly. Now she says 'Get out, I need private time,' and does everything by herself, even washes her hands, which I think is pretty good for two years seven months."

THE TWENTY-FIRST-CENTURY FEAR OF POTTY TRAINING

In the early 1950s, before the widespread use of plastic diapers, 90 percent of American children were potty trained by the age of eighteen months. By 2001 the average age of potty training rose to thirty-five months for girls and thirty-nine months for boys. A decade later, most parents assumed that three-year-olds will still be in diapers, teachers and medical doctors report a rising number of incontinence problems among preschool and school-aged children (as one doctor puts it, "toileting troubles are epidemic in Western culture and, alarmingly, on the rise"), up to 30 percent of children ages two to ten are chronically constipated, and, as any parent who takes her child to the playground can attest, it has become increasingly common to see four-, five-, and sometimes even six-year-olds still in diapers.

American parents have come to fear potty training, worrying they will inadvertently harm their child in the process. Tiffany Vandeweghe, a mom of four in San Diego, was told by her son's preschool teachers not to rush. "It will happen when he's ready," they counseled, sensible-sounding advice echoed by pediatricians, psychologists, and child development specialists across the United States.

This American aversion to potty training has largely originated with one man: Dr. T. Berry Brazelton, who invented the concept of child-led potty training. Brazelton's idea is that children—not parents—should decide when they are ready to use the toilet. Though parents today aren't as familiar with him as parents were two decades ago, Brazelton is one of the giants in pediatrics, a well-respected and much-loved doctor who is among the most famous pediatricians in America. His CV spans twenty-five single-spaced pages, revealing that he has authored or coauthored more than 100 scientific papers, 140 book chapters, and 35 books, which have been translated into more than twenty foreign languages. Instead of dictatorial potty training using punishments and rewards, Brazelton believes parents should have a relaxed attitude toward potty learning and simply follow a child's lead.

"I began to realize if you allow children to find their own way to the toilet they would do it," he tells me when I visit his office in Boston. Not based on scientific study or an expertise in urology, Brazelton's idea came from his observations in his practice. "If you allow them to watch and learn by imitation and finding their own way, they began to feel like, when they finally did it, 'Wow! I just did it. Aren't I great?' " he elaborates. "And so standing back and allowing the child to take the lead was eminently successful, and that's what led me to my theory children will train themselves if you give them plenty of time. And many children aren't ready till they are three or four."

While Brazelton's notion of child-led potty training sounds smart and easy, delaying potty training actually makes it harder for children to learn to stop using their diaper as a toilet and start using the bathroom. As the current epidemic of toileting troubles shows, delayed potty training harms children, who are having more accidents, withholding bowel movements, and contracting more urinary tract infections than ever before. It also makes more work, more cleanup, and more headaches for parents. At the same time, child-led potty training has been a fantastic boon for the disposable diaper company that has paid Brazelton hundreds of thousands of dollars to promote their products over the years.

THE ADVENT OF "DISPOSABLE" DIAPERS

As a kid, I remember loving Mr. Whipple, the harried, bespectacled supermarket clerk in the TV commercial who admonishes customers not to squeeze the Charmin while absentmindedly squeezing it himself (because Charmin is "irresistibly soft"). In 1957 Procter & Gamble bought the Charmin Paper Company. After the purchase, the company searched for new products to add to their paper line. An ingenious, energetic, and creative inventor worked for Procter & Gamble, Victor Mills. Mills was known for thinking outside the box: he revolutionized soap production by devising a simpler, continuous process to make Ivory Soap in lieu of the expensive time-consuming batch-by-batch method; he came up with the idea of using milling machines to grind flour so fine to make Duncan Hines cake mixes a consistent texture; and he also thought up a processing method to make every Pringles potato chip a geometrically perfect saddle shape (by grinding the potatoes into a slurry, pressing, and then baking them).

It was Victor Mills who developed the earliest version of Pampers, which hit the market in Peoria, Illinois, in 1961: a plastic rectangular product packed with creped tissue that fastened with pins. These diapers were bulky, expensive, and so gimcrack that one dad remarked that if you as much as tipped the baby the contents of the diapers would fall out. The paper core of the diaper, though absorbent, held the urine against a baby's sensitive skin, causing a terrible rash. Replacing the creped tissue with cellulose only made the diapers bulkier.

But over the next twenty years the plastic diaper improved: In 1968 Kimberly-Clark (which makes Huggies) introduced Kimbies, which were shaped to fit a baby's legs. In 1976 Procter & Gamble elasticized the leg openings to keep the diapers from leaking. And inventors at both Johnson & Johnson and Dow Chemical came up with the idea to replace the cellulose in the diaper's core with a superabsorbent polymer, white plastic balls no bigger than sugar crystals that were made from petroleum and could hold many times their weight.

Since the mid-1980s both Procter & Gamble and Kimberly-Clark have made their diapers with superabsorbent polymer at the core, the same material found in tampons. But initially there was a problem. In September 1980, Pat Kehm, a twenty-five-year-old woman in Cedar Rapids, Iowa, died of toxic shock syndrome after using Rely, a Procter & Gamble tampon made with SAP (in this case superabsorbent synthetic chips of superabsorbent polymer as well as carboxymethylcellulose derived from wood pulp). Pat was one of more than a thousand young, menstruating women who had become severely ill or even died after using Rely. Despite the fact that the company knew as early as 1975 that women were suffering from debilitating health problems after using free test samples of Rely, safety studies were ignored or repressed, scientists who exposed the dangers silenced, and consumer advocates belittled. Procter & Gamble instead spent more than $10 million to send 60 million sample packets to American households. But about two weeks after Pat's funeral, Procter & Gamble could no longer ignore the mounting evidence against the tampons and was forced to spend $75 million to take Rely off the market.

Six years after it recalled Rely, Procter & Gamble introduced a new disposable, Ultra Pampers. Many parents were unhappy with the product, complaining their babies' bottoms were breaking out with severe rashes and that slimy gel was coming out of the diaper, adhering to their infant's skin. Reported reactions including oozing blood, fever, vomiting, and staph infections, and were so bad that at least one hospital withdrew the Pampers. When an independent laboratory analyzed the diaper they discovered that Procter & Gamble was actually making Ultra Pampers with carboxymethylcellulose, the same material that had already been linked to toxic shock.

For some families, diapers that could be thrown in the trash were a welcome convenience. But just as many households were happy to continue using cloth. Procter & Gamble, and their competitor Kimberly-Clark, faced the problem of how to convince parents to switch from using an inexpensive product that only had to be purchased once and could serve for multiple children (to finally end up under the sink as a rag) to using a onetime product that was expensive, had to be purchased every week, often did not work well, was full of chemicals and perfume, and created the new problem of needing to dispose of voluminous and smelly additional waste. It took more than two decades of aggressive marketing campaigns and redesigns for plastic diapers to become the norm in the mid-1980s. Even once the majority of American consumers started using plastic diapers in greater quantities, Procter & Gamble still had to advertise and market the product vigorously in order to keep babies swaddled in Pampers and Luvs (introduced in 1976) instead of a competing brand.

DOCTORS SHILLING PRODUCTS

Procter & Gamble has been ingenious in its marketing strategies. One innovative way to promote the brand is to create consumer loyalty by pretending to offer unbiased parenting advice to anxious new parents looking for answers. Procter & Gamble's consumer research revealed that American moms responded most to advice given to them by doctors. "Pediatricians are most valued and obtain a high rating of 4.5 on the 5-point usefulness scale," reads one confidential internal company document. "Three out of five women (57 percent) give the highest rating of 5 to the usefulness of pediatrician's information and advice." Therefore, the more pediatricians the company could associate with their brand, either by advising parents to use their products or simply offering advice "sponsored by" Pampers, the more plastic diapers and other products they could sell.

Enter T. Berry Brazelton.

At the side entrance to an auxiliary building of the Boston Children's Hospital at 1295 Boylston Street a locked door and an intercom stand between the journalist and the entrance. Wide corridors, fluorescent lighting, and dropped ceilings make the place seem ominous. But Brazelton's current office, in a third-floor suite at the Brazelton Touchpoints Institute, which is also locked, has a warmer feel. Almost life-size color photographs of children hang from the walls: a boy wearing a cobalt blue turtleneck and rust-colored jeans with suspenders; a toddler with café-au-lait skin in a pink tunic with purple butterflies; a big-headed baby in a red sweater looking pensive. Brazelton's spacious office is filled with the dozens of accolades and awards he's garnered over the years, and the walls are hung with pictures of him posing with presidents and celebrities.

Personable, polite, and enthusiastic, Brazelton has an almost photographic memory of the names of the children he has cared for. As he became increasingly famous, parents of his former patients would get in touch with him. He would respond with notes written in his compact old-fashioned cursive hand inviting them to stop by and see him, promising them copies of their old medical records.

Born in Waco, Texas, on May 10, 1918, T. Berry Brazelton came from a close-knit southern family. He tells me when we meet that every weekend when he was growing up his parents would take him to his grandmother's house for Sunday lunch with all of their siblings and children. When Berry was eight or nine years old he was assigned to take care of all the cousins so the grown-ups could eat and talk in peace. In order to take care of eight small children of different ages, the youngster realized he had to learn to get inside their brains. "I had to watch them and observe them and help them control themselves, and help them interact with each other and so forth. That showed me how important it was to observe, be involved, and to understand what was going on inside each child's head. From that time on I wanted to work with children, and I thought the best way to do that was to be a doctor."

Brazelton graduated from Princeton University in 1940 and earned his M.D. from Columbia University just three years later, in 1943. In 1969, he was hired as an assistant professor at Harvard University and published his first book: *Infants and Mothers: Differences in Development.* He wrote three more books in just four years and quickly moved up the ranks at Harvard, becoming an associate professor in 1972 and a full professor in 1986. Brazelton's interest in observing babies and learning from their behavior, combined with his ability to reassure worried parents and inspire young medical students, made him a natural teacher.

In 1998 Pampers developed a large diaper (Pampers Size 6), suitable for children over thirty-five pounds. Procter & Gamble is such a giant that it describes itself as "a force in the world," selling products to 4.4 billion of the world's 7 billion people. "Our market capitalization is greater than the GDP of many countries," the company boasts. If you look at the small print, chances are your household contains products—toothpaste, toilet tissue, laundry detergent, air freshener, disposable razors, even batteries—made by brands owned by Procter & Gamble.

In 1999 Professor T. Berry Brazelton became the face of Procter & Gamble's plus-size diaper. "Don't rush your toddler into toilet training or let anyone else tell you it's time—it's got to be his choice," a smiling Brazelton tells parents.

One Boston pediatrician, Robert Sege, was shocked to see his colleague on prime-time TV: "I was pretty surprised to see him hawking a product," Sege told a reporter for the *Boston Herald.* John Rosemond, a conservative child psychologist and syndicated columnist, called Brazelton's promotion of large-size diapers "a fairly blatant conflict of interest" and decried the idea of delayed potty training as harmful to children and undermining of parental control. Beverly Beckham, a columnist for the *Boston Herald,* suggested we "flush Harvard baby doc's diaper pitch," exposing that Brazelton's "shilling for Pampers" has led to four-and-a-half-year-olds who refuse to be potty trained. "This child and all the others like him are the multibillion-dollar disposable diaper industry's dream," Beckham wrote, "and Brazelton is in the industry's pocket."

Though it didn't come to the public's attention until the late 1990s, Brazelton's relationship with Procter & Gamble actually started years earlier. From 1983 until the last one in 2000, Brazelton was the host of the Lifetime cable television show *What Every Baby Knows.* The show, which aired daily in the early afternoon, usually featured an office visit with a family facing a particular issue, followed by a roundtable discussion with parents, and concluding with the central family's story and a solution to the problem. As a TV personality, Brazelton became to babies what Carl Sagan was to "billions and billions of stars": an enthusiast, a champion, an expert. "During the 1980s you had the Baby Boomers becoming parents, wanting to do a good job, and running into roadblocks," remembers the host's producer and director Henry ("Hank") O'Karma, who wrote many of the episodes. "They had a tremendous appetite for information. We could not go into a restaurant or walk through an airport without being besieged by a parent. They would tell him their troubles with toilet training, and look to him as a reassuring source. He understood parents' struggles." The main advertiser and corporate sponsor of *What Every Baby Knows* was Procter & Gamble.

Then in 1996 Procter & Gamble invited Brazelton to be the chairman of the Pampers Parenting Institute, an online parenting resource developed to strengthen the brand's credibility, communicate directly with consumers, and influence buying choices through "the right third-party influencer," according to internal documents.

"Procter & Gamble came to me and asked if I would work with them," Brazelton explains to me about his decision to do the television advertising. "They offered me a reward for doing it."

At first he hesitated. It was one thing to host a television show funded by advertising from a diaper company. It was another to take money directly from that company in exchange for promoting a specific product. "It took me a long time to decide to do it," he told a *New York Times* reporter in 1999. While personal arrangements between doctors and corporations, in which physicians or researchers would be paid for their services often for consulting, lecturing, and research, were commonplace at that time, Brazelton worried that accepting so much money from Procter & Gamble might be a violation of his ethical position as a professor at Harvard. Many of his colleagues thought it was wrong of him to align himself with a corporate sponsor. But Brazelton was eager to take the opportunity to convey the idea he had developed decades before that toilet training pressure on young children did not facilitate toilet training and could cause unnecessary harm. He believed the larger diaper sizes would facilitate a child-centered approach to toilet learning and ultimately decided the reward was too great to refuse. Besides, Brazelton tells me frankly, since he already had tenure, Harvard could not fire him even if he was in violation of their policies. "I was eager to do it," he explains, "and they [Procter & Gamble] funded our television show for thirteen years."

At ninety-four, T. Berry Brazelton is as charming, outgoing, and avuncular as ever, though perhaps a little shorter and more distracted. Toward the end of our interview I mention I've noticed more and more doctors are endorsing specific baby products. I ask Brazelton what he thinks of that.

"I don't like that," he responds without hesitating.

Me: Why not?
Dr. Brazelton: I think it's tying yourself to a product that you may not know
    enough about. I think it's dangerous.
Me: What's dangerous about it?
Dr. Brazelton: It may not be what you think it is.
Me: Do you think doctors do it for the money? Or they do it for the
    recognition?
Dr. Brazelton: They do it probably for the money.

DELAYED POTTY TRAINING CAUSES HARM

Linda Sonna, Ph.D., a psychologist and the author of twelve books about parenting, points out that starting potty training late and adopting a hands-off approach is actually associated with both physical and psychological problems: bladder infections, frequent urination, bedwetting, low self-esteem, social rejection, and behavioral problems.

One child who experienced social ostracism firsthand is Zoe Rosso, a three-year-old from Arlington, Virginia, who, according to the *Washington Post,* liked to bake brownies with her mom and make up elaborate worlds with plastic animals. On December 3, 2010, Zoe and her mother were escorted to the door by the preschool principal. Zoe was suspended from the $835-a-month Montessori preschool because she defied school policy by wetting her pants eight times in one month.

Daytime incontinence and bowel control problems in three-, four-, and five-year-old children have become more common in the past decade than ever before. Urologists are now studying the problem to figure out why. When Dr. Joseph G. Barone starting seeing an increasing number of children with daytime wetting problems, he collected data from 215 children, aged four to twelve. Barone was surprised by the results. The data revealed that children whose parents initiated toilet training *before* thirty-two months had far fewer problems with "urge incontinence" (bladder and bowel accidents).

Other research shows similar results: School-age children with bladder dysfunction—including daytime urinary incontinence, bedwetting, and recurrent urinary tract infections—are more likely to have started toilet training at a *later* age.

"I have always been a proponent of the child-oriented approach to training, telling parents not to rush training and let their child be the guide," Barone admits to a reporter for *Urology Times*. "I thought that starting a child 'too soon' would more likely result in day wetting compared with waiting and starting the training process later. Our data indicate this may not be the case."

Though they may have started with the best intentions, the plastic diaper industry, the doctors who promote them, and the hospitals that give parents "free" diapers have created long-lasting problems for children whose parents have been duped into thinking it's normal for a three-year-old to use their pants for a potty. They've made it more difficult and more stressful for parents to potty train their children, and they've turned what in many countries around the world is still a seamless process into a money-generating industry. Though very few parenting experts believe we should go back to the more rigid, even coercive training methods common in the 1950s and before, the science shows us clearly that Brazelton is wrong and delaying potty training is not in the best interests of the child.

"Potty training was not such a problem sixty years ago," says Jean-Jacques Wyndaele, a Belgian urologist who is an expert on this issue. "Now it has become a problem. That's surely not because of the child."

PHYSIOLOGICALLY READY BETWEEN TWELVE AND EIGHTEEN MONTHS

Dr. Wyndaele, a professor of urology, chair of the University of Antwerp's Department of Urology in Flanders, Belgium, and a urologist in practice for thirty-six years, has been studying the physiological and psychological effects of potty training in healthy children for more than a decade. His team first took an interest in delayed potty training when they noticed that more children in Belgium were having incontinence problems in kindergarten than ever before. In one ongoing study, which has not yet been published, they found that kindergarten teachers now report having to spend up to one third of their time dealing with urinary and fecal accidents in the classroom.

Wyndaele contends that, as with all human behavior, there is a wide range among babies and every baby is unique, but that he and his team have found consistently that potty readiness begins between twelve and eighteen months.

"Neurologically and physiologically children are ready between one year and one and a half years," Wyndaele tells me in a phone interview. "Of course, there's a big variety because every child is different, but it's very rare that a child would be older than two years before being ready." While the style of potty training seems to play a role (one study of Belgian schoolchildren found that stricter methods result in more incontinence problems later on), Wyndaele has found that a delay in potty training itself causes myriad problems, including anxiety for the child and even higher incidents of disease transmission, like hepatitis B.

Wyndaele's research has revealed that a large part of the problem is the kind of diapers parents are using: "We've found that diapers that have the most absorbency are the worst for potty training," Wyndaele says. "Potty training is about awareness and sensation. If you break the link and take away the sensation, you have a problem. The diapers have not helped. They have broken this link. The feeling that 'Something has to happen' and 'Now I'm wet' is no longer there for the child. Because they are so good, so wonderful, and they absorb so much, you can pee many times in the diaper without getting wet."

Wyndaele's solution is to leave disposable diapers behind as soon as a child shows readiness. "You should leave them out when you start training. Use cloth diapers or don't put anything on."

AMERICAN PARENTS SPEND $27 MILLION A DAY ON DIAPERS

Since the advent of the plastic diaper in 1961, the longer babies remain in diapers, the more money is made by the industry. Plastic diaper manufacturers have an enormous incentive to discourage parents from potty training.

The numbers tell the story: Since the average age of potty training has risen so precipitously, on any given day in America there are approximately 13 million children in diapers. More than 95 percent of these children are swaddled in plastic. The average cost of one plastic diaper is about 25.5 cents (a brand-name diaper costs about 28 cents, a store-brand diaper 23 cents), which means Americans are spending an average of $27 million *per day* on diapers, which amounts to a staggering $9.8 billion a year. A family that uses plastic spends approximately $2,400 just on the diapers for that child. And that's not counting the disposable wipes, diaper rash cream, cost of increased garbage volume, gas for emergency diaper runs and for trips to the store to stock up on diapers, and other expenses like the Diaper Genie (a disposal pail lined with plastic bags that is designed to mask the unpleasant odors of the diapers until they are thrown in the trash).

THE HALO EFFECT

The tactic of aligning a brand with as reputable and credible a "third-party influencer" as possible is one of corporate America's standard selling tactics. Though her background was in international relations, Leslie Becknell Marx spent two years at Procter & Gamble's corporate headquarters in Cincinnati (which employees call the "G.O." for general offices), first as a brand assistant assigned to a team that sold Downy fabric softener and later promoted to assistant brand manager for Bounce dryer sheets.

Leslie and I sit in her living room talking about Procter & Gamble while her husband, Adam, her son, Aaron, and the nanny hang paper moons and stars on the walls for Aaron's second birthday party. She tells me she left corporate America in 2002 seeking more meaningful work, which is how she came to her current position as the minister of the Rogue Valley Unitarian Universalist Fellowship in Ashland, Oregon. After five minutes of talking to Leslie I realize that "P&G" is a lingo-laden place. Much of the shorthand is familiar to business school students but I have to interrupt Leslie to ask for a translation every five minutes.

Though Leslie worked in the "soap" side of the business, much of the Procter & Gamble sales strategy is the same. According to Leslie, brand managers spend countless hours talking about "sampling," "trial," and "repeat." "You want people to try your product because if they haven't ever tried it then they don't know if they're going to like it . . . so in marketing you talk about 'trial' and 'repeat.' Trial means people buy your product for the first time. Repeat means they buy it again and again and become loyal customers. So one of the questions in marketing is: Do we need to focus on trial (getting more people to try our product) or repeat (building more loyalty)?"

This answers a question that most first-time parents who have a baby in the hospital never think to ask: Why do American hospitals put *brand-name* diapers on a newborn instead of using generic diapers that are less expensive for new families?

Because marketers have discovered that the most effective time to get someone to try a new product is when they are going through a life-changing event, like having a baby. "You suddenly have new needs that you've never had before, and in those moments you're more likely to change your behavior," Leslie explains as Aaron toddles over for a snuggle. "So I advocated for increasing our sampling with appliance manufacturers [that is, putting P&G product samples in new appliances] because you're likely to be hitting someone at a life-changing event. Why does someone buy a new washer or dryer? Because they just bought a new house, because they just got married, because they just moved away from home."

If Leslie's team put Downy in Sears's Kenmore appliances, Sears would in effect be lending its name and reputation to Downy and vice versa. "There's an implied endorsement by the manufacturer, just like if you get a sample from the hospital, there's an implied endorsement by the hospital," Leslie explains.

This endorsement is so effective that marketers use religious terminology (think of baby Jesus in a Pampers diaper) to describe it: the halo effect. "P&G has a pretty easy time finding organizations willing to provide samples of our products because our products tend to be the best in their categories. So there's this dual halo effect. If I'm buying this dryer that has a free Bounce sample in it there's also an implied endorsement from Bounce of this dryer, and vice versa," Leslie details. "We each get this light shining on us, this benefit from associating with this other esteemed organization or business."

Over the waistband of Aaron's jeans I see the brand of diapers he is wearing: Huggies.

When I reach him by phone, Kai Abelkis, sustainability coordinator at the Boulder Community Foothills Hospital in Colorado, tells me he uses only cloth diapers on the fifteen hundred or so infants born there each year and is troubled by this kind of branded marketing to new parents. Kai points out that there is a blatant conflict of interest when American hospitals give new parents free samples of name-brand diapers: "We don't give away Pampers or any other freebies," Kai tells me. "We have a policy against it. When you get something from a hospital, there's an underlying blessing on that product. We believe that each family should make their [own] decisions. . . ."

DISPOSABLE CHEMICALS

Boulder Community Foothills Hospital has good reasons for using cloth diapers instead of disposables. Though they are widely used in the United

States, disposable diapers, are not the healthiest choice. It was when Shawna Cummings, thirty-seven, was pregnant that she decided to use cloth diapers for her baby. She was concerned about the environment, she knew cloth diapers would save her family money, and she didn't want chemicals or plastic next to her son's skin. Even more compelling, Shawna had a friend whose dog had gotten into the diaper pail, eaten part of a disposable diaper, and died. "I know toddlers will get into anything and eat anything," Shawna says. "The idea of having something that toxic around my child was just horrifying."

Diaper manufactures work hard to counteract the viscerally negative reaction to the idea of a baby having his most intimate parts in contact with plastic all day, every day, for years. But not only do disposable diapers contain several different plastics, but conventionally bleached plastic diapers also contain trace amounts of dioxin, a carcinogenic by-product of the paper-bleaching process. Over the last twenty years some five thousand scientific studies have highlighted the harmful effects of dioxin.

Jay Bolus, a researcher who has been working on environmental and health issues for fourteen years and is currently vice president of technical operations for MBDC, a company that helps businesses evaluate how sustainable their products are, puts it most concisely: "The chlorine bleaching process makes this really toxic stuff," Bolus says when I interview him by phone. "Dioxins can be toxic and persistent, stick around in the environment for a long time, and accumulate in our bodies." Dioxins can cause skin lesions, alter liver function, negatively affect the human immune system, disrupt the reproductive and endocrine systems, and cause cancer. The amount of dioxins a human can tolerate without detectable health effects is minuscule: only 70 *trillionths* of a gram (smaller than one grain of sand) per month for every kilogram of body weight. According to the WHO, dioxin exposure has the most serious consequences on the developing fetus. A newborn, with his rapidly developing organs and nervous system, is among the most vulnerable to dioxins. Plastic diaper enthusiasts are quick to point out that any potentially harmful by-products or substances are contained *inside* the diaper. But if baby diapers contained different toxic substances, like asbestos or lead, wrapped in a plastic casing so that it was not actually touching a baby's skin, there would be public outcry.

Though plastic diapers have improved considerably since the early 1980s, anyone who has used one knows that the stuff inside the diapers sometimes comes out. The small white beads of gel that appear on a baby's skin when a diaper has been on for a while contain the improved superabsorbent polymer that has enabled plastic diapers to be so compact yet work so well. Best-selling writer Malcolm Gladwell praises single-use diapers in uncharacteristically hyperbolic prose as "more than a good idea" but rather "something like perfection."

A consumer who would like information about the contents of a plastic diaper and who calls a customer-service representative at Procter & Gamble to inquire what ingredients are in their diapers will not be met with enthusiasm or transparency. Instead, the spokesperson will become very nervous on the phone, refuse to give her name (but insist you give yours), remind you that the conversation is being recorded, put you on hold four times in forty minutes to check with her boss about what she is "at liberty to disclose," and tell you that you are asking about "sensitive and proprietary" information—even after you've explained that all you want to know are the names of the ingredients in the product that touches your baby's skin every day, all day long.

At the end of the conversation, a customer still won't know all the ingredients in a Pampers, because some of these ingredients "cannot be disclosed." Translation: Consumers do not have the right to know exactly what their infants are wearing on their privates. The customer service representative, after a good deal of awkward silence and hemming and hawing, will tell the consumer that, among other things, the "base ingredients" present in Pampers' Easy Ups, Baby Dry, Swaddlers, and Cruisers, as well as all the Luvs products, include petroleum, polyethylene, processed wood pulp, "absorbent gelling material," polypropylene, and perfume (to mask the unpleasant chemical odors of the other ingredients).

You could then spend the next ten days trying to understand the components of these different plastic polymers and realizing that plastic diapers are made of petrochemical products derived from petroleum, a nonrenewable resource that has caused wars and environmental devastation, and which, as it gets more scarce, oil companies must go to greater lengths to extract, from hydraulic fracking to deep-water drilling in the Gulf of Mexico. But instead of wasting all that time on hold, on the Web, and in the library, a parent may simply decide that she knows enough already: These diapers are made from petroleum-derived plastic, chemically treated pulverized trees, and synthetic perfume. To think this wrapped-up package of plastic is a healthy (or even benign) option for a baby is like believing that McDonald's food is natural and wholesome: It is an illusion based on the power of corporate advertising and the willingness of American consumers to be co-opted by a cult of putative convenience.

PAMPERS DRY MAX DIFFICULTIES

Even the most enthusiastic plastic diaper users have found themselves in conflict with big-business interests. In May 2010, the U.S. Consumer Product Safety Commission launched an investigation of Pampers Dry Max because parents started complaining that the diapers were causing bleeding and rashes, some so severe that they looked like chemical burns. That month more than 4,500 parents joined a Facebook page called "Pampers bring back the old CRUISERS/SWADDLERS." Within a matter of months the group had more than 10,000 members (it has since been taken down).

Mandy Fonck, a mom writing on another Facebook group, "RECALL PAMPERS DRY MAX DIAPERS!" with more than 3,000 members, reported that after just an hour of wearing Pampers, her son got a blistery rash so severe it started to bleed. Another mom, Rebecca Boxer, wrote that her daughter got a rash up to her belly button that disappeared when she switched to a different brand. Jenniffer Brown had a similar experience: "I used the Swaddlers w/dry max—after one day my 3 week old daughter had a bright red burned look to her skin [that] matched everywhere the inside of the diaper had touched. Even on her hips where the diaper wraps around," she wrote on Facebook. Procter & Gamble denied any problem existed. Instead, the company accused unhappy parents whose babies had severe rashes of deliberately promoting misinformation, secretly working for competitive brands, or being undercover cloth diaper advocates.

"We've been accused of many things: being irresponsible, uninformed parents, rival diaper company employees, cloth diaper advocates, and emotionally attached to an old, outdated product just to name a few," wrote April Weber, a stay-at-home mom of three and a loyal Pampers customer who was one of the parents who spoke out against the diaper. "We have been the victims of many pieces of hate mail from Pampers supporters and frighteningly enough Pampers employees." Ultimately the Consumer Product Safety Commission did not find a link between the diapers and the rashes. But Procter & Gamble agreed to settle a class-action lawsuit filed against them (without admitting there was anything wrong with their product) by paying $2.7 million in attorney's fees, including a label about diaper rash on both the diaper packaging and their website for two years, funding research in pediatric skin diseases, and giving $1,000 to each of the named plaintiffs.

LONG-TERM RISKS

There are negative long-term consequences of plastic diapers on a baby's health that are not as visible, or as fleeting, as diaper rash. A boy's penis and testes are on the outside of the body in order to keep them at a temperature cooler than 98.6°F. But plastic diapers keep the genital area hotter than nature designed it to be. A study of forty-eight healthy children, published in *Archives of Disease in Childhood,* found that scrotal skin temperatures were significantly higher in boys who wore disposable diapers than in boys wearing cloth. The researchers from the Netherlands posited that the overuse of disposable diapers may be one of the reasons for the decline in male reproductive health, and concluded that disposable diapers keep a boy's privates too hot: "The physiological testicular cooling mechanism is blunted and often completely abolished during plastic nappy use," the authors wrote.

Plastic diapers have also been linked to asthma, an illness that now affects millions of children. A study published in *Archives of Environmental Health* found that mice exposed to several brands of plastic diapers experienced eye, nose, and throat irritation, as well as bronchial constriction similar to that seen in asthma attacks. Chemicals known to be respiratory irritants, including toluene, xylene, ethylbenzene, styrene, and isopropylbenzene, were off-gassing from the diapers, and were thought to be the cause of respiratory distress. The cloth diapers studied did not cause the mice breathing difficulties.

"You want to think about what you're putting on the most vulnerable skin—a newborn baby's skin," says Dr. Deborah Gordon, a family physician who has been practicing for thirty years and specializes in preventive medicine. Gordon believes we should not be using plastic diapers on infants. "Cloth diapers are absolutely healthier and better for the baby," she says. Kai Abelkis at the Boulder Community Foothills Hospital in Colorado agrees. "We'd rather put a natural product around a child's private area than something that has chemicals and petroleum," Kai says. "We believe that cloth diapers are a better option."

There's no doubt that disposable diapers keep a baby dry: the superabsorbent polymers in plastic diapers can now hold up to three hundred times their weight in water, and disposable diapers work well to wick away moisture. But while this seems like a technological advance, how long do we really want a baby urinating into the same diaper? Parents of newborns in single-use diapers have no way of knowing if their baby is properly hydrated (unless they see uric acid crystals on the diaper's surface, which is an indication of severe dehydration). Though advertising has made parents view diapering as a burdensome chore, changing a baby's diaper is a way for the mom, dad, or other caregiver to enjoy and interact with the baby. Finally, the efficiency of disposable diapers has a long-term downside that all but ensures that parents with babies in plastic will be changing more diapers for a lot longer than parents with babies in cloth: A baby or toddler cannot understand the association between peeing and wetting, making it harder to learn to use the potty when the time comes.

FOR THE LOVE OF CLOTH

Heather McNamara, born in Bethpage, Long Island, in 1973, was the oldest of six children. By the time her three youngest brothers came along Heather was old enough to help her mother with their diapering. Though by the early 1980s the diapering trend had shifted almost entirely toward plastic, Heather's family used old-school cloth diapers: Chinese prefolds (a rectangular piece of cloth that is triple layered) and rubber pants. As a six-year-old, Heather liked practicing different diapering folding techniques, and she didn't mind sloshing the soiled diapers in the toilet before putting them in a soaking bucket.

"It just was the way we did things," Heather remembers when we talk on the phone. "Origami folds were fun for a six-year-old." Heather's father worked for the New York telephone company and her mom stayed home. They didn't have the money to buy plastic diapers. But her mom's sister, whose husband was a doctor and whose family lived in a more upscale town in Connecticut, used disposables.

Since Heather was little there has been a veritable revolution in cloth diapering. Most parents no longer use the old-fashioned flat diaper, a single rectangular layer of cotton cloth folded into place. Instead they use prefolded cotton diapers with some kind of waterproof cover that closes with Velcro or snaps. In addition to this tried-and-true method, there are now many convenient and versatile new diapering systems to choose from. With adorable patterns of leopard skin, cowhide, and baby giraffes, cloth diapering has become cute (manufacturers report that fashion is a major selling point to new parents, more so than concerns about health or the environment).

When Heather was deciding what kind of diapers to use, the decision was both about finances and familiarity. "What we knew was cloth. It just didn't make sense for us to spend money on something we were just going to throw away," she tells me. "I don't think we even viewed it as something that was not mainstream." Heather is a middle school teacher and her husband, Tim, teaches writing and rhetoric at various universities in San Diego. The couple lived in an eight-hundred-square-foot apartment on Bancroft Street that did not have a washing machine. They asked friends and family to give them cloth diapers, and Heather's parents bought them a portable washing machine they could roll up to the kitchen sink. They hung the diapers to dry outside, which is easy in sunny San Diego.

"It was exhausting having a baby, but there was no real hurdle with us with cloth. To me the bigger hurdle would be to go out shopping to get stocked up on diapers."

INFANT POTTYING

Heather kept her babies diaper free when she could. She learned to read their signals and became expert at "catching" their poop, even when they were infants. "I think most cloth diaper users potty their infants," Heather tells me. "Every time an infant poops in the potty that's one less diaper to clean."

Unlike in the United States, more than 50 percent of the world's children are potty trained by the time they turn one. In Niger, West Africa, plastic diapers are so uncommon they are a status symbol used only by the wealthiest families, and in some orphanages that receive them as donations. A traveler to Ghana, Togo, Benin, Ivory Coast, Burkina Faso, Senegal, the Gambia, Mauritania, and dozens of other countries on the continent will quickly realize there are very few, if any, African babies wearing diapers, and very few babies having accidents. Indeed, to a Nigerien the idea of a toddler in diapers is disturbing. People in many other cultures react with surprise, dismay, hilarity, and sometimes disgust when they see how Americans and other rich nations increasingly delay teaching their children to urinate and defecate in the toilet.

When she was an exchange student in Japan in her twenties, Christine Gross-Loh, author of *The Diaper-Free Baby: The Natural Toilet Training Alternative,* saw something she would never forget. The hosts she was living with had twin baby granddaughters. When the three-month-old babies came over for the day, their *obaasan* (grandmother in Japanese) put them down for a nap on the bed. When they woke from their nap, the *obaasan* held each baby girl over the toilet. She made a *pssss* sound and the baby—responding to the noise—peed.

The young American woman was astonished. She also felt totally uncomfortable about what she saw. Offering an infant, especially one that was just twelve weeks old, the chance to use the potty went against everything she had ever learned. Having grown up in America, she worried it was coercive, maybe even cruel, to potty train a baby.

But several years later, when Christine started having children of her own, she realized that what the *obaasan* had done with the twins was far from unusual: It is what billions of parents and caregivers do with babies. She discovered that in at least seventy-five countries around the world, including Russia, China, India, and Greenland, vast numbers of parents respond to their newborns' cues and keep their infants dry and diaper-free much, if not all, of the time.

Christine's research made her realize that the resources to acquire diapers, then to clean or dispose of them, are a recent luxury. It has only been in the last few centuries—a millisecond in evolutionary time—that a small percentage of our species has chosen to cover infants' bottoms with any sort of diaper, intending that the infant soil it. The move to keeping children indoors on surfaces one wouldn't want soiled may be part of the motivation. When parents are outside moving around all day, and infants are with them, the infants can eliminate on the ground outdoors wherever they are, as other primates do. Most women in Africa today carry infants in a loop of cloth tied to their backs. The mother goes about her business without the cloth becoming dirty because she can feel and recognize her infant's squirm of discomfort when he needs to go. The mother reacts by unwrapping the baby and letting him relieve himself in an appropriately out-of-the-way place outside, or over a toilet if she is indoors.

When their son was born, Christine and her husband noticed—as many new parents do—that Benjamin would pee or poop whenever his diaper was off. He responded to the physical cue of no longer being diapered or swaddled by bedewing his parents. But still Christine was skeptical, so much so that when her mother (who grew up in Korea) bought Benjamin a potty just after his first birthday, she got indignant. "I thought I, a hip, modern parent, knew better than she, and that 'better' now meant waiting until he was two or three, not starting with a preverbal thirteen-month-old! I even thought that early pottying could be harmful somehow," Christine writes in her book.

But instead of returning the potty, Christine tried sitting her son on it. Baby Benjamin, to his mother's astonishment, peed right away. She sat him on the potty throughout that day and the next, whenever she thought he needed to go. Sure enough, Benjamin went.

"I began to realize that he had been waiting for me to understand that he wanted to go to the bathroom outside of a diaper. He had been watching all of us using the toilet and was eager to join in." It was that experience that made Christine realize that Benjamin had been eager for an alternative to peeing and pooping in a diaper. She realized that helping Benjamin use the potty was doing the opposite of harming his self-esteem: It enhanced his sense of well-being and helped him be more comfortable. Christine and her family lived in Japan for five years and still spend time there each summer. "Seasoned grandmothers tell me, 'Isn't it nice for your baby to enjoy the good feeling of being dry?' "

BABIES KNOW MORE THAN WE THINK

Melinda Rothstein, a mother of three in Dover, Massachusetts, remembers sitting in her living room looking at her newborn's tiny face. It was late summer and the pink-and-white flowers on the dogwood trees outside the window had already fallen to the ground. Samuel tightened his mouth, looked uncomfortable, and started to grunt and push. It was so obvious he was pooping that Melinda, who opted for disposable diapers because she couldn't find a cloth diaper service in her area, considered holding Samuel over the toilet instead of having him soil his diaper. A Web strategist for Johnson & Johnson, Melinda banished the thought as soon as it came into her head, deciding it was "too weird" even to mention to her husband.

Eight weeks later, Melinda met Christine Gross-Loh at a La Leche League gathering in Newton. Drawn to Christine, a more experienced mom (Christine had two sons by then), partly because she was carrying her five-month-old in a colorful Korean-style baby wrap that Melinda had never seen, Melinda invited her over.

"I have to take the baby to the bathroom," Christine whispered almost apologetically when she first walked in the door. Melinda was intrigued. That day Christine loaned Melinda a book by Ingrid Bauer, the person credited with coining the rather unfortunate phrase "elimination communication" (EC), which is often used in the United States to describe infant pottying. Melinda read Bauer's *Diaper Free: The Gentle Wisdom of Natural Infant Hygiene* in one day and bought Baby Samuel a little potty. The book has a decidedly woo-woo bent: Bauer sleeps her unclad infant on washable lamb's wool, makes her own baby carrier so she can wear the baby naked against her skin, and encourages him to urinate outside to better understand the cycle of life. But the woo-woo bent didn't temper Melinda's interest. When she sat Samuel on the potty for the first time and he peed, Melinda was hooked. She switched from disposables to all-in-one cloth diapers (where the cover and the absorbent diaper inside is all one unit—a washable, reusable, cloth version of a disposable), but Samuel defecated so consistently on the potty that she has no memory of his soiling them.

Though Melinda is quick to explain that potty *training* is not the point (when your baby becomes interested in food and you respond by feeding them, you are not "food-training" them or expecting them to eat on their own), Samuel was dry at night by eleven months and in training pants by thirteen months. Melinda's second child, Hannah, pooped in the potty for the first time when she was seven days old and went mostly diaper free with her mom after that (Melinda's husband was more comfortable keeping her in diapers in the early months of her life). By seven months Hannah was wearing svelte German-made baby underwear. By then Melinda had become such an advocate for infant pottying that she cofounded a nonprofit support group for moms called DiaperFreeBaby in 2004 and was interviewed in a *New York Times* article about the practice. The article features a picture of Baby Hannah, her blue eyes wide and her ears sticking out, sitting happily on the potty, steadied by her mother.

"People don't think that babies understand things about their bodies. But they do," Melinda tells me over the phone. "I'm convinced that their perspective on life is much broader than we think. A baby who is EC'ed understands that when I am brought into the bathroom and my bottoms are taken off and a sound is made, this is a good time to pee . . . There's this disbelief that babies are capable of signaling when they have to pee or poop, or of going when given the opportunity. But what about all the other babies in the world that do that from birth? Are they physiologically different from us?"

As a new mom, Tiffany Vandeweghe had read the Brazelton books and was afraid to teach her sons to use the potty. Tiffany's oldest was out of diapers at twenty-seven months but at four years old her second son was still struggling. When Tiffany first learned about infant pottying she thought it was absurd, something for hippies who lived in Hawaii or for moms who stayed home full-time. But she had had so many problems with her second son (who would hold his feces and have painful bowel movements and messy leakage) that she decided to try it with her third. When Odella was three weeks old she woke up from a long nap dry. Tiffany held her over the tub and made the noise. The baby peed right away. "I'll never forget that moment. It was like, 'Oh, my gosh, now I have to do this.' It was so much easier than I thought, and so rewarding." Her third and fourth babies started wearing underwear around six months old and both were dry at night and naptime very early, using the potty independently by the time they were about a year old. Now it's hard for Tiffany to imagine *any* baby wearing diapers.

Like Melinda, when our oldest daughter was an infant my husband and I always knew when she needed to poop. Her face would get red and she would start grunting. We would laugh and hold her on our laps with her legs cocked in what we liked to call the "poopy position" to make it easier for her to defecate in her diaper. It never occurred to us that we could have taken off the diaper and held her over the toilet. It wasn't until I was an experienced mom with my fourth baby and a new mom told me she had friends who had successfully done it that I even realized infant pottying was possible. Unless you actually try it, it's easy to dismiss infant pottying as too hard or too messy or simply too weird. But we started responding to and teaching our littlest to poop in the potty when she was seven weeks old. I was surprised to find it was so much easier, cleaner, and cheaper not to change poopy diapers (from seven weeks on she pooped in a diaper less than twice a month). Maybe it is just coincidence but our lastborn was also our happiest and our only baby never to have diaper rash.

Diaper debates weigh the advantages and disadvantages of plastic diapers versus cloth and make little mention of the third alternative: to forgo diapers completely. Maybe because no one—not the plastic diaper giants or the cloth diaper advocates—makes money from parents who choose no diapers at all. Infant pottying aside, every child learns differently, and there is a wide variation of when children are ready to use the potty independently. Our oldest daughter learned around her second birthday. With very little help from us, and to our astonishment, our secondborn was completely out of diapers by the time she was eighteen months old. She was very independent, eager to be like her older sister, and she happily took herself to the bathroom. My son, though he began peeing in the toilet when he was a year old and was accident-free by age two, needed diapers at naptime until he was four and at night until he was six.

Toilet learning is a process. For several years after a child has full bladder and bowel control a parent or caregiver may still need to help wipe and wash hands. But we can all agree that there is nothing convenient, cheap, healthy, or easy about a child being in diapers for years longer than necessary.

It is wrong for hospitals to endorse diaper brands. It is wrong for diaper companies to give parents misleading advice. And, in Brazelton's own words, it is wrong for doctors to promote products they may not know enough about. Gentle, kind, patient potty training should take place when a baby is physiologically ready. We know from Dr. Wyndaele's research that a baby is ready between twelve and eighteen months, not two or three years of age, as the disposable diaper companies who profit off us and the doctors paid to promote their products would like us to believe.

Amount Americans spend on plastic diapers: $27 million/day

Average cost of plastic diaper: $0.25

Average cost of high-end, one-size-fits-all cloth diaper: $20

Cost of diapering a child in disposables: $2,400/child

Cost of diapering a child in cloth: $480/child

Average number of plastic diapers needed for 1 baby for 1 year: 3,000

Average number of cloth diapers needed for 1 baby for 1 year: 24

Cost to practice infant pottying: $1.00 (for a mixing bowl to use as a potty)

Errol Matherne: A Better Way to Care for Your Kids

Twenty-nine-year-old Errol Matherne, who lives in San Antonio, Texas, checks on his five-year-old, two-year-old, and eleven-month-old every morning. He works nights and can't go to sleep until he assures himself that his children are safe. He used to be completely against cloth diapers, but now he and his wife own a baby store that specializes in natural products and holds free monthly workshops on how to use cloth. It was when his oldest son, Tristan, was just a few weeks old and Errol was about to be deployed that his thinking about diapers totally changed.

Right after he was born, my son Tristan, our oldest, got a very violent and bloody diaper rash. His penis and testicles were red and inflamed and raw. He had bloody blisters in between his legs, thighs, and butt. He was in so much pain he couldn't sleep. When we wiped him he was in pain. It even hurt him to put ointment on him. We tried everything. We bought every over-the-counter diaper cream, the kinds you get from Babies "R" Us or Walmart: Desitin, Burt's Bees, Boudreaux's Butt Paste. We went to the doctor and got a prescription for Nystatin cream. We tried switching diaper brands and tried every name brand and off-name brand of disposable diapers. Nothing worked.

We were totally desperate to figure out something to make our son feel better. It's very hard to see your baby in pain like that. I was getting ready to be deployed to Guantánamo Bay, Cuba, to be a guard for detainees. Before I left, one of the girls in my wife's moms' group showed her a cloth diaper. Brooke came home and said she thought we should try that.

I snarled at it. I got the big lip about it. I didn't like the idea of washing poopy diapers with your clothes. Using cloth diapers just seemed like way too much trouble. Instead of just throwing a diaper away, you actually had to deal with it. I thought it was gross. And I thought it would be too much of a burden on my wife while I was gone.

But two days after Brooke put Tristan into cloth diapers, his rash completely cleared up. Now my son wasn't screaming in pain anymore, he was finally comfortable and could sleep through the night. I didn't care what fixed it. I was just so glad the rash was gone.

When I got back from deployment and Tristan was a little older, I started changing diapers. I was surprised that everything with cloth diapers is actually very simple. They aren't old-fashioned or weird like my friends think. They're actually designer and cool-looking. I like all-in-ones—simple diapers where everything is all there and you don't have to worry about an insert with a cover. You get the kids and you get out the door to give Mommy a break and you just want it to be simple. That's my preference as a dad. Now I have two girls in diapers and my son still wears a nighttime diaper, and they are all in cloth.

Using cloth diapers is a better way to take care of your kids. With all the stuff they are putting in the diapers, the chemicals and everything, you don't really know what harm they are doing. Does anyone ever read the warnings on the labels? Disposable diapers can actually cause your child to become infertile. Cloth is also so much more economical: My baby daughter is wearing the diapers that my son wore five years ago. If you are only planning on having one kid, maybe it doesn't matter. But being able to buy twenty-four diapers one time only and never have to buy diapers again is a big thing I like. I don't have to go to the grocery store and buy diapers for the kids. And I don't have a giant trash can literally full of crap in my laundry room.

We put our middle daughter in a disposable diaper once when she was a newborn and she broke out into a rash two days after she started wearing them. When we traveled when our son was little we put him in a disposable diaper because we thought it would be easier. Two hours after he started wearing it, even though the diaper was dry, he broke out into a rash. So we travel with cloth diapers as well. Our youngest daughter has never worn a disposable diaper. Disposables are horrible.

Before I had kids I was a hard-ass military guy. Now my wife and I run our own business, and we sell natural baby products. Our son is what got us into it. We don't want people to have to go through what we went through. Our biggest product is cloth diapers. Maybe Tristan's rash was unusual but think about it, if the chemicals in disposable diapers caused my son to break out like that, what are those chemicals doing to every baby, even if the baby doesn't have a visible reaction?

CHAPTER 9

BOOST YOUR BOTTOM LINE:
Vaccinating for Health or Profit?

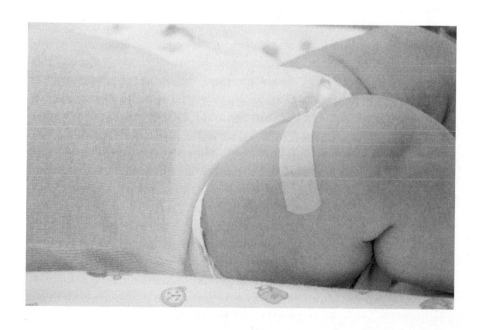

Vaccinations save lives. They represent a tremendous step forward in medical history—a way to jump-start the immune system so it can recognize, build an immunological memory of, and better fight off disease. Because vaccines are such an important part of public health, in the United States today a baby is given his first inoculation, against hepatitis B, in the first or second day of life. At one to two months, American babies get a second shot for hepatitis B, and at two months they receive vaccinations against rotavirus, diphtheria, pertussis, tetanus, *Haemophilus influenzae* type b (Hib), pneumococcal disease (PCV), and poliovirus. Immunizations against eight different diseases, usually given in several separate shots, are injected into the fatty tissue of the thigh of a baby with floppy muscles who is not too young to smile but still too young to hold his head up.

That's just the beginning.

At four months of age an American baby receives another round of booster shots against rotavirus, diphtheria, pertussis, tetanus, Hib, pneumococcal, and polio. At six months, there's yet another round of shots: This time possibly including the yearly flu shot, which a baby will get between six months to a year of age and then *every year afterward*.

By the time he is one, a baby in America—if his parents choose to follow CDC guidelines—will have received twenty-one injections against ten diseases. By the time he is eighteen months old he will have received as many as twenty-six injections against fourteen diseases. The CDC now recommends that children receive no fewer than fifty injections against sixteen illnesses by the time they are eighteen years old—more than *four times* as many injections (including oral polio) as when my generation was growing up in the 1970s and *double* the number of injections given in Norway and other European countries. And there's a strong possibility that even more vaccines will soon be added to the infant immunization schedule: In April 2011 the FDA approved another vaccine, Menactra, Sanofi-Pasteur's vaccine against several strains of meningococcus bacteria for use in infants. Now the CDC is considering whether to add to the infant schedule two doses of this vaccine, to be administered at nine months and twelve months.

"I don't know any rational person who could truly be antivaccine, because the benefits are so profound," said Louis Cooper, M.D., professor emeritus of pediatrics at the College of Physicians and Surgeons of Columbia University and former president of the American Academy of Pediatrics, to a group of some sixty-five listeners at a Federal National Vaccine Advisory Committee–sponsored community meeting on vaccine safety.

Vaccine proponents—including Martin G. Myers, M.D., a professor in the Department of Pediatrics and Preventive Medicine at the University of Texas Medical Branch at Galveston, former director of the Department of Health and Human Services' National Vaccine Program Office, and coauthor of a book about vaccine safety who also attended the meeting— argue that vaccines are fundamentally safe. They believe that the risk of side effects has been grossly overreported by the media, that parents' easy access to unfiltered (and sometimes inaccurate) information about vaccines on the Internet is at odds with the health of America's children, and that the risk of side effects or long-term immune system damage from vaccines is negligible compared to the public health crisis we would face if vaccine-preventable diseases were to reinfect the general population because large numbers of parents have stopped vaccinating their children.

It's a basic tenet of modern medicine that childhood vaccination is essential for a healthy America. But even parents who believe wholeheartedly in vaccines, and who usually follow their doctor's advice without question, are beginning to wonder about the current vaccine schedule and if public health officials have adequately considered the long-term ramifications of injecting so many pharmaceutical products so often into babies at such a young age. Parents are questioning if we can really keep adding vaccinations to the childhood schedule without subtracting anything. They are starting to ask why children in America receive so many more vaccines than children in countries where health is not a for-profit business and where infant mortality rates are significantly lower. They conjecture that there may be a tipping point when so many vaccines administered all at the same time overload the immune system and cause more harm than good. When they see that a disease like polio has been completely eradicated in the Western Hemisphere and, in fact, in 98 percent of the world, and that the only way it is contracted in America has been from the vaccine itself, they begin to distrust that current schedule is actually of benefit to American babies.

"When you get the doctors away from charts and business and records and liability they can say that human beings are wonderfully capable of healing themselves," one doctor tells me. "I can tell a mom who comes in that 'Yeah, I feel like we're really turning our kids into pincushions. There are so many more vaccines than when I was a child. I had chicken pox, and I survived, we really don't need the chicken pox vaccine.' But I can't say that in public." Though they often go along with the system for fear of being ostracized by their medical peers or sued by patients, many of the nation's best and most conventional pediatricians, family practitioners, and other medical doctors themselves have deep reservations about vaccines. It's a well-kept secret that doctors, nurses, physician's assistants, government officials, vaccine developers, and others who work in the health care industry, and whose job it is to publicly urge parents to vaccinate, often delay vaccines for their own babies, pick and choose vaccines, or decide not to vaccinate at all.

Amy Miller, who graduated with a degree in nursing from Humboldt State University's School of Nursing in 2000, worked in a family practice where one of her main jobs was vaccinating babies. But she chose not to vaccinate her infant son Cassidy or her twin daughters born seven years later. "One instructor didn't vaccinate her kids either and totally agreed with me off the record, but on the record and in class she taught the students that kids would die if they weren't vaccinated," Miller tells me. "She felt she would have lost her job if she said what she really believed."

As we talk outside on a bright but chilly December day, a high-ranking government official, who has authored several scientific papers on the benefits of vaccines, confides off the record that he and his wife decided not to vaccinate their own newborn against hepatitis B. The International Medical Council on Vaccination is a newly formed group of medical doctors, registered nurses, and other medical professionals who, after thousands of hours of research, have concluded that vaccines are neither as safe nor as effective as the pharmaceutical companies that manufacture them and the government officials that promote them have led the American public to believe. The late Bernadine Healy, M.D., health editor of *U.S. News & World Report* and former director of the National Institutes of Health, publicly critiqued the current schedule, expressing her deep reservations about the lack of safety studies of vaccines as administered today.

"How can these vaccinations that we give to tiny babies not affect their immune systems?" wonders Michele Pereira, a nurse and mother of two from Yreka, California. Although Michele administered vaccines to children and is married to a doctor, she and her husband stopped vaccinating their now six-year-old daughter after she started getting continuous upper-respiratory infections and showing signs of developmental delays. When Michele asked her pediatrician if the vaccines could have been a trigger, he said absolutely not. A scientifically minded person and a medical professional, Michele felt frustrated by the doctor's certainty. How could he know the vaccines had not contributed to her daughter's problems if he had no idea what was causing them? Both Michele and her husband believe that though vaccines have a place in modern medicine, we are giving too many vaccinations too soon.

Mainstream media reports suggest that childhood vaccination is a black-and-white issue. On one side are the crazy parents who insist, with no proof, that vaccines cause autism and irrationally oppose all vaccines. On the other side are rational doctors with the science to back up their belief in vaccines. But the truth is very different. Parents *and* doctors are becoming increasingly concerned that the current CDC schedule is not in the best interest of our children's health. They question whether some vaccinations on the CDC schedule—including the hepatitis B shot at birth when neither parent is hepatitis B positive, the rotavirus vaccine, and the chicken pox vaccine—are really necessary. They are concerned about both short-term and long-term side effects from vaccines, pointing to an increasing body of scientific evidence that shows that some vaccines are harmful and that the risks of having side effects is sometimes higher than the risks of contracting the illness against which the vaccines are supposed to protect.

A VACCINE AT BIRTH FOR A SEXUALLY TRANSMITTED DISEASE

Just hours after our baby was born at Crawford Long Hospital in Atlanta, a nurse bustled in with a tray. "Time for her hepatitis B vaccine," she said in a chipper voice. The nurse caught me and my husband unprepared. Having worked on the literacy component of a child survival campaign in one of West Africa's poorest countries, meeting malnourished mothers who birthed eight children only to have three of them die in infancy, and seeing paralyzed polio survivors begging at stoplights, I had more than a healthy respect for vaccines. To prepare for my work overseas, I had been vaccinated as a young adult against diseases most Americans have never heard of. But I had never heard of vaccinating a newborn—especially against a sexually transmitted disease.

Hepatitis B is a blood-borne virus that can cause liver damage and liver failure. It is spread through sexual contact, by sharing unclean needles, and from transfusions of tainted blood. Hepatitis B is not common, affecting only between 0.1 and 0.5 percent of the population, the greatest reported incidents occurring in young adults. Intravenous drug users, men who have sex with men, and promiscuous heterosexuals, as well as people traveling to countries with high rates of hepatitis B, are the populations most at risk. Because a hepatitis B–positive mother can transmit the disease to her newborn during childbirth, infants whose mothers are hepatitis B positive can also contract the disease. But every woman in America who receives prenatal care is tested for hepatitis B. If a pregnant woman and her partner do not have hepatitis B, their infant has little, if any, chance of getting the disease.

In most First World countries, the hepatitis B vaccine is not part of the routine schedule, recommended only when medically indicated. In the United States, every newborn is given the hepatitis B vaccine. When my husband and I told the nurse we needed more time to decide if the vaccine was the right choice for our baby, she narrowed her eyes in anger and her whole attitude changed, as if we were posing a health threat just by wanting to discuss the matter with our pediatrician and get more information. But it turns out many doctors believe the mandate of this vaccine for all American newborns is a mistake.

"The hepatitis B vaccine doesn't make any medical sense," argues Larry Palevsky, a pediatrician in private practice on Long Island, New York, when we meet at a coffee shop in Portland, Oregon. Palevsky is on the West Coast to give a workshop to health professionals. I've lost a run-in with some poison oak and my right eye is so swollen I look like Frankenstein's monster. Even though his plane was delayed and he barely slept the night before, Palevsky radiates good health and confidence.

A pediatrician who emphasizes nutrition and healthy lifestyle practices, Palevsky is outspoken in his critique of this vaccine. "Hepatitis B has never been a disease prevalent in neonates," he continues, telling me he was taught in medical school that the reason for the vaccine in newborns is the adults who were most at risk were not accessible to medical care, so if all babies were vaccinated in the hospital, should any of them grow up to be members of high-risk groups (intravenous drug users, prostitutes, or others engaging in frequent sexual intercourse without protection) then they could be protected from the disease. "But there was *no evidence* that you could actually be protected," Palevsky insists. "There was no evidence that the vaccine would be effective when the person was thirty-five and shooting up drugs."

"If I'm a rational person and I listen to what you've said, the recommendation that we vaccinate a newborn against hepatitis B sounds absolutely insane," I interrupt.

"Asinine," Palevsky agrees. "There's no medical indication for it. Unless you have a mother who is hepatitis B positive, then the protocol changes. But there's no medical indication to a healthy newborn who is not at risk . . . healthy newborns, infants, and small children are not using intravenous drugs, they are not having sex, and even if they are exposed to blood products, this is America, and the risk is still low."

Palevsky takes a sip of green tea, leaning forward to explain that the problem is not just that the birth dose of the hepatitis B vaccine is unnecessary, it also may be damaging: Humans are born with very immature immune systems, more so than other mammals, he tells me. "Most of the maturity of the human organism happens outside the womb. Most of the maturity of other mammals happens inside the womb. . . . The injection of material into the body is disruptive to the way the immune system functions. You are injecting foreign proteins into an immune system that has yet to even understand its relationship to the world."

Like Palevsky, Robert Sears, a pediatrician in private practice in Orange County, California, and author of *The Vaccine Book,* does not believe hepatitis B is a medically indicated vaccine. In one manufacturer's safety trials of the hepatitis B vaccine among 147 healthy infants and children, the most common adverse reactions, occurring in up to 10 percent of the children, included irritability, fever, diarrhea, fatigue, loss of appetite, and rhinitis (inflammation of the mucous membranes inside the nose). As Sears points out, while none of the vaccine's common side effects are worrisome in older children (this study was done on children up to ten years of age), high fever, diarrhea, or loss of appetite can be devastating in newborns and lead to prolonged hospital stays. That's what happened at Chaim Sheba Medical Center in Israel: The number of newborns admitted to the NICU in 1992, the first year the vaccine was administered to every newborn, was more than double the number from 1991, when the vaccine was not routine, though there was no increase in infectious causes of the fevers.

Vaccine-induced NICU stays, difficulty eating, and digestive problems are not the only ill effects of this vaccine. More than fifty scientific studies in both humans and animals, the majority of which have been published in peer-reviewed journals, have revealed myriad problems with the hepatitis B vaccine, including that it can cause lupus, as well as a fatal form of inflammation similar to Guillain-Barre syndrome, that it can put children at up to two and a half times more risk for juvenile diabetes, and that it can trigger other chronic autoimmune disorders. One 2012 controlled study in mice done by Chinese researchers found that the hepatitis B vaccine causes cell death and mitochondrial disorders.

But perhaps the most disturbing research to date was a study done on twenty male rhesus monkeys by researchers at the Washington National Primate Research Center in Seattle and the University of Pittsburgh School of Medicine. Thirteen monkeys were injected with the hepatitis B vaccine, four received saline injections, and three had no shots at all. Researchers who did not know the vaccination status of the monkeys found that monkeys exposed to the vaccine showed significant developmental delays when compared to the nonvaccinated monkeys. Published in 2010 in the *Journal of Toxicology and Environmental Health,* this monkey study (which is the only study of any kind to date to use a control group of unvaccinated subjects) led researchers to conclude that there is evidence that the hepatitis B vaccine can cause abnormal neurodevelopmental responses. Ignoring the nurse's anger, my husband and I refused the hepatitis B vaccine for our newborn. After we left the hospital we talked to our pediatrician, who actually applauded our decision. Given that an American baby with hepatitis B–negative parents has a negligible chance of getting the disease and there are serious concerns about its safety, and so many doctors are against it, more and more parents are beginning to wonder why it is mandated by the CDC.

HOW NECESSARY IS THE ROTAVIRUS VACCINE?

Rotavirus is an intestinal virus, like influenza, that causes watery, often foul-smelling diarrhea, fever, and vomiting. The vast majority of the children in America who get rotavirus (and most children do) get better without complications. The problem is not the infection itself but the possibility of becoming dehydrated due to severe diarrhea. Rotavirus is most severe in the first year but can be miserable at any age. The CDC estimates that before introduction of the rotavirus vaccine in 2006, the disease was responsible for 410,000 outpatient visits, more than 200,000 trips to the ER, and between 20 and 60 deaths among children under five annually. In 2005, before the vaccine was introduced, 1 in 400,000 children under five died from rotavirus. (In other words, the chance of an American baby under five dying of rotavirus was 0.00025 percent.)

Karen Driscoll, a mother of four in Wallingford, Connecticut, remembers when her twins got rotavirus in 1999 when they were two and a half, before there was a viable vaccine. "Brittany lost five pounds (25 percent of her body weight) and was throwing up blood by the time they finally hospitalized her for three days and two nights. Holly got checked into the same room the day Brittany checked out. Fun times," Karen writes me in an email. "I took Brittany two times to the pediatrician that week and called almost every day but they kept giving me the brush-off. That was back when you had to have prior approval from your doctor to go to the ER. I have never been so furious. We left that practice immediately after. Rob was one and still breastfeeding—he didn't get it, and neither did I. But Mark [Karen's husband] did. . . . If I could go back and prevent that from happening to them, I would have vaccinated them."

As Karen's experience shows, a bad case of rotavirus is not something to take lightly. Yet infants who are exclusively breastfed and do not attend daycare have very little risk of catching rotavirus in their first year of life, when the disease is most serious. Like the disease itself, the vaccination can sometimes have serious, even life-threatening, side effects. An earlier version of the rotavirus vaccine, Wyeth's RotaShield, was taken off the market because the side effects were so severe. The newer vaccine has proved problematic as well: During the first months when Rotateq was introduced in 2006, twenty-eight cases of intussusception occurred in infants (none were fatal but half required surgery). About 20 percent of the infants given the rotavirus vaccine will get a fever; vomiting or diarrhea will occur in 10 percent of vaccinated infants and poor feeding in 25 percent. Other reported side effects include bloody stools, rotavirus infection from vaccinated child to nonvaccinated person (because the live virus is excreted in bowel movements), and hives. During safety trials, 1 in 1,000 infants had seizures and 1 in 10,000 contracted Kawasaki syndrome, a poorly understood illness that may be an autoimmune disorder that causes inflammation of the arteries leading to the heart. A study published in June 2011 in the *New England Journal of Medicine* found that the vaccine caused a fivefold increase in intussusception in children in Mexico in the first seven days after the dose, occurring in as many as 1 in 51,000 infants. There were ninety-six cases of intussusception and five deaths due to the vaccine.

Hannah, the infant daughter of Dr. John E. Trainer III, got severe diarrhea for seven days after each of the first two rounds of the Rotateq vaccine. "The pediatrician said it was a fluke, so we went ahead with dose number two," Trainer, a family physician in Jacksonville, Florida, tells me during a phone interview. "The same thing happened: Hannah had knock-your-house-down diarrhea. The vaccine was worse than the disease." Trainer and his wife (also a doctor) decided not to give Hannah the third dose.

"I did research at Children's Hospital on rotavirus for two years," says

Lyn Redwood, a nurse who has a master's degree in public health and is married to a doctor. "It's a minor childhood infection. It's more dangerous in developing countries, but that vaccine is not necessary in the United States. We've actually found nucleic acid from pig viruses in the rotavirus vaccine that can get incorporated into our human DNA. We don't know what that can do."

Still, in the face of the severity of the disease, public health officials in the United States feel that the risk of these side effects is worth the protection the vaccine gives against the disease. Sears agrees: "During my training years the hospital hallways would be overflowing with dehydrated babies and worried parents because of this bug," he writes. "This vaccine should help us get rid of rotavirus. I consider this a fairly important vaccine." Apparently doctors and public health officials in Italy, France, the Netherlands, Germany, Ireland, Spain, Switzerland, the Czech Republic, Sweden, Norway, Iceland, Denmark, Singapore, and Japan—all countries that have lower infant mortality rates than the United States and all countries where the vaccine has *not* been adopted—do not believe this vaccine is useful.

HOW NECESSARY IS THE CHICKEN POX VACCINE?

In the United States, at his one-year birthday, a baby will receive a vaccine against chicken pox (varicella), which is another vaccine mandated in the U.S. but not found on the childhood vaccination schedule in much of Europe and the developed world. The chicken pox vaccine was introduced in 1995 as part of the CDC's schedule of routine vaccinations. This surprised many parents and doctors since vaccines have historically been developed to protect children against serious childhood illnesses, but chicken pox has always been a mild disease. Until the vaccine was introduced, chicken pox caused about a hundred deaths each year, about half of which were in children. According to vaccine safety advocate Barbara Loe Fisher, cofounder of the National Vaccination Information Center and coauthor of the book *A Shot in the Dark,* about the dangers of the whole-cell pertussis vaccine, the vaccine was first designed for children with compromised immune systems. But once Merck & Co. had developed Varivax, the chicken pox vaccine used in the United States, they needed to find a market for it.

The argument in favor of adopting the varicella vaccine was not because chicken pox was a lethal disease but because it was in the best economic interests of big business. "If working mothers could keep working instead of taking time off to care for their sick children, *business would save hundreds of millions of dollars,*" writes one pediatrician (her emphasis) who continues to be opposed to the vaccine. The vaccine seems to be working: Now that the chicken pox vaccine is required for enrollment in most public schools, there are far fewer cases in America. On one hand, this is a good thing: fewer children miss school, parents don't have to take time off work, and currently there are fewer than five deaths per year from wild chicken pox. Yet these positive benefits come with at least one unforeseen downside: Once exposed to wild chicken pox, you have immunity, which is important since the disease is almost always more serious in adults than in children. Previously, when wild chicken pox broke out in the general population, the immunity of adults exposed as children would be asymptomatically boosted by coming into contact with the virus, which helped remind their immune systems to recognize the disease. But now that children aren't getting chicken pox, that natural immunization boost is effectively gone, leading to growing numbers of baby boomers who had chicken pox as children getting shingles, a much more serious disease caused by the chicken pox virus.

"We now have an epidemic of shingles among older children and adults, which is far more expensive and actually far more painful than if we had let chicken pox remain endemic in the population, and used the vaccine selectively," Fisher says. According to the CDC, one million people a year get shingles, and there is an increasing body of scientific data showing that the numbers of adults suffering from shingles is on the rise. One study showed a 90 percent increase in the number of adults infected with the herpes zoster (shingles) virus in Massachusetts from 1999 to 2003 as the chicken pox vaccine become more widespread. Other studies have estimated increases as well. One reason European countries like England, which has not adopted the chicken pox vaccine, have taken a conservative wait-and-see approach is because of this concern over rising rates of shingles.

John Grabenstein, senior medical director, adult vaccines, at Merck & Co., Inc., argues that the slight recent rise in shingles—which, he says, is something Merck & Co. also found in their shingles prevention study of more than 38,500 people—is not related to the chicken pox vaccine. "Shingles was with us even before the childhood vaccine came along," Grabenstein points out. "It's not correct to say that because we are vaccinating children therefore adults are getting shingles." Instead, Grabenstein attributes the rise in shingles to the rising life expectancy among American adults. "The longer you live, the greater your risk of getting it." He discounts the idea that reexposure to the wild virus circulating in the population provides a potential boost to the immune system and is preventive. "Your exposure to your grandkids is a minor player," Grabenstein insists. "The dominant reason shingles occurs is that the immune system ages along with the body."

What everyone does agree on is this: Shingles is a potentially serious and terribly painful disease. To address the shingles problem, Merck has developed a new vaccine: Zostavax, licensed by the FDA in May 2006, and now being recommended for everyone over sixty. Barbara Loe Fisher finds this troublesome. "You have a situation that was created by vaccination and you have a vaccine to counteract that situation," she contends.

Dr. Trainer has questions about the cost of Zostavax and how well it works. "It's a coin toss of efficacy," he says. "If I give it to a sixty-year-old, they are fifty percent likely to benefit from it. That means you can flip a coin and see whether it works or not. The vaccine is expensive, and the cost to the patient varies widely, depending on insurance coverage. My partner and I have not embraced the shingles vaccine."

In addition to the problem of an upsurge in shingles, the chicken pox vaccine, which is a live virus, can cause several side effects that are as or more dangerous than the disease itself. Since the chicken pox vaccine came into widespread use, reported reactions to it have included bleeding disorders, pneumonia, skin infections, severe rashes, and such nervous system problems as Guillain-Barre syndrome, encephalitis, and seizures. According to the CDC, 10 percent of the children who get the vaccine will spike a fever, and 4 percent will get a chicken pox–like rash up to a month after being vaccinated. Early findings from an ongoing CDC study have shown that the combination vaccine—MMRV—has been found to cause severe fever-induced seizures in approximately 2,100 children a year. For this reason, the CDC recommends getting two shots—the MMR vaccine and the chicken pox vaccine—separately for children under four. Two children in Dr. Sears's practice who received the vaccine have gotten benign thrombocytopenia, a blood-clotting disorder. Both were hospitalized, and each took more than two months to recover.

DAMAGED BY VACCINES

It took Sarah Lipoff three years to get pregnant. She and her husband started trying when she was thirty years old. During that time Sarah worked at a preschool in San Anselmo, California. One of her favorite pupils was a little boy, Joey, whom she had known since he was eighteen months old. Sarah spent eight hours a day with Joey five days a week for more than a year. Joey was an active, healthy, lively toddler who babbled all the time and loved to play the preschool piano. His mom, who had a high-profile job in town, had told Sarah and the other teachers that she had not had time to vaccinate her son when he was a baby. So when Joey was two and a half, she took him in for a round of catch-up vaccinations, including the measles, mumps, and rubella shot (MMR) and a flu vaccine.

The little boy who came to school at 10:30 a.m. that day was unrecognizable. His face was wan, he would not interact with other children, and he was lethargic. Sarah sat beside Joey doing paperwork while he napped. The next morning Joey was uncharacteristically quiet and withdrawn. When there was no change in his behavior, Sarah called his mom and insisted she take him to the doctor. When his mom dropped Joey back at school in the afternoon, she told Sarah that the pediatrician had reassured her Joey was having a "normal response to the immunizations," and that if he didn't develop a rash or a fever there was nothing to worry about. By the end of the week Joey was sitting in the corner, rocking. Three months later he was assessed: The neurological damage he suffered was irreversible.

When Sarah gave birth to her daughter five years later, what happened to Joey was still vivid in her mind. Sarah and her husband declined the hepatitis B vaccine and delayed the others until Ivy was eighteen months old, choosing to give their daughter one shot at a time, with plenty of time in between in order to observe adverse reactions. Their pediatrician was unsupportive and every visit felt like a battle, but Sarah continues to decline the flu and chicken pox vaccines, despite her pediatrician's insistence. "I think Joey was at a pivotal point in his development. His language capacity was amazing. His brain was on hyperdrive. It makes me sad that the doctor would deny what is so terrifyingly obvious: To overload *that* child with the chemicals in so many vaccines all at once damaged him beyond repair," Sarah tells me.

It's not just parents like Sarah who worry there is a link between the ingredients in vaccines and neurological damage among American children today. Lyn Redwood sits on the Board of Health for Fayette County, Georgia. For more than ten years, she worked as a pediatric nurse at Tampa General Hospital, the University of Vermont Medical Center, and the Children's Hospital in Birmingham. Her husband is the director of Wellstar Kennestone Hospital's emergency room in Marietta, Georgia, the third-largest hospital in the state. The Redwoods' son was diagnosed with autism in 1997. "I used to preach vaccines," Lyn tells me. "If anyone had initially told me that vaccinations had anything to do with my son's autism, I wouldn't have taken that information as credible."

Will, who had multiple ear infections as a baby and got so sick as a toddler that he was admitted to the ER for vomiting, fever, and wheezing, started showing signs that something was wrong after his first birthday. When he was about fifteen months old, he stopped eating, stopped talking, didn't seem to listen when spoken to, became very sensitive to sunlight, and lost the feeling in his fingertips. (Lyn found this out when a wooden hamper fell on his finger and so much blood pooled under the nailbed that her husband had to drain it by burning a hole in the nail. Will didn't even flinch.) He started chewing on his shirtsleeves and sticking his fingers in his mouth to make himself gag and vomit.

At first the doctors thought Will's loss of speech was due to hearing loss from fluid buildup in his ears from the ear infections. But even after tubes were put in his ears to drain them there was no improvement. So Will was referred to a neurologist who ran a series of tests. Though he didn't mention it to Lyn and her husband, the neurologist made a note in Will's chart that Will had behaviors and characteristics consistent with PDD-NOS (pervasive developmental disorder, not otherwise specified). "Nobody used the A-word around us. They tiptoed around it," Lyn remembers. "It was the ENT [ear, nose, and throat] doctor who said to us, 'Your son may never talk . . . he may end up being put in a residential facility.' " When Will was two years, three months, and nineteen days old, he was evaluated by Georgia State University's Toddler Project. The evaluation revealed that his developmental age was twelve months. On a test that was like a baby IQ, Will scored a mental developmental index of 51, which put him in the category of mental retardation. Though technically he had not been diagnosed, when he was three years old he was enrolled in an early-intervention program through the Fayette County public schools for children with autism. "Nobody at the time knew what caused autism," Lyn told me.

But when Will was five Lyn remembered she had saved a lock of hair from his first haircut at twenty months old. She decided to have the hair tested. That lab report changed her life. The Environmental Protection Agency has an "action level" for exposure to mercury in water, meaning that you need to do something to reduce your exposure—that level is one part per million. The mercury level in Will's body at twenty months old was 4.8. He also had more than four times the normal level of aluminum in his body (normal for the lab was 0 to 9, Will had 40.2). Lyn examined her son's medical records and discovered something else: All of Will's vaccines that could have possibly contained thimerosal, a mercury-based preservative that has since been taken out of infant vaccines (except the flu shot), did. At his two-month well-baby visit he had received the DT and Hib vaccines, which both contained 25 mcg of mercury, and a hepatitis B vaccine, which contained another 12.5 mcg. According to EPA guidelines for safe daily mercury exposure based on his weight, Will's allowable exposure for that day was 0.5 mcg. He had received 62.5 mcg, an exposure to mercury, a known neurotoxin, approximately 125 times the EPA's safe allowable daily exposure. It was then that Lyn and her husband realized that the mercury in the vaccines, combined with the thimerosal-preserved immunoglobulin she had been given during pregnancy to prevent RH-negative incompatibility (and possibly some environmental toxins, like the mercury amalgam in her dental fillings and the old coal-burning power plants in Georgia, emitting high levels of mercury into the air), are what damaged her son's brain.

Sue Latour's daughter's brain damage happened much more dramatically. In 1981, Kimberly was nine weeks old when she had her first round of vaccines. "The day she had her shot, the doctor's remark to me was, 'My goodness, she is so bright!' " her mom remembers. Immediately after the shot Kimberly started high-pitched crying, like a wild animal. She had developed encephalitis (swelling of the brain) and brain seizures. Kimberly, who is mentally retarded, holds her head awkwardly to one side when she walks. Her body and face are mangled, and she needs around-the-clock care. In the 1980s, as the vaccine schedule in the United States started getting more aggressive, so many American children, like Kimberly, suffered severe vaccine-induced side effects (including death) that parents started filing lawsuits against vaccine manufacturers. These manufacturers appealed to government authorities to protect them against liability, threatening to stop manufacturing all vaccines if they were to be held financially responsible for the side effects of their products. In 1986 the National Vaccine Injury Compensation Program was created by an act of Congress as a no-fault nonadversarial way to pay for the care of children injured or killed by vaccines.

In the 1970s fewer than 3 children in 10,000 were diagnosed with autism; in the 1990s more than 30 children per 10,000 were diagnosed, a tenfold increase. Though some argue that part of the dramatic rise in autism is linked to changes in how we define or identify autism, the most detailed and careful study to date has shown that is not the case. When

I first started researching this book, 1 in every 150 children was on the spectrum. Today about 1 in every 88 American children has a neurological disorder classified by the CDC under the umbrella of autism.

Almost every mainstream news article in the United States summarily rejects the idea that vaccines can cause autism or other neurological damage. Although scientists do not yet understand why or how (possible culprits include vaccine preservatives like thimerosal, which is mercury-based; the trace amounts of formaldehyde, which is a known carcinogen, found in some vaccines; the adjuvant aluminum, which is a neurotoxin that can cause seizures in primates), we know that vaccination causes autism in some children. Neurologist Manuel Casanova theorizes that vaccines given at the wrong time to already susceptible children may trigger abnormal brain cell migration in the same way ultrasound exposure does. In June 2012, the Italian Health Ministry concluded that the severe neurological damage suffered by fifteen-month-old Valentino Bocca was triggered by the MMR vaccine. Three eminent Italian doctors independently examined Bocca's medical history, all concluding that there was no other scientific explanation for what caused Valentino's brain damage. Once a thriving toddler, Valentino cannot speak or hold a pencil.

In 2008, the U.S. courts awarded $1.5 million in damages to Hannah Poling, whose father is a neurologist who also has a Ph.D. in biophysics and whose mother is a registered nurse and an attorney, as well as a yearly sum of more than $500,000 to help pay for her care. A panel of medical evaluators from the Department of Health and Human Services found that the nine shots Hannah received at one doctor's visit when she was a nineteen-month-old and developing normally resulted in brain damage (the once healthy redheaded little girl now needs twenty-four-hour care and may never be able to live independently). What scientific arguments showed the connection between Hannah's autism and vaccines? We don't know because the government sealed the files related to her case.

A DEARTH OF ADEQUATE SAFETY STUDIES

"As a parent I researched the safest car to drive, the best car seat to put my child in. I made organic baby food and froze it in ice-cube trays," Lyn Redwood tells me. "But I didn't research vaccines because I thought somebody else did that for me. I was wrong. My husband, who's a doctor, and I made a mistake in believing the vaccine safety studies were adequate, a mistake that will stay with us for the rest of our lives. My best advice is to do your own research."

That research reveals that there are two problems with our vaccine safety studies that most new parents are not aware of:

1. To date there have been no large-scale safety studies of vaccines in humans conducted with a control group of entirely unvaccinated babies. Public health officials claim it would be unethical to conduct studies where children remain unvaccinated because of the danger not vaccinating children would cause. Vaccine skeptics point out that there are now hundreds of thousands of parents choosing to delay or forgo vaccinations. Though it may not be possible to conduct double-blind studies, the gold standard in scientific research, it is certainly possible to include a control group of unvaccinated children in the research.
2. Though multiple vaccines are administered to infants at the same time, vaccine safety studies (which are usually sponsored by vaccine manufacturers with a vested interest in proving the vaccines are safe) are conducted on a vaccine-by-vaccine basis. We have no way of knowing what the effect of multiple vaccinations is on the human organism because it has not been studied scientifically.

"I think it's really hard for people who have actually given vaccines to think something you did could have caused injury," says Vicky Debold, a registered nurse who has a Ph.D. in public health and firsthand experience with infants who have died from vaccine-preventable diseases, and whose son was damaged by vaccines. "Vaccines are something that I willingly participated in for my son. The natural defense as a mother and a nurse is that I didn't do anything wrong." The more Debold researched the issue, the more skeptical she became. When she went to the library to find articles that proved the efficacy and safety of simultaneously receiving multiple vaccinations, she was shocked to find almost none. "The more I looked, the less I found," Debold says. "As a nurse, I had just assumed that everything we had been asked to do was based on good, solid science. But the gaps in the science are, frankly, what we're still dealing with now."

The fact is, large-scale vaccine safety studies have never compared the immune systems of completely unvaccinated children to vaccinated children, nor have we studied the possible long-term damage that may be done to the immune system by multiple vaccines. We simply don't know what the effect of giving so many vaccines at once is on the immediate or long-term health of a child. When John Iskander, M.D., M.P.H., chief science officer at the CDC's Immunization Safety Office, was asked at a CDC-led community meeting about childhood vaccination held in Ashland, Oregon, if vaccines could cause long-term immune dysfunction, he answered honestly that we don't know, and that the issue is a concern to doctors and government officials, and something that needs more study.

Though the FDA has been discussing in closed-door meetings the problem that vaccine technology has outpaced our ability to predict adverse events, more vaccines are being added to the infant immunization schedule without testing how the human infant's immune system responds to multiple vaccines given at the same time. In 1999, then Center for Biologics Evaluation and Research Viral Products Division director Peter Patriarca, M.D., said, "[T]he technology used to make these vaccines actually exceeds the science and technology to understand how these vaccines work and to predict how they will work," explaining that federal budget constraints often contribute to the lack of adequate safety testing. At the same meeting, his colleague, then National Association of County and City Health Officials executive director Tom Milne, cited an informal study done by the Washington State health department of physician office immunization practices, which found that "in every office visited there were significant quality assurance problems," including "wrong doses, wrong routes of administration, wrong kid [or] missed opportunities for vaccine."

More than a decade later we are dealing with the same problems: A Department of Health and Human Services investigation, published in June 2012, revealed that the majority of providers (76 percent) exposed vaccines to inappropriate temperatures for at least five cumulative hours, almost a third of providers stored expired vaccines with viable vaccines, "increasing the risk of mistakenly administering the expired vaccine," and *none* of the providers met the government guidelines for appropriate oversight of the vaccine administration, even after follow-up visits and education from investigators.

"Our national vaccine program started out with all the right intentions in terms of preventing horrible infectious diseases," Lyn Redwood tells me, "but in the last two decades, when the vaccine manufacturers were given release from any liability from their products, our children became cash cows. . . . Now corporations can introduce new products and get them mandated for every child with no legal responsibility if something goes wrong. The growth in the vaccine industry has been tremendous. Now we're targeting minor childhood illnesses with vaccines that in my professional opinion we don't need. The vaccines are mandated before we have adequate safety testing. And often the industry-sponsored and government-sponsored tests aren't done thoroughly enough. We are not using a placebo group of nonvaccinated children. What we're really doing is playing Russian Roulette with Mother Nature."

EARLY AND OVERVACCINATION IMPLICATED IN IMMUNOLOGICAL DISORDERS

Dr. Kenneth Saul, originally from Cleveland, talks faster than a New Yorker, gesturing with his hands for emphasis. He's been a pediatrician for twenty-eight years and is so dedicated to his patients that he works seven days a week, sees sick kids after hours, and makes himself available by phone or email to worried parents, even if it's ten o'clock at night. Dr. Saul tells his patients honestly that the birth dose of the hepatitis B vaccine is probably unnecessary if you are not hepatitis B positive, that he thinks it is reasonable to space out vaccines if you know the risks, that a baby who is not old enough to crawl or walk probably doesn't need a tetanus shot yet (Saul has seen one case of tetanus in the past twenty-eight years, in an unvaccinated farmer who had a tractor accident in the dirt—he recovered without incident), and that he is not very concerned about the risk of polio unless a family is traveling overseas. Though he likes to poke fun at Bob Sears and vaccine safety proponents, Saul readily admits to me that he has noticed that the nonvaccinated or "under" vaccinated children in his practice tend to be healthier. But the infectious disease experts and CDC colleagues to whom he's mentioned this observation dismiss it as "impossible."

In fact, Paul Offit, a pediatrician at Children's Hospital in Pennsylvania, author of several books about vaccines and perhaps the best-known and most vehement vaccine champion in the United States, has theorized that an infant can be vaccinated against ten thousand illnesses all at once with no negative repercussions.

Unlike Offit, other doctors and researchers worry that early and aggressive vaccination is damaging to a baby's immune system. Heather Zwickey earned a Ph.D. in immunology and microbiology from the University of Colorado Health Sciences Center, considered one of the best immunology programs in the country. She was then hired by Yale University, where she spent three years doing postdoctoral research and teaching at Yale Medical School. She does highly technical research on human cells and bandies around words like "immunomodulation" and "attenuator peptides." Zwickey is now dean of a research institute in Portland, Oregon, and professor of immunology at the National College of Natural Medicine, an accredited four-year medical school that trains naturopathic doctors.

It's a cloudy November morning, the sky a blanket of gray when I drive to Zwickey's home on a tree-lined drive up a steep hill in southwest Portland. Moss clings to the cottonwood and alder trees, and the fog is so thick that you can see droplets of moisture in the air. Zwickey is pro-vaccine. She herself worked as a vaccine developer—as a graduate student she did basic science that contributed to the development of a vaccine for tuberculosis. Zwickey explains that the most interesting part of the human immune system is how it interacts with the nervous system and the endocrine system, working together to fight disease and keep humans healthy. Flames dance in the fireplace and I see a cascading waterfall out the enormous bay window in the living room. "There's a benefit from vaccination," she tells me. "But as an immunologist I don't think we've taken into account the physiology of a human with our current vaccine schedule."

Zwickey explains that a newborn's immune system can't mount an effective response to diseases (or vaccines) because it is protecting the baby's brain, which would be damaged by a full-fledged immune reaction the way an adult would react to a virus or bacterium. So newborns rely on their mother's antibodies, which they get in breast milk, to give them the immune cells and proteins needed to combat infections. It is not until the brain is more developed—probably between nine and twelve months of age—that a human baby can mount an effective immune response. Until then, the baby has only mild, general immune response (the scientific name for this is a TH2 response), the sort we associate with allergies, but which doesn't tailor any of the special white blood cells (called TH1 cells) to respond to a specific bacterium or virus.

After twelve months the baby's brain is better able to handle a stronger reaction, and the immune system's full range begins to come online. A baby's short-lived and immature immune response is the reason we have to give so many doses of any given vaccine in the first few months of life. Any vaccines that doctors give before twelve months of age need another dose given at or after twelve months, because the immune system didn't really *learn* anything from doses given before then, Zwickey explains. This dovetails with what other medical professionals have told me. In nursing school, Michele Pereira's instructors informed students that from an immunological standpoint it would be better to give vaccines to children when they are a bit older. But since parents less reliably bring children over a year old to the pediatrician, the best way for public health officials to ensure high levels of vaccinated children is to vaccinate them as early as possible. Zwickey says the same thing. Early vaccination makes sense from a public health perspective but it is not immunologically optimal. It is only after about twelve months that the immune system is sufficiently mature to have a memory. That is, a baby can develop antibodies in response to a vaccine, and these antibodies will stay in the bloodstream without needing booster after booster. I find my face going numb as Zwickey, who is articulate and patient when I make her backtrack over the technical bits, explains to me why the way we are aggressively vaccinating infants is not the best way for infants to be protected against disease. "Everyone who studies human immunology knows that the TH1 response doesn't come up until the end of the first year of a human baby's life," she says matter-of-factly.

The timing of childhood vaccines is not the only drawback. After twelve months, when the body can mount a specific response, there's another problem. In a normal infection, the bacteria or virus would be living in your body and excreting something that makes you sick. Your body would recognize that a foreign substance was making it sick, so it would react to get rid of it by mobilizing a whole system to attack whatever foreign protein it doesn't recognize in the system. TH1 cells are specifically keyed to lock on to foreign proteins to form a chemical bond to kill the bacteria or virus. The garbage (the used-up T cells and the killed bacteria or virus) is then excreted—sometimes with great force into the porcelain goddess (if you're lucky and your child doesn't miss). With a vaccine the response is similar: The body finds the toxin in the vaccine and looks for a foreign tissue to glom on to. But because the bacteria or virus has to be disabled (the scientific word for this is "attenuated," which means weakened) in a vaccine, so as not to give a child the very disease the vaccine is intended to prevent (which happened so much with polio and pertussis whole-cell vaccines that both were taken off the market and reengineered), the body may not react to them at all. So vaccine makers need to put something in the vaccine that the body recognizes as foreign in order to provoke an immune response. This extra something is called an "adjuvant."

Aluminum is adjuvant of choice. Most attenuated viruses and bacteria are too weak to stimulate a strong immune response. So in order to make a stronger response, aluminum is added to vaccines. The body reacts to aluminum because it recognizes it as a poison, but the immune system can't attack aluminum. Instead it treats the foreign aluminum like an allergen, and makes a TH2 response. If babies have a lot of aluminum in their bodies, they will be making a large TH2 response, and that response will spread to additional things the baby is exposed to that are actually harmless, like pollen or cat dander. So aluminum in vaccines makes it more likely that a baby will develop allergies. A recent study done in Canada showed that when vaccinations were delayed until the babies' systems were mature enough to mount the TH1 response, children developed allergies at significantly lower rates. Another study has shown that children exposed to wild measles virus have lower rates of allergies than children who are vaccinated against measles.

But the most worrisome part is that the immune system can sometimes mistakenly identify *the body's own cells* as foreign, causing the body to attack itself. When the body inappropriately attacks its own proteins, a child develops an autoimmune disorder. There has been an exponential rise in autoimmune disorders among children since the 1950s. One study found that cases of type 1 diabetes among children under five increased fivefold between 1985 and 2004. Graves' disease, once unheard of in children, is becoming common. Other autoimmune disorders on the rise in American children include asthma, allergies, Crohn's disease, and atopic dermatitis. One 2008 epidemiological study found that infant immunization leads to a statistically significant increase in type 1 diabetes in children. Some leading immunologists, like Jean-François Bach, professor of immunology and head of the immunology unit at Necker Hospital, Paris's most renowned children's hospital, hypothesize that increased immunizations, which have in turn led to a decrease in the circulation of infectious diseases for the body to respond to, may be responsible for the exponential rise in autoimmune disorders.

Dr. Larry Palevsky shares Zwickey's concern that we are doing long-term damage to our babies' immune systems. "Fatal allergies and autoimmune diseases are exponentially on the rise in our country," Palevsky says. "That's not anecdotal. You can't look at this rise and not look at the fact that we are instigating a change in the immune system that could potentially be the catalyst for the development of these diseases."

Although she fully supports some kind of national vaccination program, Heather Zwickey has a further concern that early immunizations can and do cause neurological damage in some children. "The nervous system and the immune system go hand in hand," Zwickey explains. "Physiologically we're designed to be protected by breast milk until we mature enough to mount an effective response without harming our brains. Like naturally occurring infections, vaccines can lead to dangerously high fevers in babies. If you have a vaccination, or any infection that makes a baby spike a high fever in the first year, it can lead to all sorts of brain disorders, including schizophrenia, bipolar disorder. It's dangerous to a developing brain."

Vaccine manufacturers are aware of an infant's weak immune response. "The inserts in the vaccine packaging say that children under one can't develop an adequate response," Zwickey points out. "The public health rationale is to link vaccine doses to well-baby visits. So there's a very large financial piece. I don't disagree with the public health officials here: There is a large percentage of parents who, if they weren't taking the child in for a well-baby visit, wouldn't take another day off work to bring the child back just for a vaccination."

BOOST YOUR BOTTOM LINE

When I call Dr. Saul, who is based in southern California, for an interview, he invites me to a talk he's giving in Bend, Oregon, to pediatricians about meningitis. Though Saul is told by the organizers that the talk is overbooked, that the other pediatricians wouldn't appreciate an outsider attending, and that they would have to get permission from headquarters, he suggests I come anyway and stand in the back if they allow. Whether or not that works, he says, we can still meet the next morning for an in-person interview.

I salvage a business suit from my days in corporate philanthropy, channel my inner Michael Moore, drive four hours to Bend with the baby and her older sister to look after her, walk in late, and sit down at an empty seat (there are plenty). Immediately a Novartis employee taps me on the shoulder. "Are you Dr. K?" he asks. "No," I answer. "I'm Dr. Margulis" (I have a Ph.D.). They can't find my name on the roster. I explain I've been invited by the speaker. "Just sign in here," the rep says kindly. "We're so glad to have you with us."

The sponsors of his talk, Novartis, the multinational pharmaceutical company based in Basel, Switzerland, that makes vaccines against influenza, meningococcal, and polio, did more than $50 billion in sales in 2010. In 2011 net sales of $2 billion came just from vaccines. Held at the Oxford Hotel, the hippest most eco-friendly hotel in Bend (I coincidentally wrote an article about it for the *New York Times* when it first opened), the talk includes a three-course meal with free alcohol and a choice of curried halibut, local free-range chicken with polenta and Granny Smith apples and fennel, or steak; with flan or chocolate cake for dessert. During the talk Saul pokes fun at Bob Sears and Jenny McCarthy, the actress, writer, and *Playboy* centerfold who has brought media attention to the vaccine-autism debate after her son became autistic following routine vaccination, and who is president of Generation Rescue, a nonprofit autism advocacy group that challenges the current vaccine schedule. Privately, I find out the next day, he shares many of his parents' unanswered questions and concerns about vaccines.

As we walk on a path strewn with goose droppings and lined with ponderosa pines along the Deschutes River, Saul insists pediatricians don't push parents to vaccinate because they make money from doing so. The markup on the vaccines is usually about 10 percent, he says, which means that for a vaccine that costs the insurance company or the patient $100 the pediatrician may make as little as $10 in profit. If an insurance company or a client refuses to pay, the pediatrician loses money. "That's why it angers so many pediatricians when there are some patients who accuse the doctors of giving the vaccines just to make more money, because they don't," Saul says. "They don't make more money on the vaccines."

The problem with this argument is that it isn't exactly true. As we'll see in the next chapter on well-baby visits, pediatricians spend a good deal of their time doing well-baby checkups, and one of the biggest reasons parents bring their babies in for these checkups is so their babies can be immunized. Although Saul may be right that the resale value of the vaccine may not provide much extra income to the practice, it is these routine immunizations that bring patients back for repeated office calls and "are the bread and butter of [the] specialty," as another pediatrician puts it.

Saul then tells me that a small subset of enterprising pediatricians have actually found a way to turn large profits through selling vaccines in his area, starting something called "buying groups." They go to each practice in the area and warn the doctors they can only survive if they join the buying group in order to get a bulk discount on vaccines. The buying group then approaches the drug companies to negotiate a discounted price. Buying groups have the most bargaining power when there are several companies making vaccines against the same disease.

But the bulk discount negotiated by a buying group comes at a price: The buying group has to agree to boycott all the products from the other pharmaceutical companies in exchange for that bulk rate. "The bad part is that if a pediatrician feels he wants to give a certain product, he can't," Saul says. "Money is dictating decisions in medicine. I was against it but

I was outvoted on the local board of the AAP. The group felt it was our one chance to stand up to the drug companies. As much as we abhor this system, it's a necessary evil, because we can't afford to keep losing money on vaccines." The pediatrician who heads Saul's buying group is a savvy businessman: "He's getting a cut on the negotiation. I have no idea how much. . . . He actually took me out for coffee to squelch rumors that he's getting rich."

As other industries feel the effects of the recession in the United States, the vaccine industry continues to expand, adding new vaccines to the market and lobbying the government to approve vaccines for younger children. The New Jersey–based Merck & Co., Inc., one of the largest pharmaceutical companies in the world, manufactures vaccines for hepatitis A and B, Hib, human papilloma virus, MMR, pneumonia, chicken pox, and rotavirus. In 2010 Merck saw their sales increase to $46 billion in 2010, $3.5 billion of which were vaccines. Since 2003, while the American economy has gone through one of the worst slumps since the Great Depression, Merck's sales of vaccines have more than tripled. Among Merck's best-selling products are prescription medications for allergies and asthma, conditions that immunologists like Heather Zwickey and other researchers have found to be linked to early vaccination.

In 2011 Britain-based pharmaceutical giant GlaxoSmithKline reported profits of £3.5 billion (about $5.5 billion) on their vaccine sales alone. Their top-selling vaccine, a combination of hepatitis B, diphtheria, tetanus, pertussis, and polio, yielded the company £688 million (over $1 billion) in sales. Sanofi Pasteur, the vaccine division of Sanofi Aventis Group (a pharmaceutical company based in Lyon, France), made $2.595 billion in revenue in 2010. While pediatricians may lose money on it, vaccination is a high-stakes game: These big pharmaceutical companies earn more than $13 billion in yearly profits through convincing parents in the United States and around the world that vaccinating (as opposed to exclusive breastfeeding and eating healthy whole foods) is the most essential component to ensure their children's health.

In order for pharmaceutical companies to convince parents to use their vaccines (in their terminology, "to capture market share"), they must first convince pediatricians. I was told off the record by a drug representative ("salesman" would be more apt) who specializes in adult medicine that part of the persuasiveness of a drug pitch hinges on bringing catered lunches and logo-branded gifts to the doctors' offices, which this rep does for the same clients about twice a week. My raised eyebrows and look of surprise (it was the word *catered* with its connotation of the kind of delectables we were served in Bend that got my attention, though I've since been told by other doctors that "catered" is more like soups, bread, salad, an entrée, and drinks from Panera Breads than the free-range grass-fed seared sirloin I was imagining) elicited a shrug. "It's not my money," the rep said. This particular employee's motivation for accepting a job hawking penile dysfunction and urological disorder drugs? "I heard you could make good money doing it, have a company car, get great bonuses, and only have to work three or four days a week."

Looking to talk to someone on the record, I tracked down Nick Servies, a thirty-something drug rep who, according to his public LinkedIn page, has been working for GlaxoSmithKline for more than five years. I was told Nick specializes in selling pediatric immunizations in Oregon and contiguous states. Nick is an "executive immunization specialist" at GSK, an impressive title that makes him sound like he has extensive qualifications. Nick is "proficient as a top revenue generating performer in all aspects of the sales process" (that is, of convincing doctors to sell GSK vaccines), including "brand positioning, product education, and customer relationship building." When I finally got Nick on the phone, I explained I was hoping to better understand how companies like his sell pediatric vaccines directly to pediatricians, how they are trained about vaccines, what kind of "literature" they provide the medical establishment with, and what other perks they use to persuade doctors to choose their brand over a competitor's. Nick said that in order to get permission for any kind of interview I would have to submit a list of questions in writing to his company's legal team. Otherwise it would be "illegal" for him to talk to me. He was cordial and apologetic. "It's not my company," he explained. "It's the government that doesn't want us talking to the public."

In 2012 the AAP, a nonprofit organization of some 60,000 pediatricians dedicated to the health and well-being of American children, received almost $7.4 million from for-profit corporations, taking at least $1.6 million from the vaccine industry. Since the 2004–2005 fiscal year, corporate donations to the AAP have reached $47,896,763. Vaccine manufacturers, including Merck & Co., Pfizer, GlaxoSmithKline, and Sanofi Pasteur, are among the AAP's top-paying corporate sponsors. In the past eight years, these and other vaccine manufacturers gave more than $9.4 million. In 2010 Merck & Co., which manufactures the MMR vaccine, donated at least $500,000. Pharmaceutical firm Wyeth (since bought by U.S. drug giant Pfizer for a record $68 billion), which makes Prevnar—the brand of the pneumococcal vaccine widely used in the United States (the Pc vaccine is currently recommended by the AAP for children ages two, four, six, and fifteen months)—was one of the AAP's top corporate donors in 2008, giving at least $750,000. Paul Offit, who wrote an article published in AAP's peer-reviewed journal, *Pediatrics,* excoriating Dr. Bob Sears's alternative vaccination schedule, has an often undisclosed financial conflict of interest: coinventor of the RotaTeq vaccine, he is also a coholder of the patent on it.

Can pediatricians really look rationally and objectively at the current vaccine schedule when their parent organization, so to speak, is receiving literally millions of dollars from the pharmaceutical industry? Can they really look rationally and objectively at the current vaccine schedule when they accept industry handouts, including gifts, prescription medication, money for travel, and food and are themselves paid by the pharmaceutical companies to give educational seminars about vaccines?

FORGOING VACCINES

Rebecca Mehta and her husband had decided to wait to vaccinate their son. Rebecca knew Osha, her good-natured, drooly, and completely bald son, wouldn't be exposed to diseases like polio because the family had no plans to travel abroad. There have been no cases of wild polio in the United States for more than fifteen years and the last reported case of the wild virus was in 1993. After that and before the live polio virus vaccine was taken off the market in 2000, children in the United States became infected with polio only from getting vaccinated, resulting in about eight cases of paralysis a year. Though Rebecca's husband's family was from India and India was one of the six countries where wild polio still existed (the others, according to the World Health Organization, were Nigeria, Afghanistan, Pakistan, Egypt, and Niger), Rebecca wasn't worried.

Parents like Rebecca who choose to forgo all vaccines (and doctors like Larry Palevsky who support them) believe that yes, polio was once a terrifying disease resulting in terrible paralysis and death, but that our bodies have evolved to coexist with hundreds of viruses and bacteria that were once harmful. They point out that other endemic diseases like scarlet fever and cholera stopped causing harm to Americans without the intervention of vaccines. They make a rational decision based on their assessment of the risks, knowing that the vast majority of those who become infected with the poliovirus will *not* get sick or show any symptoms and that fewer than 1 percent of polio infections result in paralysis. They also realize from reading the statistics that in a country like the United States, where the risk of getting polio is so small it cannot be quantified, the vaccine has become more harmful than the disease.

In Sears's practice in Orange County, California, about 20 percent of his two thousand patients are not immunized at all. "It's most often parents with a higher level of education choosing not to vaccinate," Sears told me when I interviewed him for a magazine article, adding that this is not because these parents are smarter, but because they are more likely to question and research everything having to do with their children— including the food they eat and the car they drive.

The data backs this up. In one study of pediatricians facing vaccine-hesitant parents, suburban physicians dealing with wealthier, better-educated parents reported more vaccine concerns and refusals. According to Sears, parents in his practice choose not to vaccinate because they would rather take the natural risk that their child may catch an infectious illness than the risk posed by vaccines. These parents decide that they do not want to intramuscularly inject their child with something that is not part of the natural course of life. "It's not a natural exposure to the disease," Sears explains, stressing that though he does not agree with this point of view, he understands it. "It's chemicals, it's artificial. And by giving it to their child, they are taking a risk. Even though it's a smaller risk than the disease risk itself, these parents are less comfortable with that kind of artificial risk."

The possibility of contracting *Haemophilus influenzae* type b (Hib), a bacterial infection that can cause brain inflammation, blood infections, bone infections, and pneumonia, is much more likely than of getting polio. Five children became sick from Hib in Minnesota in the winter of 2008 and one unvaccinated baby died. But that does not negate the risk from the vaccine itself. The known side effects listed from trial experiments by Merck & Co. of the PEDVaxHIB vaccine include high-pitched crying and crying that lasts for more than four hours, fever, the inability to eat, vomiting, and irritability. In addition to those "mild" side effects, two very serious reactions from the Hib vaccine have been reported: Guillain-Barre syndrome (which causes muscle weakness and paralysis) and serious Hib infections. In December 2007, Merck had recalled more than one million doses of the Hib vaccine because equipment used in its manufacture was found to be infected with *Bacillus cereus,* a bacterium associated with food poisoning and diarrhea. Most parents don't realize that the Hib bacteria is common in the human body. It usually remains in the nose, ears, and throat; complications like pneumonia from Hib are very rare.

Rebecca's decision to delay vaccines ultimately wasn't based on research. When she thought about vaccinating her tiny baby, Rebecca had an instinctual feeling that putting chemicals into his body at such a young age would be the wrong choice. She felt protective of Osha and willing to risk that he might catch a disease naturally. Besides, she was breastfeeding her baby and eating healthy food herself. And she knew Osha wouldn't be around any older children—who are usually the vectors for more common childhood illnesses like chicken pox, whooping cough, and measles—because she was a first-time stay-at-home mom who didn't have friends with school-aged kids.

"I don't think I'll ever vaccinate him," Rebecca told me later. "If that's something he wants to do when he's older, he can decide himself."

THE BACKLASH AGAINST PARENTS WHO DELAY OR FORGO VACCINES

In November 2008, parents in Maryland were threatened with jail if they could not show proof that their children were vaccinated. Some doctors believe so strongly in the current vaccination schedule mandated by the CDC that they consider it child abuse not to vaccinate. Still, there is evidence that the majority of American pediatricians are actually sympathetic to parents who delay vaccines. In one survey in Washington State, 61 percent of pediatricians reported supporting parents who wanted to use an alternative schedule. In another survey, of Midwestern pediatricians, 79 percent reported accepting families who chose to forgo some vaccines. Still, nonvaccinating parents who try to have their children treated by mainstream health care providers often face difficulties.

When Osha was six months old, Rebecca brought him to the doctor in Boulder, Colorado, because he had a cold.

"I think he might have an ear infection," Rebecca said as the doctor looked over her paperwork.

"His vaccines aren't up to date," the pediatrician responded in a perplexed voice, flipping up the last page as if there might be something on the clipboard that he had missed. "What hospital was he born in? Is there a reason he's not vaccinated?"

Holding her brown-eyed boy a little closer, Rebecca started to explain why she and her husband had chosen to delay Osha's vaccines. But the doctor wasn't listening.

"You're just one of those moms who doesn't care if your baby lives or dies," he muttered under his breath. "Really, you should be seeing another doctor in this practice," the doctor added in a louder voice, looking at her angrily. "He makes moms like you walk through the graveyard to look at babies who have died from not being vaccinated."

Despite the vehement emotional reaction of some doctors to nonvaccinating parents, the AAP generally advises against reporting vaccine-hesitant parents to Child Protective Services, reserving that option for when there is a disease outbreak. "Continued refusal after adequate discussion should be respected unless the child is put at significant risk of serious harm (as, for example, might be the case during an epidemic)," the AAP's Committee on Bioethics writes in a 2005 article in *Pediatrics*. "Only then should state agencies be involved to override parental discretion on the basis of medical neglect."

Doctors' attitudes seem to vary by region. In Ashland, Oregon, an upscale tourist town of some twenty thousand inhabitants known for its Shakespeare festival, organic food co-op, and mild weather, most pediatricians assume parents will individualize their child's vaccine schedule, and they spend time talking to the parents about their choices. But in some parts of Pennsylvania parents choosing to selectively vaccinate report that it's almost impossible to find a pediatrician to treat their kids. When Tasha Pittser moved from Las Vegas, Nevada, to Gwynedd Valley, Pennsylvania, and needed a pediatrician for her four- and seven-year-olds, she was turned away from five different practices. Her two healthy children ("They've never had anything more than a cold," she wrote me in an email) were missing polio, MMR, hepatitis B, and chicken pox. The practice they finally found would not treat her family today: though Tasha's children are grandfathered in, they've stopped accepting selectively vaccinated kids.

Jake Marcus, a lawyer who lives in the suburbs of Philadelphia, never found a supportive pediatrician for her three sons. One pediatrician responded with open belligerence, threatening to report her to Child Protective Services when Jake mentioned she might want to delay some of the vaccines: "I would never allow my patients to vary from the schedule!" the doctor cried. "You have to vaccinate! And if you don't you should be reported to Children and Youth!" Realizing she had upset the doctor, Jake, seven months pregnant with her third child, began to waddle away. "I said 'Okay,' and I turned to walk out," she told me. "[The doctor] followed me to the parking lot, yelling at me as I walked to the car. 'How dare you not vaccinate your children! Do you know what kind of risks you are putting your children in! You pose a threat to the rest of the community!' " Jake met with five pediatricians and called fifteen family practitioners before finding a doctor.

Many pediatricians in Thousand Oaks, California, the wealthy suburb of Los Angeles where Dr. Kenneth Saul practices, fire parents who are looking for an alternative vaccine schedule. Saul, who believes this is wrong, explains that his colleagues reject these families partially for financial reasons. "Some doctors are very adamant that it has to be the standard way or else. I understand their point of view but I don't agree with it," Saul tells me. "They feel that the people who won't listen to them about vaccines won't listen to them about other things, like if their child came in with appendicitis and they recommended surgery, they wouldn't do it. That puts them at legal risk. But there's a bigger reason. The economics of medicine has changed so much that a lot of doctors have taken a tougher stance about doing anything that's not compensated. They say, 'I'm not going to spend fifteen uncompensated minutes discussing the pros and cons of vaccines' because it costs them too much money. They don't want what they call 'that kind' of patient in the practice— someone who is constantly going to want to dialogue about everything they recommend—because they see those lengthy discussions as wasted time for no compensation."

After Saul's talk in May 2010, officials from Novartis had a conference call with him to talk about me. Their biggest fear was that I was a mole for another company, looking to steal trade secrets. Although Saul reassured them I was an independent journalist writing a book, he has not been invited to speak on vaccines since. "I have asked if I have been blacklisted," he writes me in an email, "and they deny it and blame the change on budget cuts, but I suspect that is only a partial truth."

HIGHER INFANT VACCINE RATES CORRELATED WITH HIGHER INFANT MORTALITY

The United States mandates twenty-six vaccine doses for infants under one year of age—more than any other country in the world. Yet we also have one of the highest infant mortality rates among industrialized countries. A 2011 study published in *Human & Experimental Toxicology* found a correlation with aggressive vaccination and high death rates among infants, leading researchers to conclude:

These findings demonstrate a counter-intuitive relationship: *nations that require more vaccine doses tend to have higher infant mortality rates.*

Efforts to reduce the relatively high US IMR have been elusive. Finding ways to lower preterm birth rates should be a high priority. However, preventing premature births is just a partial solution to reduce infant deaths. A closer inspection of correlations between vaccine doses, biochemical or synergistic toxicity, and IMRs, is essential. All nations—rich and poor, advanced and developing—have an obligation to determine whether their immunization schedules are achieving their desired goals [their emphasis].

In March 2008, when Hannah Poling's family was awarded more than a million dollars by the government to pay for her care, Julie Gerberding, M.D., then head of the CDC, insisted, "the government has made absolutely no statement about indicating that vaccines are the cause of autism." In December 2009, Gerberding was named president of Merck Vaccines. Our comparatively poor infant health outcomes combined with this kind of revolving door between pharmaceutical companies and the U.S. government makes it no surprise that increasing numbers of parents no longer trust the American government's public health campaigns.

When I asked health officials in Iceland and Norway if childhood vaccination in their countries was mandatory, the question surprised them. The surgeon general of Iceland, himself a pediatrician, told me childhood immunization is entirely voluntary and most Icelanders follow the government's recommendations. The same holds true in Norway. Babies in Norway and Iceland, where breastfeeding rates are also much higher than in the United States, receive *no* vaccines before they are three months old. In countries with less aggressive vaccination schedules, lower infant mortality rates, and health care systems not tied to profit, parents know doctors and government officials have their best interests at heart. In America even the doctors most vocally championing the CDC's current schedule are choosing an alternative route for their own infants. At the same time, parents who selectively vaccinate or forgo vaccines completely are accused of being selfish, hiding in the herd, and causing disease outbreaks. This media vilification of parents obscures the real problem: evidence is increasing that our current vaccine schedule, while a boon for pharmaceutical companies with a captive market for their products, is not in the best interests of our babies' health.

Merck & Co.'s revenue for 2012: more than $13 billion

Amount Merck & Co. gave to AAP in 2010: $500,000–750,000

Donations to American Academy of Pediatrics from the vaccine industry during an eight-year period: more than $9.4 million

Money paid by the government to parents of vaccine-damaged children since the National Childhood Injury Act was established in 1986: more than $2 billion

Money paid by British government to parents of vaccine-damaged children between 1997–2005: $5 million

Money awarded by the court on August 28, 2010, after a British panel of medical experts reviewed a case where a baby suffered brain damage from MMR vaccine: $146,000

Cost of not getting vaccines: $0

Michelle Maher Ford: Choosing Not to Vaccinate the Second Time Around

Michelle Maher Ford, an insurance agent in Culver City, California, is tall and slender with dark hair and gray eyes. At forty, she has four children from two marriages. Her oldest daughter, Chelsea, who is completely vaccinated, just turned eighteen. Her three-year-old curly-haired blue-eyed twins and their ten-month-old blue-eyed baby brother have not been vaccinated. As an infant Chelsea was sick with allergies and ear infections, even though she was breastfed longer than Michelle's other children. Her three younger siblings, who aren't vaccinated, have never needed an antibiotic or even Tylenol.

When Chelsea was born I was twenty-two. I didn't even know that vaccines were optional. It was one of those things where your baby's born and in the delivery room they sweep her out of your arms and start poking her. "What are you doing? What's that for?" I asked the nurse. It was so sudden, I wasn't told that would be happening, and I didn't know why. I was told, "It's just the usual vaccines." I said, "But she's brand new, she doesn't need them yet." That was the extent of my protest. I didn't really do my homework.

Chelsea spiked a really high fever one day after her shots. She was screaming in the night. I called the doctor's office the next day. The doctor said, "It's probably just because of her shots," and told me not to worry. I felt reassured because the high fever was a normal reaction to the shots. I really trusted the doctor.

I breastfed Chelsea for nine months but she was always getting ear infections. She would be on antibiotics for ten days, the infection would clear up, and then within a week to ten days she'd be pulling on her ear, she'd have a fever, and we'd go back to the doctor to try a new antibiotic. She was too young to verbalize it but I could see her ears hurt her. By the time she was a year and a half she was immune to most antibiotics. There were only a few antibiotics they gave to children and she had had all of them numerous times. She was a really unhappy little angel. When I look back, I feel really guilty. I am confident that her weak immune system was because of all the shots.

I used to be a fast-food girl, I used to drink soda all day long. I didn't pay that much attention to what I was eating. I never drank water, it was either coffee or Diet Coke. But my new husband and I knew we wanted to have children so I thought I better get my health in order. I started exercising. I gave up soda. I still drink coffee but only two cups a day instead of a pot. I rarely eat candy or sugar. Maybe I'll have a piece of cake at a birthday party. I've become much more aware of eating organic.

While I was pregnant with the twins I asked my OB about vaccines and she said, "Oh, you have to do them!" I said, "You have to do them? I was reading it was optional." She said, "No, you'd be crazy not to!" And I said, "I'm actually thinking about not vaccinating the twins." "Oh, boy," she answered, "I'm not a pediatrician. You should talk to your pediatrician. You should go to the CDC website."

So I did. The more I read on the CDC website, the more I was convinced not to vaccinate the kids. It was like one big advertisement. It became very apparent to me that the current vaccination schedule in this country is a moneymaking machine. I'm not someone who believes in conspiracy theories. I laugh at that idea. But the research I did led me to the conclusion that it is almost a conspiracy. Millions of people a year are vaccinated. It creates illness, which creates customers. One of the books I read mentioned one of the side effects of the vaccines is chronic ear infections, like my oldest daughter had.

Vaccines are one size fits all. Everyone gets the same dose, everyone gets the same vaccines. When I told the pediatrician after the twins were born that we decided not to vaccinate, I had to sign every waiver in the world. The waivers scare you. The guilt is on you that you're not vaccinating. Even after all the research I did I almost caved in. I was being scared into vaccinating. The doctor was not in favor of our decision but ultimately he honored it.

So we didn't vaccinate our twins. And we also didn't vaccinate baby Russell, who's ten months old. I'm constantly amazed at how healthy they are. If they get a cold, they're over it really fast, within twenty-four to forty-eight hours. I've had the leftovers from the flu for weeks now and I'm jealous that I don't get over stuff as fast.

For the twins and for Russell we have the same pediatrician. At their two-month checkup, he said, "If you won't be doing vaccines, I won't be seeing you for a while. Just call me if they get hurt." He was really matter-of-fact.

When my daughter was an infant I was a junkier eater. Maybe she wasn't as healthy because I wasn't as healthy. That's very possible. I'm certainly open to that idea. I'm not a trained doctor or a nutritionist, it's not my field, but as a mom I think I'm right that the vaccines are highly dangerous, highly toxic, and they should not be administered so early.

If people are going to vaccinate their children, they should wait until their brains have finished developing. We've seen such a huge increase in cognitive impairment. You read these stories of people who say, "My kid was fine, bright and alert, and the night of the vaccine they had a fever and they woke up the next day and nobody was home." It's so common to hear those stories. Neurological damage has become so much more prevalent, I think it's caused by the early vaccines.

There are hundreds of thousands of parents (and even more who feel they can't talk about it because they'll be deemed a weirdo) who have seen their children damaged by vaccines. It's a pretty hefty debate. When I was consenting to Chelsea's series of vaccines I could never really get over seeing her look at me as they were shooting the needle into her leg or her arm, and looking at me like, "You're letting this happen to me. You're supposed to be protecting me, and you're letting them hurt me." Every time I was consenting to her shots, I felt like something wasn't right. I just didn't have the skills or the confidence as a young parent to really press the issue.

I'm hoping my kids get the chicken pox. I had the chicken pox. What's wrong with the chicken pox? I know the measles can cause complications but in the vast majority of cases—for my father and grandfather—it is a routine childhood illness, even when it makes you really sick. What's wrong with the measles? The mumps? What's wrong with becoming naturally immune?

Calling parents who choose not to vaccinate selfish is an easy way for other people to not take responsibility for the choices they made that could damage their child. It's easy to say someone else is a freeloader. I want my kids to get exposed to disease so they can be immune to it, but I want them to get exposed naturally. Medicine has its place. If you break a bone you need a doctor to set it properly, but our bodies know how to fight disease.

CHAPTER 10

SICK IS THE NEW WELL:
The Business of Well-Baby Care

When I took my oldest daughter to her six-month well-baby visit at a big pediatric practice on Peachtree Boulevard in Atlanta, the doctors were running behind. At first the baby stood on my lap, holding my hands and bouncing up and down, smiling at the other patients. But as it got closer to naptime, she started to fuss. As the minutes ticked by, her whines became more strident. I walked her around, even taking her outside to look at the cloudless sky. Nothing worked. Used to napping in her crib, Hesperus started to sob from fatigue. When it was finally our turn to be seen, I was also close to tears.

"What's wrong with the baby today, Mom?" the thirty-something pediatrician asked, genuinely concerned. I explained we had waited for almost an hour and Hesperus was long overdue for a nap. The doctor clicked the light on her otoscope, a handheld device the width of a thick marker with a funnel on the end, to peer into the baby's ears.

"I think Miss Hesperus has an ear infection!" she announced, clicking the light off with her thumb and rolling her chair to the counter to scribble a prescription. My face felt numb. Hesperus had never needed medication—or been sick—before this visit.

"Has she been tugging on her ears?" the pediatrician asked.

"Not that I've noticed," I admitted.

"I've prescribed an antibiotic," the doctor handed me a scrip, "and here are some free drops . . . for the pain."

I filled the prescription at the practice's pharmacy, which entailed another twenty minutes of waiting, paid the copay for the visit and the additional copay for the medication. But I waited to give my daughter the antibiotic. After a much-needed nap, the baby was her happy self again. I found out later that if our pediatrician had waited twenty minutes and looked in her ears again when Hesperus was calm, she would not have seen *any* sign of infection. Why? Because my daughter did not have an ear infection.

Middle ear (*otitis media*) infections are more difficult to diagnose correctly than you might expect. When the doctor looked into my daughter's ear with the light on the end of the otoscope, she was looking at the eardrum. Behind the eardrum is the middle ear. The doctor made the diagnosis of an infection based on observing the eardrum. Normally the eardrum is white, a semitranslucent membrane covering what looks like a tiny arm. This is actually a bone (the malleus or hammer), which picks up sound vibrations. An eardrum may be red or pink because of an infection. Hesperus's eardrums looked bright red. But an eardrum may be red or pink for another reason. In this case Hesperus's eardrum was red because she'd been crying.

Since the pediatrician was already running forty-five minutes behind, she needed to get us out of the office as quickly as possible. Pediatricians, like everybody else, want to please their patients. Our pediatrician was delighted to solve a problem, even though the problem she diagnosed didn't actually exist. In the United States ear infections are one of the most overdiagnosed and overtreated infections among American children, who collectively receive more than thirty million courses of antibiotics for ear infections each year. The only problem with the baby was one the doctor's office had caused.

ASSEMBLY-LINE PEDIATRICS

Parents in America are currently advised to bring their babies for well-child visits at least seven times in the first year of life: a few days after birth, at one month, two months, four months, six months, nine months, and one year. The average American doctor spends about 18.3 minutes with a child under three during a well-child visit. During the visit, a baby is given a thorough physical examination by the doctor and routine immunizations, usually by a nurse. A baby's skin, eyes, ears, nose, mouth, heart, lungs, abdomen, hips, and genitals are all checked. Pediatricians are also expected to discuss whether a baby is meeting development milestones, answer questions from parents, and look for red flags of problems in the home. According to one study, well-child care accounts for 57 percent of all office visits for children from birth to age one, which means the average pediatrician spends more of her time with babies doing well-child care than anything else. Well-baby visits and routine immunizations are the backbone of pediatrics.

Why are pediatricians so rushed? Why is it normal, especially in bigger practices, for parents to be left in the waiting room for over an hour? And why are pediatricians answering questions based on their own biases rather than medical expertise—about pacifier use and sleeping arrangements—problems they have *not* been trained to solve in medical school?

We need a crash course in medical economics to understand that this has as much to do with pediatricians—even though most have the best intentions, enter the profession because of a love of children and medicine, and are as frustrated and demoralized by the system as so many parents are—as it does with the way insurance companies reimburse doctors. I ask Dr. Edward Schor, a policy analyst and pediatrician, who has written about the systematic problems with well-child visits in *Pediatrics* and other professional journals, to help make sense of what's going on.

A father of three, Schor has worked in pediatrics for forty years and has a particular interest in promoting children's health and development. Schor explains that pediatricians, like everybody else, are trying to make a living. Though most of us really want to think of our doctor, especially our baby's doctor, only as a medical expert whose job is to minister to the sick, a pediatrician in private practice is actually a business owner first and a care provider second. That's an uncomfortable truth that American parents, eager to view their doctors as altruistic authority figures and not money-conscious body mechanics, would rather not admit.

The average pediatrician earns significantly less than the average obstetrician, about $168,650 a year before taxes. While pediatricians are on the lowest end of reimbursement among physicians and they make less than every other medical specialty, they still earn more than three times the salary of the average American: about $14,050 a month, or $462 a day. Though most doctors don't specialize in pediatrics to make money, they do leave medical school with the reasonable assumption that, after years of training and hard work, they will be well remunerated for their work. They also leave medical school saddled with debt. According to the American Medical Association, the average debt of a medical school graduate in 2010 was $157,944. Though some students manage to pay for school out of pocket, the majority of graduates (78 percent) have a debt of at least $100,000.

In order to make a comfortable living, pediatricians have to do everything they can to maximize profits. This isn't as easy as it might seem. The licensed contractor who replaced our corroded water pipe hires workers when he needs them and lays them off when work slows down. With a home office and only one other full-time employee, his office costs are minimal. In contrast, pediatric practices generally have a lot of staff—nurses, nurse practitioners, administrative assistants, physicians' assistants, and often a full-time or part-time administrator whose entire job is to collect unpaid bills from insurance companies and patients. They need a big space and have to pay high heating, electricity, water, and rental bills. Since they have had very little, if any, training in running a business, doctors' offices tend to be inefficiently managed, Schor and other pediatricians tell me, which adds to their already high overhead.

"Anywhere between sixty and sixty-five percent of everything that comes in goes to overhead," says Brandon Betancourt, practice administer for Salud Pediatrics in Algonquin, Illinois, who writes a blog about pediatric office management. "So if you make one hundred thousand dollars, sixty-five thousand goes to overhead. That's pretty standard across pediatrics." However, one place pediatricians save money over other specialties is on malpractice insurance, which can cost anywhere from $10,000 (or less) to $50,000, depending on the geographic area and the number of years of expertise (more experienced doctors pay more than younger ones). Brandon says the biggest expenses in his small practice, which has two pediatricians and one nurse practitioner on staff, include personnel, prepurchasing vaccines (which accounts for up to 25 percent of their expenses), and rent (which costs more than $5,000 a month). Only about 30 percent of the practice, Brandon tells me during a phone interview, is wellness appointments and 70 percent same-day walk-ins.

Both Brandon and Schor share the same insight with me: To make money, a pediatric practice needs to be an expert at "coding." This is the convoluted system of converting medical terms into a numerical code that has been developed by the American Medical Association. It's so lengthy and complicated that the AMA publishes a yearly book to help doctors with coding. They also need to have someone in the office skilled at negotiating rates with insurance companies.

That brings us back to the problem. Whether you are a confident third-time parent bringing a healthy baby to a well-child visit, or a first-time mom with a dozen concerns, insurance companies reimburse the same amount for a well-baby visit. "There's actually no allowance for individualizing well-child visits from the reimbursement. If you take a child in for a *sick visit,* if it's an easy problem—a rash or a cold—the office uses a certain billing code. If it's more complicated, you can code it for the complication or for more time," explains Schor. "*But there's only one code for a well-child visit.* If a doctor spends five minutes with a patient or twenty minutes, she's paid the same. This creates a huge incentive to move patients through as quickly as possible, so they can support their practice and their staff."

From an insurance standpoint, well-child visits are only worth a lump sum of reimbursement. Financial considerations on the part of the pediatric practice—not what's best for the growing baby—are also why most parents see *the doctor* at these well-child visits. A visit with the doctor yields the practice the highest remuneration. Doctors are the most educated, experienced, and highly trained people in a pediatric practice, but the truth is they have little, if any, training in basic parenting practices and preventive medicine.

Pediatricians themselves are the first to admit that the non-medical information they provide to parents is outside of their knowledge and expertise. "Sometimes we would pick up the *Family Circle* in the waiting room to see if there was anything in there," confesses Dr. Maggie Kozel.

When Kozel became a pediatrician she knew how to thread a tube down the nostrils and into the lungs of a twenty-eight-week-old premature infant, so small its whole body could fit into the palm of her hand. She knew how to administer blood-clotting factor to children with hemophilia, a rare blood disorder. And she knew how to analyze the calcium metabolism of a sick newborn. But though she was trained to speak and act with great authority, she realizes, looking back, that she—and her fellow pediatricians—had no special expertise when it came to the majority of parental concerns. Even as she gained experience by becoming a parent herself (Kozel has two daughters), she had little training in the seventy topics that a pediatrician is now expected to review with parents during a well-child visit.

"Eighty percent of what I did could have been handled better by a nurse or nurse practitioner," Kozel confesses. "But there would have been no way to pay for it."

AS FAST AS AN AUCTIONEER ISN'T FAST ENOUGH

The AAP publishes guidelines for infant care in a series of educational materials called Bright Futures, which makes recommendations for what pediatricians should cover during well-child checkups, topics such as nutrition, TV watching, poison control, and car-seat safety. "Of course, the opportunity for a physician to cover those things well is nil," says Kozel, who speaks with a hint of the working-class accent she acquired growing up in Point Lookout, New York. "I'm talking as fast as an auctioneer. My goal isn't really educational because there's no way a parent can absorb that much information in that kind of rushed setting. You pray that a parent won't ask a question, because that will really throw off the time allotted for the visit. You talk as fast as you can, you check things off, and then they're on their way."

Sharon Rising, a certified nurse midwife and former instructor at the

Yale School of Nursing, agrees that while the AAP guidelines are laudable, they are rarely, if ever, put into practice. "Pediatricians look at me and say, 'Yeah, right, like there would be time to do this,' " Rising says. "There just isn't. The guidelines are completely unrealistic. Every doctor agrees that the content should be there, but that it's just not possible."

These days Kozel does a mean imitation of a mole. She scrunches up her nose, curls her fingers into claws, bucks her teeth, and makes the classroom of fourteen- and fifteen-year-old high school students squeal with delight. She's teaching chemistry, not mammalian biology, but Kozel likes to keep things interesting. Though it was this same love of being silly and flare for theatrics that made her a beloved pediatrician, her career in medicine is over. Kozel left pediatrics after practicing for seventeen years. "It was very discouraging spending most of my day doing things I didn't go to medical school for," she explains.

STANDARDIZING APPLES AND ORANGES

Every well-baby visit begins with charting a baby's height and weight against a standardized curve. But because pediatricians are usually so rushed, this quantitative evaluation is often done without taking the particular context of the baby's family into account, and without consideration to the problematic nature of the growth charts themselves.

Jennifer Rosner and her husband, Bill, were told their baby, Sophia, was "falling off her growth curve." Their pediatrician in Los Gatos, California, had Sophia return for extra checkups to be weighed. He was especially concerned because the baby's head circumference was small and she was slight for her height. When Sophia was nine months old, the family moved to Northampton, Massachusetts. There the pediatrician instructed Jennifer and Bill to start feeding Sophia high-fat foods. Jennifer was breastfeeding and also pumping to give Sophia extra milk. She started adding cream to the bottles so her baby would ingest more fat and gain more weight. When that didn't work, their pediatrician referred them to a pediatric gastroenterologist and a neurologist.

Jennifer always imagined herself making healthy homemade food. Instead she and Bill combed the supermarket aisles looking for anything fattening that Sophia might be willing to eat: Key lime pie? Cheesecake? Rice pudding? Their tiny baby (though she was born in the 50th percentile for weight, Sophia was in the under 1 percent category on the growth charts by the time she was a few months old) would take a spoonful of whatever Mom and Dad were offering, scrunch up her nose, arch her back, and turn away. The stress made Bill and Jennifer famished. They ate every leftover bite of the fattening food their daughter had no interest in.

The specialists they saw in Springfield and Boston (in Massachusetts) and in New Haven, Connecticut, told them that if Sophia didn't gain weight she could suffer severe neurological damage. All three pediatric gastroenterologists they consulted examined her growth charts and insisted the baby needed a feeding tube, which would either be threaded down her nose (a nasogastric tube) or surgically implanted directly into her stomach (a G-tube). Jennifer balked, knowing that with a feeding tube a child bypasses the experience of feeling hungry, tasting food, and even swallowing. She worried an intervention this drastic could mean that Sophia would permanently lose the connection between hunger and eating.

One doctor gave them an ultimatum: If Sophia didn't gain seven ounces in one week, it was time to insert the supplemental feeding tube. Jennifer sat at the kitchen table, holding her wisp of a baby on her lap, and sobbed. Sophia already had hearing aids to deal with genetic deafness. Another intervention and the possibility of surgery seemed unimaginable. But that wasn't the only reason Jennifer and Bill were hoping to find a noninvasive solution: even though they were first-time parents and they trusted their doctors, they both believed Sophia was developing normally. She was eager and alert, responding with remarkable intelligence to her hearing training. As she learned to recognize the sound of a dog barking, Sophia would point to her little plastic gray-black dogs and then to their shaggy German shepherd.

Though it is standard medical practice at a well-baby visit to chart a baby's height and weight, these seemingly innocuous grids can actually be harmful. Until 2006 the growth charts most commonly used in the United States were made from data collected by the CDC's National Center for Health Statistics on almost exclusively white, bottle-fed babies in Yellow Springs, Ohio, from 1929 to 1975. Moms who breastfed did so for only the first few months. We know now that breastfed babies grow differently than bottle-fed babies, tending to gain weight more slowly over time. But pediatricians in the United States (and worldwide, since the World Health Organization adapted the American charts for use overseas) have used these charts to tell millions of breastfeeding mothers over the years that their babies were failing to thrive and they should switch to formula.

Though the CDC recognized the problem in the late 1990s and issued revised growth charts in 2000, these charts also lumped breastfed and bottle-fed babies in the same category and used predominately bottle-fed babies to plot average growth over the first year. "Women who were breastfeeding their babies who didn't stay on the chart were told to supplement," laments Jay Gordon, a pediatrician in private practice in southern California who has more than thirty years of experience. "This problem has not gone away."

Recognizing that suboptimal breastfeeding negatively impacts health globally, the WHO spent six years collecting data to accurately represent the standard growth of breastfed babies in the United States, Ghana, Norway, Oman, Brazil, and India. The new WHO charting gives a baseline for babies based on the way humans have evolved to grow, not how they have already grown when they drink artificial milk. But although this new information was released to health care professionals in 2006, the majority of American pediatricians and family doctors have yet to incorporate it into their practices. Since so many doctors get "free" growth charts from formula manufacturers, one wonders how quickly—if ever—these new charts will be put into use in the absence of a government mandate. In the meantime, doctors continue to compare babies to a faulty national average and use this information as a reason for unnecessary intervention.

"For my clinical practice the specific weight grid is irrelevant," argues Jeffrey Brosco, a pediatrician, medical historian, and professor of clinical pediatrics at the University of Miami. Brosco's office switched to electronic medical records two years ago and Brosco admits he "has no idea" where the growth charts in the computer come from. But he also thinks it doesn't matter. "How much a baby weighs at any given checkup is more related to whether the baby just had a bowel movement, the scale and how it's calibrated, and who's taking the weight. I don't find the information from a single visit to be particularly clinically helpful," Brosco says. "I look for growth over time. Where they stand on the chart is less important than the overall well-being of the baby and the mother."

Sophia didn't gain a single ounce that week but Bill and Jennifer, waistlines burgeoning, decided to wait on the surgery. The doctors finally realized nothing was wrong with Sophia. Jennifer, who is forty-four and has curly brown hair, a lopsided grin, and hazel eyes, is only 5 foot 1; her husband exaggerates his height at 5 feet 7 inches. They both come from small families. Jennifer's mom, also 5 foot 1, has weighed 98 pounds her whole adult life; Sophia's great-grandmother was 4 foot 11 and weighed 90 pounds. The people in Jennifer's family also have small head circumferences.

Today Sophia is eleven and almost as tall as her mom. After school she likes to listen to a book on tape and make jewelry. She's still a picky eater, her mother tells me, but she eats a wide variety of foods. Shrimp, mangoes, and coffee ice cream are among her favorites.

"There's this range of normality that they are imposing on all children. Sophia fell outside the margin," Jennifer says. "But that was only because we have a small family. If during this whole process someone had just said 'Let's look at your family,' if any doctor had been reasonable and looked through our basic genetics . . ." Jennifer grows thoughtful. "She was a healthy baby, alert, engaged. If someone had just looked at the actual markers for wellness, rather than a piece of paper, we would have been saved a huge amount of stress."

But when parents ask pediatricians to consider genetics, common sense, and what's in a child's best interests over quantitative measurements, they are often met with defensiveness, hostility, or worse. Janie Oyakawa and her husband, who was born in Hawaii and is half Japanese, are both big-boned. The Oyakawas and their five children (ages eleven, ten, six, three, and eighteen months) are bigger than average. Janie, an occupational therapist with extensive medical training, describes herself as "fat but healthy"—her family exercises and eats well, she says. But since her ten-year-old is sensitive about her weight, Janie told the pediatrician she did not want her weighed on the public scale in the office hallway. The pediatrician in Frisco, Texas, insisted weighing was mandatory. Janie held her ground, saying that if a medical indication from the checkup indicated her daughter needed to be weighed *after* the physical exam, she would consent. A week later Janie received a letter in the mail that her family had been dropped from the practice for "refusing treatment."

While Jennifer and Janie's stories may be striking, they are, unfortunately, not unusual. "The use of standard growth charts is an example—and American medicine is filled with them—of quantitative nonsense superseding qualitative sense," warns the late Robert S. Mendelsohn, M.D., in his book *How to Raise a Child . . . In Spite of Your Doctor*. Mendelsohn points out that the use of growth charts has led pediatricians to prescribe hormone therapies to alter the height of girls deemed too tall. A popular practice when Mendelsohn wrote his book in the 1980s, estrogen therapy to prevent girls from growing too tall is less common today. Instead, doctors are prescribing growth hormones for otherwise healthy short children, especially boys, to keep them from becoming short adults.

KEEN TO INTERVENE

When their son Lochran was eight weeks old, Elizabeth Hunter and her husband, Alan, were concerned because their baby seemed constipated. His stomach seized up and he would arch his back as he strained to poop. When he finally managed to go, the apricot-colored curdlike poop would shoot like a projectile across the room, spraying everywhere. Though they knew it is unusual for breastfed babies to have constipation or diarrhea, and that a breastfed baby can go for two weeks without having a bowel movement and be perfectly fine, Lochran's parents felt something was wrong. Their pediatrician in Birmingham, Alabama, ordered a laxative: MiraLAX.

MiraLAX, advertised as having "no harsh side effects," is marketed as a miracle laxative, one that works when all others fail (hence the name). But the long list of side effects reported by adults who have given it to their children, or taken it themselves, includes severe gas, dehydration, jaw pain, diarrhea, bloating, and even heart palpitations. The National Institutes of Health lists nausea, bloating, cramping, and gas as side effects of the drug, as well as two rare but potentially serious problems: diarrhea and hives. Elizabeth tells me that she and her husband didn't like the idea of putting such a tiny baby on MiraLAX, made from polyethylene glycol, which is also an ingredient in antifreeze. They asked the pediatrician if there had been long-term studies done about the safety of using this drug in infants. He said no.

Elizabeth instinctively felt it would be better to figure out *why* baby Lochran seemed to be having trouble pooping instead of giving him such a harsh medication. A week later she took him to a La Leche League meeting. The facilitator suggested the baby might be sensitive to the protein in the cow's milk Elizabeth was drinking. She stopped eating dairy products the next day and Lochran's constipation went away.

PRODUCT PLACEMENT AT THE PEDIATRICIAN'S OFFICE

My father has a white magnet on his fridge with the contact information for my younger sister's pediatrician in Wellesley, Massachusetts. In a repeating pattern in capital letters around the border of the magnet are some bizarre word combinations for the uninitiated: ENFAMIL, LACTOFREE, PROSOBEE, NUTRAMIGEN. These are names of infant formulas. Your pediatrician's office is "a treasure trove of free stuff," enthuses Vincent Iannelli, a pediatrician in private practice in the Dallas area. Formula manufacturers and pharmaceutical companies give pediatricians much more than magnets—they stock their offices with branded growth charts, formula samples, pacifiers with the company logo, coupons, notepads, clipboards, magnets, business cards, pens, coffee cups, stethoscopes, and other swag. Some doctors keep containers of the free formula stacked on a shelf, often next to the baby-weighing station.

Pharmaceutical companies also give doctors free samples of prescription and over-the-counter medications and "educational" materials with the company logo at the top (Elizabeth's pediatrician handed her a pamphlet on how to treat a feverish child with the logo of Johnson & Johnson, the company that makes baby Tylenol). Using the doctor as the middleman is an effective way for multinational companies to reach parents of babies and young children, capitalizing on the trusted relationship between a pediatrician and his patients. They also target advertising directly to the smallest children. In 2006 Walt Disney advertised its Little Einstein DVDs on the paper liners of examination tables in two thousand pediatricians' offices.

Iannelli informs parents that they can get free formula; antibiotics; allergy and asthma medications, including Claritin, Clarinex, Zyrtec, Allegra, Flonase, Nasonex, Rhinocort Aqua, Flovent, Pulmicort, Foradil, Serevent, Advair, Xopenex, and Singulair; coupons for ADHD medication; acne, eczema, and ringworm dermatological medication; cough and cold medications; and Advil, Tylenol, and Motrin. "Other samples I have recently received in my office include MiraLAX for constipation, Ovide for head lice, and Triple Paste for diaper rashes," Iannelli writes. With all these visits from drug representatives, one wonders when Iannelli's office has time to practice, let alone write about it. In the same "educational" series on About.com, Iannelli touts the benefits of specific prescription medications he no doubt has been freely supplied, advocating for the use of acne medication, including antibiotics, for teens and allergy medications for young children.

The company salesmen who ply pediatricians and other doctors with coffee, doughnuts, and "free" lunches are called "detailers." In 2004, drug companies spent more than $20 billion on them, doling out about $16 million worth of free samples. Detailers establish friendly rapports with the doctors and their staff, keep offices stocked with educational materials that conveniently double as company propaganda, and even offer free medication to doctors and their families for their own personal use. This is not a gesture of goodwill on the part of the drug companies. It is money well spent. When parents get free samples from pediatricians or see advertising in the doctor's office, they are more likely to buy those same products in the future.

"For as long as I've been in business, the drug representatives have been coming to my office," says Dr. Jay Gordon. "They bring us free samples, they buy us lunch. They offer to take me out to dinner whenever I want. If there were a conference and I wanted to go, I could say to them, 'I'm having a bad year,' and they would pay to fly me there *and* give me an honorarium. They fly doctors all over the world for free, very often just to give a ten-minute speech."

Despite peer pressure from colleagues to accept the swag, Gordon closed his doors to drug companies five years ago. He refuses to attend dinners sponsored by formula manufacturers or take freebies from pharmaceutical companies. Gordon believes it is unethical and professionally undermining for medical professionals to accept "free" products from drug companies and hand them out to patients. It can also negatively impact a child's health. After all, it is much easier to recommend MiraLAX than take a nutritional history to find out why a child is constipated, much faster to prescribe an allergy medication than to problem solve with the parent what allergens are prompting a toddler's symptoms, much simpler to give a "free" can of formula to a mom than for a doctor to take the time to help her breastfeed.

NICE-LOOKING DRUG REPS

But accepting the "free" stuff is alluring and unfortunately very few pediatricians have the integrity to say no. The detailers who used to come to Gordon's office were attractive, likable, and friendly: "The rep for Zithromax was an ex-marine. My office staff loved him," Gordon says. "The female reps are almost exclusively good-looking women. The male reps are good-looking guys."

It took Gordon years to realize how these friendly salesmen were undermining his patients' health. "What they are doing is biasing us, and often giving us information that is completely inaccurate," Gordon says. "They are influencing us unduly to use their drugs, which may be more expensive and less effective, by giving us free pens and free trips. It influences all of us, even if it influences us subtly." Last year Gordon received more than thirty invitations to have dinner at upscale restaurants at the expense of the drug companies. He declined them all.

"I overlooked it for years because it was convenient for me to have samples around the office, and free medication for family members who needed it," he admits. "I didn't think hard enough about it. Once you give it any thought at all, you realize that it was always immoral and should be illegal."

PEDDLING PEDIALYTE

When a flu bug was going around, the nurse instructed Elizabeth to give her son Pedialyte. Elizabeth asked the nurse if she could give Lochran homemade chicken broth instead. The nurse looked confused.

On the bottle of grape-flavored Pedialyte is a bunch of Concord grapes with water splashing off of them. But there is nothing in this product that resembles fruit in its natural state. In fact, there is no fruit of any kind in Pedialyte. Instead, grape-flavored Pedialyte, which costs about $6 a quart, contains a host of unpalatable ingredients, including artificial flavor, sucralose (a noncaloric artificial sweetener that is made by chemically reacting sugar with chlorine), acesulfame (another artificial sweetener that some scientists believe to be carcinogenic), FD&C Blue #1 (a petroleum-based dye that has been found to cause allergic reactions, tumors, and possible nerve-cell damage), and Red #40 (another petroleum-based dye that causes allergic reactions and has been found to be contaminated with a carcinogen).

The Pedialyte labels boasts: "#1 pediatrician recommended brand." How could pediatricians—or any doctors—recommend this toxic brew of literally nondigestible substances (one reason the artificial sweeteners are not considered harmful at low doses is because they are so foreign to humans that the intestines do not absorb them) for small children over chicken soup or a homemade rehydration drink?

The simple answer is they are rightfully worried about dehydration. In the early 1990s, I worked on a child survival campaign in Niger, West Africa. I've seen firsthand how dehydration is a leading cause of death among children in the developing world. It's devastating. Dehydration can also be a serious concern for any child anywhere who gets a stomach bug. So community health workers in Niger and other poor countries instruct parents to breastfeed as often as possible and give their small children sips of one liter of boiled water mixed with one teaspoon of salt (no saltier than tears) and one tablespoon of sweetener. The cost of one quart of this drink in the United States: less than 16 cents.

Abbott Laboratories, the pharmaceutical giant that makes Pedialyte as well as Similac, has given more than $5 million to the American Academy of Pediatrics in the past eight years! In the 2011–2012 fiscal year alone,

Abbott was one of the top corporate donors to the AAP, listed in the over $750,000 category. Like other pharmaceutical companies, Abbott spends enormous amounts of money on product placement in pediatricians' offices. No wonder Pedialyte is the #1 pediatrician-recommended brand.

WHY TAKE A WELL BABY TO A PLACE WITH SICK CHILDREN?

Elizabeth started questioning why she should take her chubby blue-eyed infant who had five extra chins and a drooly smile to the doctor and pay the $30 copay at all. She noticed that each well-baby visit created problems. "Getting a checkup just to make sure nothing's wrong and paying a premium for it seems to invite the opportunity to look for problems," she tells me, "or to inflame problems that would have resolved on their own."

But it was the time she sat in the doctor's office with a healthy Lochran, when the waiting room was filled with sick patients, that disturbed her the most. Even though she scheduled the first appointment after lunch, Elizabeth had to wait an hour and a half. One wan seven-year-old with dark circles under his eyes played with his mother's iPhone as she watched a video on the waiting room screen. He looked so pale and sick Elizabeth wondered why he wasn't home in bed. Another child, a toddler, was coughing so hard Elizabeth could see the droplets in the air.

The situation felt absurd: Why was she bringing her healthy child into what she calls "a clambake of germs"? By the time they were called in, Lochran was completely undone. So was his mom. It was then that she decided it did not make sense to follow the advice of a doctor who saw her child for ten minutes once every three months. Pregnant again, Elizabeth recently switched from a pediatrician to a chiropractor trained in nutrition and homeopathy. Elizabeth appreciates her chiropractor's whole-child approach and how she takes the time to examine Lochran's skeleton, digestive system, diet, and even to talk about his personality. "She will come to my house at three a.m. if it's an emergency," Elizabeth wrote me in a follow-up email after her second son was born. "For injuries like broken bones I would take the boys to the ER, but for sickness I just don't trust the pediatricians."

A PATIENT-CENTERED MODEL OF WELL-BABY CARE

Laura Wise, a family practice doctor based in Oakland, California, understands Elizabeth's concerns. She thinks the biggest downside to well-baby care—which is a problem of medical care in America in general—is that it is too fragmented. Wise believes in order for checkups during the first year of a baby's life to really promote a baby's health and well-being, the mother's health needs to be considered as well. "The mom and baby in the first year of life are so inextricably linked in their well-being," Wise tells me in a phone interview as she leaves her office on the way to the gym. "If the mom is depressed, it impacts the baby. If the baby is colicky and doesn't sleep well at night, it has a huge impact on the mom."

Wise herself was isolated as a new mom. When her daughter was born in Petaluma, California, nine years ago, Wise knew very few other first-time moms. None of her med school friends had had children yet, and she found it lonely to be home all day with the baby without family in the same town or a community of friends. Despite being a medical doctor, she felt unmoored. "I didn't know what I was doing," she admits. "I got postpartum depression. I had doctors and could get medication, but I didn't have any support."

Then, in 2006, six weeks after her second child was born, a colleague who was going out on maternity leave asked Wise to run a group for her. In this model of group care, pairs of moms and babies come together for a two-hour appointment facilitated by two people, one of whom is a health care provider, usually a pediatrician, family doctor, or a nurse practitioner. Well-woman care (including family planning, mental health, and achieving weight goals) is combined with well-baby care (including safety, immunizations, and developmental assessments). Unlike in the traditional medical model, the group appointments always start on time, patients have a full two hours with the provider (who is there the whole time, although she does the physical exams on each baby in a space apart), and moms learn as much from each other as from the health professionals guiding the group.

When parents bring concerns—a three-month-old baby who has stopped pooping frequently, for instance—they hear from other parents who are experiencing the same issue, and often learn that it's not actually a cause for concern. The group wisdom also gives women a chance to try less invasive alternatives to clear up medical problems, Wise says. In one of her Spanish-speaking groups a baby had eczema and another mom suggested trying a gentle moisturizing soap. If Wise had been alone with the patient, she might have written a prescription for a topical steroid cream. But the next month the mom came back with a rash-free kid.

"What I wouldn't give to be receiving care that way," said a mom of a two-month-old baby in Silver Spring, Maryland, who goes to the same church as Sharon Rising, CenteringParenting's founder. Laura Wise agrees. "Having a group of moms at the same stage would have been so helpful to me," she admits. "If my doctor had taken the time to understand my background, and what I brought to mothering, it would have made a huge difference to my daughter's well-being. I was not outwardly high risk but if you knew me, you would know that I had a lot of risks for postpartum depression. There was so much that I didn't know. It's kind of intense. If someone had just given me a Moby wrap, and said, 'Just put your baby in here and walk around and you'll have a much better day.' That's what happens in a parenting group, someone gets a good carrier and they share it with someone else. It's transformative."

First started in 2006, there are now thirteen health centers in the United States where women and their babies can receive CenteringParenting postnatal care. In one randomized controlled study with ninety-seven mom and baby pairs, participants had better attendance, recalled getting more education and advice, and generally reported higher satisfaction than with traditional care. Lead researcher on this study, Ada Fenick, a pediatrician with more than fifteen years of experience who teaches pediatrics at Yale School of Medicine, says almost twice as many of her medical students requested to receive training to do group well-baby care this year than last. "These residents are learning a lot from the moms in their group, and about how to talk to people. I think it's wonderful," Fenick says. "And the parents are getting a lot out of it. For me it's very exciting. You've got the moms teaching each other; every now and then you look over at the doctor, who's nodding, because that's just what she would have said."

Despite the positive results, in the Oakland area, Wise has found it challenging to get her colleagues to embrace group care. "We are trained to take care of people in the exam room, and we are trained to be or act like experts," Wise explains. "The idea of a facilitated group, which can be chaotic and draws on the knowledge of the people in it, is uncomfortable for health care providers who haven't been trained that way." But, Wise believes, it's a battle worth fighting. The biggest difference between this kind of well-baby care and a more traditional model is that it changes our current pediatric paradigm by embracing "the radical idea that patients have more knowledge about their health than anyone else."

Average cost for visit to pediatrician: $150/visit

Average amount pediatrician spends on overhead: 65 percent

Salary of average American: $43,000

Salary of average pediatrician: $168,650

Average debt of a medical school graduate: $157,944

Amount of antibiotics consumed for ear infections by American children each year: 30 million rounds

AAP reported revenue for the 2009–2010 fiscal year: $87,194,548

Amount drug companies spend on doctor goodwill: $20 billion/year

Amount drug companies give in free samples: $16 million/year

Stephanie Precourt: Parents, Not Doctors, Know Best

Stephanie Precourt, a thirty-five-year-old mother of four who lives in Valparaiso, Indiana, is the kind of person who likes to follow the rules. She faithfully took her first two children to well-baby visits. But when her pediatrician ignored her son's life-threatening illness, Stephanie started to think differently about how best to ensure her children stay healthy.

We met with the pediatrician before my son, Noah, was born in 2001. We thought he was fabulous. He was young, like us, and he had a new baby as well. It seemed like we really clicked, and I felt like we were friends.

The extent of my preparation for becoming a parent was *What to Expect When You're Expecting.* I assumed it was law that you go to every well-baby visit and you do every shot and you do everything the doctors tell you. I honestly had no idea that there was an option to do it any other way. We felt we knew the doctor very well and were very influenced by his authority. But it also seemed like we were at the doctor's office all the time. A few days after our son was born, then two weeks later, then when he was a month old, then at four months. And we saw the pediatrician in between visits too. When it wasn't for a well-baby visit, it was for ear infections. My second son, Carter, was born a year after Noah and we did the same with him. Both boys had all their vaccines on the recommended schedule, and both were sick all the time.

When Noah was two and a half he had a grand mal seizure. I thought he was being electrocuted. He collapsed on the floor rigid and shaking. We called 911 and went via ambulance to the hospital. The pediatrician told us it's very common for a child to have one seizure, and it was probably nothing to worry about. We were hoping it was a fluke. A week later Noah had another.

The doctors we went to in northwest Indiana were stumped. They had no idea what was wrong with our son. They prescribed Topamax, an antiseizure medication. As soon as we started giving it to him, Noah went from having had only three seizures over the course of two weeks to having hundreds a day. He had to wear a helmet and sit strapped into a high chair all day long so he wouldn't fall and hurt himself. He stopped talking. He lost his memory. "The medication isn't working. He's getting worse," I told both the local neurologist and our pediatrician. "Maybe you need a higher dose," they said.

We were lucky to get an appointment at Children's Memorial Hospital in Chicago with Dr. Linda Laux. When she heard what was happening with our son, she came in on her day off to examine him. Within minutes she diagnosed him with Doose syndrome, also known as myoclonic-astatic epilepsy (MAE), a type of epilepsy. A lot of people think it's genetic but no one on either side of our family ever had childhood seizures. It was very rare in 2003 but over the years MAE has become more common. For children with Doose syndrome, antiseizure medication often makes them worse.

Dr. Laux told us there's been a good response of kids with Doose to a medically prescribed diet, which is started in the hospital, and suggested we try it. I called our pediatrician to ask his opinion. "Nobody does that. It doesn't work. It's too hard," he scoffed. "Can you imagine not giving him Goldfish crackers or anything he wants?" I never got that conversation out of my head because it felt like the doctor was ridiculing me. My child, who was a normal little boy before all this happened, was having these seizures every three minutes, and nobody knew why. Instead of trying to help us figure out what was wrong with his patient, the pediatrician laughed at the alternative another doctor suggested we try. We were desperate. Maybe they saw me as Annoying Mom. I understand if he was irritated with me for calling his office so often. But it was an emergency.

We checked into the hospital and started the diet. A dietician had to teach me what to do. They weighed Noah's food to the nearest gram. He had to fast for the first two days. It was really hard. But the ketogenic diet worked. Within four days Noah was seizure-free. His case was very unusual. Usually it takes more like three to six months to see improvement. He was on the diet for two and a half years with no medication of any kind. We were very strict about it and never wavered from it. In all that time he had one seizure.

I want people to like me. I'm an Indiana girl who was used to following the rules. It was really hard for me to realize that the people we had put so much trust in with our first baby didn't know what they were doing when we really needed them. We're so thankful we took our blinders off with that doctor. If we had stayed with him, and listened to him, I don't think Noah would be alive. I asked the first neurologist we went to, "What's Noah's future?" and he said, "Probably an institution."

My third son, Gray, was born in 2005. We had a different pediatrician by then. We met with him before Gray was born and told him our story. He was very understanding. We told him we were having Gray in the hospital but we didn't want any interventions. We were not going to give Gray any vaccines, we didn't want the eye salve or the vitamin K shot. He was fine with that. After Gray was born we went to the well-baby visits. Gray was very tiny. His head size was always in a negative percentile.

When Gray was four months old I took him in because his nose was running. The doctor put him on antibiotics for a "sinus infection." When the antibiotics didn't work he gave me a free sample of Zyrtec. "Can we just find out what's wrong with him? Maybe he's allergic to our pets?" I asked. But the doctor said it would be better to try the drug first. So I called an allergist in our area on my own. They did a blood test on him and found out Gray was allergic to dairy, egg, and nuts. When the tests came back the allergist recommended putting Gray on a really expensive hypoallergenic infant formula. I wanted to keep nursing. As soon as I stopped eating the foods he was allergic to, Gray was fine. I nursed him for almost three years.

That was the last time I took a child to a well-baby visit. I was feeling more confident in myself and I was tired of going to doctors who didn't really get it and would never admit they were wrong. I had no interest in talking to them anymore. I felt like the well-baby visits had no benefit. I felt like every time we went to the doctor we were at more risk of being exposed to whatever the kids had in the office. By then I didn't feel afraid that I was going to get in trouble for not going. Before that I had a lot of fears. I was afraid I would get reported to Child Protective Services. I had this false sense that these doctors had authority over me and I didn't want to upset them or get in trouble with them. I didn't want them to look down on me. But by then I felt confident enough to realize well-baby visits were a waste of time.

Before Gray was born I had joined a local natural parenting group. I started surrounding myself with other moms they don't talk about in parenting magazines. I started to feel less alone. I had other people who were on my side. We ended up having our fourth child, Ivy, at home. Ivy never saw a pediatrician and she never got sick. When she was a newborn, the homebirth midwife (who was practicing illegally because lay midwifery is illegal in Indiana) would come every few days to check on me and check on the baby. She believed a new mom shouldn't leave the house for the first thirty days. But if you follow conventional doctors, three days after giving birth new moms have to lug themselves and their baby to a doctor's office.

If there's an emergency I don't hesitate to go to the doctor. I was just at the ER with Carter because he had an allergic reaction to a nut in a cookie. I'm very thankful for medicine in that way. But the well-baby visits often lead to more interventions than you really need. I remember how much they worried me about Gray's head size, which didn't matter at all. It took me a really long time to realize I'm the one who knows my child better than anyone else. The doctor should be used as a guide but never as the person who has complete authority over your child or their care. What they say should be taken as a suggestion, not as the end all be all. They're just wrong too much of the time.

CHAPTER 11

SO WHERE DO WE GO FROM HERE?

Robbie Goodrich's wife, Susan, died of amniotic fluid embolism less than twelve hours after giving birth to a baby boy at Marquette General Hospital. Robbie, a forty-four-year-old professor of history at Northern Michigan University, was devastated. He took a leave of absence from his job. Consumed by grief and despair, he did his best to care for his newborn son, Charles Moses Martin Goodrich, and his two-year-old daughter, Julia.

In the hours after Susan died Robbie tried to find his son breast milk, knowing it was the best option and what he and Susan wanted. The hospital did not have any, but one of the nurses helped him order pasteurized breast milk from a bank in Kalamazoo, which cost $5 an ounce and would take two days to arrive.

Then Robbie received a phone call.

In a shy, almost apologetic way, his friend mentioned she was nursing her own child and offered to nurse Moses. Her offer sparked a movement. For an entire year a group of more than twenty Michigan moms, organized by Susan's best friend, Nicoletta Friare, came to Robbie's house to nurse his son or drop off pumped milk. They often brought their children, and sometimes their spouses. Some, like Carrie Fiocchi, who was twenty-nine years old and had given birth to her first child just six weeks before, didn't know the family when they started but were so upset by news of Susan's death that they wanted to help.

Though he will never get his mother back, Moses is a smiley, energetic preschooler these days. "He's a healthy, happy, well-adjusted boy," his father says, "who has always known a mother's love."

America is a nation of caring people, creativity, and great ideas. We are on the cutting edge of disease research, information technology, and the pursuit of new ways of thinking. Our society is based on free speech and freedom of choice. We offer our young people exceptional higher education and have the best universities in the world. We find creative solutions to seemingly unsolvable problems.

Yet we have one of the highest maternal mortality rates in the industrialized world, we offer medically unnecessary 4-D ultrasounds at the mall, we wash our newborns with baby soap that off-gasses formaldehyde, we vaccinate at birth against a sexually transmitted disease, we aggressively advertise products that undermine breastfeeding to pregnant women, let corporations give bleary-eyed new moms breastfeeding and potty training advice, and we put starch, food coloring, sugar, and unhealthy additives into food products made for babies. We continue these practices despite the science, best evidence, and common sense that show us they are harmful. We spend more money than any country in the industrialized world on health care, but every day we are failing to care safely for new moms and young babies. Ironically, it is scientific studies done at our outstanding research facilities that often form the basis of protocols that protect mothers and infants in other developed countries that enjoy better outcomes.

As parents it is our job to create a better, safer world for our children. We cannot blindly trust the medical establishment or the government to have our best interests in mind. We need to educate ourselves and make the best decisions we can for our families. When things go wrong, we need to share our stories with as many people as possible, contact the media, write blogs and articles, inform the politicians who represent us, even hire lawyers if we have to, and insist on change. We need to demand that competent journalists pay attention to our concerns, do their homework, interview experts on both sides of an issue, and read the scientific studies themselves or point out when the science simply has not been done.

As I have tried to show on these pages, the unfortunate result of our profit-driven medical system is that financial gain, not good health, is the motivation behind much of what happens to women and their babies in America. That hidden knowledge can be a powerful tool for every person in the United States, whether you are already a parent or hope to become one. Instead of allowing the system to continue, we can vote with our feet and our wallets. We can choose to go elsewhere. We can seek out people to care for us and our babies who have our best interests in mind: we can get our prenatal care from a homebirth midwife instead of an obstetrician, say no to the birth dose of the hepatitis B vaccine, devise an alternate vaccine schedule (or choose to forgo vaccination), combine infant pottying with hand-me-down cloth diapers instead of keeping our children's genitals wrapped in plastic for three or four years, and offer our babies whole, wholesome foods at no extra cost to us, instead of processed mush in plastic packaging.

In the early spring of 2011, two men pushed their baby girls in prams around Oslo's modern new Opera House while tourists scrambled across its sloped, pedestrian-friendly roof. Audun Lysbakken, the minister of children, equality, and social inclusion, and Knut Storberget, the minister of justice and public security, were both on paternity leave at the same time, taking advantage of Norway's family-friendly policies. In Norway parents get forty-seven weeks of parental leave (almost eleven months divided between mother and father) at full pay. If either parent wants to spend longer with their infant, they can opt for fifty-seven weeks (about thirteen months) at 80 percent of their salary. Even if she plans to go immediately back into the workforce, moms in Norway are *required* to take three weeks off before the birth and six weeks off afterward, so they have time to rest and bond with their new baby. Fathers in Norway also get time to care for and establish strong ties with their infants: they are allotted twelve weeks of parental leave at full pay. These weeks specifically set aside for the mother and father are included in either the forty-seven-week or fifty-seven-week options. Wanting to spend more time with his baby, Lysbakken opted to take four months of leave instead of three.

Just weeks before my visit to Oslo an unexpected attack by a white supremacist, purportedly upset by Norway's immigration policies, killed seventy-seven people and destroyed an entire block of government buildings. Security is tight around the government buildings and I am late for the appointment at the Ministry of Children, Equality, and Social Inclusion because of the maze of police tape and checkpoints. But Lysbakken's deputy minister, Kirsti Bergstø, ushers me warmly into her office, gesturing to a tray of grapes, orange wedges, almonds, and dried fruit. Bergstø and Arni Hole, director general of the department of family affairs and equality (who herself has five children), tell me Norway is repeatedly voted among the top five places in the world to be a mother (in 2010, 2011, and 2012 Save the Children's annual reports on maternal affairs placed Norway in the #1 spot) because of the country's generous leave policies, emphasis on gender equity, free universal health coverage, and pro-family emphasis.

"We invest in children," Arni Hole says, explaining that corporations cannot advertise to children or youth, moms who want to stay home longer receive a government stipend for taking care of their children, and the country does everything it can to support young families. The birthrate in Norway is higher than some countries in Europe but not as high as the government would like it, 1.95 in 2010, about the same as France's; the hope is that the generous government policies will encourage more Norwegians to have more children.

As I mentioned earlier, when I asked Iceland's surgeon general if their national vaccine program was mandatory, the question astonished him. Nearly 100 percent of Icelandic parents follow their current schedule.

Perhaps partly because they are a nation of rule-followers, but also because parents believe that the government schedule, which contains far fewer vaccines than ours, ensures their children will be in optimal health. In countries with universal health care the choice about what medicine to give to children is not driven by profit. In fact, since the government pays for vaccines and medication, it is in their best interest to keep costs down by doing only what is of proven benefit. The discerning American parent can look to Europe's vaccine schedule (where rotavirus, hepatitis B, and chicken pox have not been adopted in most countries) to find a more rational, health-based way to vaccinate. Countries in Europe that enjoy better maternal and infant health can be models for ours.

Ideally, improvements in the United States would happen at the government level. That's where the change really needs to take place. But in the meantime young families cannot afford to wait. We can change the system ourselves, starting right now. We can share information, form moms' groups, bring our neighbor a nutritious meal and take her older children to the park after she has had a baby, sign up for group prenatal care and group well-baby visits, nurse a newborn when his mother cannot, walk away from health care providers who mistreat us, insist on gentle practices, opt out of having our children in a sterile hospital setting and instead have our children at a birthing center or at home. It's time for every American to stop being duped by corporate interests and for women to take back control of our pregnancies, our birth experiences, and the precious first years of our babies' lives.

In Korean culture a baby's first birthday is celebrated with great fanfare. The birthday baby wears a colorful traditional Korean dress and sits in front of a low table piled high with rice cakes, apples, persimmons, and other treats. Everyone watches as the baby chooses a symbolic item thought to predict her future. If she reaches for a calligraphy brush she might become a writer; if she reaches for the coins on the table, a businesswoman. In American culture the first year of a baby's life similarly culminates in a big birthday party with friends and relatives, lots of gifts, and a cake with just one candle.

We didn't realize it until our second child came along, but this celebration is as much for the parents as for the baby. It is a tremendous achievement to survive the crazy whirlwind of the first year of your baby's life and to navigate everything from breastfeeding to changing diapers to dressing a baby whose muscles are so floppy he can't yet hold his head up to figuring out the new family dynamics now that a third person has been added to it. Babies teach us as much as we teach them. As they grow we gain confidence in our ability to parent and, especially, to know when something is wrong. "Read your baby, not the books," one Kenyan grandmother advised her granddaughter who had been living in England and found herself overwhelmed by all the advice she had gleaned from books and so-called experts. Though our tendency is to defer to the doctors and others in the medical establishment, as well as to the businesses that make products to sell to our babies, it is we parents, and our babies, who actually know best.

Abbreviations

AAP: American Academy of Pediatrics
Nonprofit trade organization for 60,000 pediatricians in the United States and Canada.
ACOG: American College of Obstetricians and Gynecologists
A nonprofit trade association representing the interests of American obstetricians and
gynecologists An affiliate nonprofit business organization, the American College
of Obstetricians and Gynecologists is a 501(c)(6).
CDC: Centers for Disease Control and Prevention
A U.S. agency based in Atlanta, Georgia, that studies and monitors communicable
diseases as well as a broad range of other health-related issues.
FDA: U.S. Food and Drug Administration
A federal agency responsible for protecting the public health in America by assuring
the safety of drugs, food, cosmetics, and other products that affect human health.
HMO: Health Maintenance Organization
An organization that provides a system between health care providers and insurance
companies or individuals that centrally manages patient care through a dedicated
network and a set of criteria for referral from a primary provider.
M.D.: Medical doctor
From the Latin *Medicinae Doctor,* a degree granted from accredited medical schools
after a student has completed four years of undergraduate studies followed by
four years of medical education and successfully passes the U.S. Medical Licensing
Exam.
NHS: National Health Service (Britain)
Established in 1948 and funded by national taxes, Britain's NHS provides free universal
health care to residents of the United Kingdom.
PPO: Preferred Provider Organization
A group of providers who have contracted with an insurer to provide services for a
discount, in return for a subscription fee.
R.N.: Registered nurse
A nurse who has graduated from a college or university nursing program and has
successfully passed a licensing exam.
RVU: Relative Value Units
Formula by which Medicare determines the pay for physicians for particular tasks, now
also used by some hospitals to determine pay for nonphysicians by comparison
to the physicians' tasks.
WHO: World Health Organization
A United Nations agency for public health, it currently spends about $4 billion per year
promoting global public health, of which one quarter comes from UN member
states, the rest from outside donations.

Glossary of Terms

Adhesion Band of scar tissue, caused by surgery, infection, or trauma, that adheres to an internal organ or other body part. In infant boys the foreskin has not yet fully separated from the glans (head) of the penis. Full separation often does not occur until age ten or eleven. Circumcision before the foreskin can retract requires forced separation, which can cause adhesions (pitting or scarring) of the glans. Abdominal adhesions are a common complication of Cesarean birth.

All-in-ones (AIOs) A cloth diaper attached to its waterproof cover, an all-in-one is like a washable disposable diaper, combining the convenience of plastic diapers with the reuse and environmental superiority of cloth.

Amniocentesis Prenatal test in which a large needle is inserted through the mother's abdomen and uterus into the amniotic sac to withdraw fluid containing cells from the fetus for testing. Normally, the sac seals and the fluid regenerates in a day or two. Just under 1 percent of amniocenteses result in miscarriage. Other risks include preterm labor and amniotic fluid embolism.

AFP (Alpha-feto-protein) screening Common blood test for a protein that can reveal several potential birth defects; often combined with other blood tests (AFP3, AFP4, "Quad," etc.).

Amniotic fluid The liquid that surrounds the unborn baby in the mother's womb during pregnancy. Amniotic fluid serves many functions, including cushioning the fetus, protecting it from injury, and helping the lungs develop as the fluid is swallowed (inhaled) and released (exhaled). Amniotic fluid is continually regenerated and reabsorbed, so the amount of fluid in a mother's uterus actually changes from hour to hour. In a normal pregnancy the volume increases as pregnancy progresses until about week thirty-six, when it starts to decrease.

Amniotic fluid embolism An uncommon but often fatal allergic reaction in pregnancy in which amniotic fluid or fetal tissue get into the mother's bloodstream when the amniotic sac and uterine veins have ruptured. Once almost unheard of, amniotic fluid embolism has been linked to Cytotec (misoprostol), an ulcer medication that is used off label to cause abortion and to induce pregnancy.

Amniotic sac The membranes that contain the amniotic fluid surrounding the fetus.

Anesthesiologist Physician specialized in anesthetizing patients for surgery, including local (skin-numbing), spinal (such as an epidural), and general (unconsciousness).

Apgar A test given to newborns to assess their condition immediately after birth and again five minutes later. Virginia Apgar was an anesthesiologist who developed the test to see how the baby was affected by anesthesia given to the mother. Breathing, heart rate, color (rosy or blue, for blood oxygen), muscle tone, and startling response when disturbed are each measured on a simple scale of 0 (none), 1 (some), or 2 (optimal), then the five are added up to give a total score from 0 to 10.

Autologous Transferred or derived from the same individual, as when cells taken from a newborn's cord blood are later given back to the same person to treat a disease.

Bilirubin A yellowish molecule in the blood, produced when hemoglobin is broken down as blood cells are replaced. Normally processed by the liver and excreted in urine (made yellow by bilirubin) and stool (likewise made brown by bilirubin). An excess of bilirubin due to physiological problems can be dangerous, but at normal levels it has the benefit of acting as an antioxidant.

Breech birth A baby is breech when the bottom, rather than the head, is in position to emerge first. Though most fetuses will turn to face head down, 3 to 4 percent of babies are in the breech position immediately before birth. Some breech babies turn to face downward during the birth practice. Breech positions are classified as frank, complete, or footling.

Cavitation Microscopic bubbles or pits (cavities) caused when a surface is struck by high-speed vibrations, like ultrasound waves, which make the surface undulate microscopically.

Cerebral cortex The cerebrum is by far the largest part of the human brain, occupying most of the interior of the skull and giving the brain its characteristic shape; the cortex (Latin for "bark") is the furrowed surface of the cerebrum, where most of human cognition takes place.

Certified nurse midwife A registered nurse (R.N.) with advanced graduate training including specialization in midwifery, CNMs deliver babies in hospitals, birthing centers, and women's homes.

Chorionic villus sampling (CVS) A method of prenatal testing by sampling tissue from the villi, rootlike extensions, the fetal placenta links into the part of the placenta supplied by the uterus. Chromosome abnormalities in the fetus are tested to identify Down syndrome or other problems. Unlike with amniocentesis, the amniotic sac is not punctured, and CVS can be done earlier in the pregnancy. Nearly 1 test in 100 will cause miscarriage.

Circulating nurse Surgical nurse who preps the operation, then orbits the doctors and nurses during surgery, monitoring their needs and the patient's and exchanging things between the sterile zone and the outer part of the operating room.

Circumcision Elective surgery, originating in prehistoric religious and cultural practices, in which the foreskin of the penis is removed so that it no longer hoods and protects the end of the organ. Though procedures differ by doctor or traditional cultural circumciser, the middle of the tube of skin covering the penis is cut out, and the resulting ends of the skin growing together, shortening the penile skin enough to keep the head revealed. The mucous membranes at the end of the penis dry out and become hardened with keratin, the protein composing fingernails.

Colostrum The yellowish or clear liquid rich in protective white blood cells and antibodies that sometimes leaks out during pregnancy and is secreted from the breasts for several days before the milk comes in. Sometimes referred to as the first milk, colostrum has a laxative effect and helps a newborn establish healthy gut biota and pass meconium.

Complete breech A baby presents as complete breech when the bottom and feet lie against the birth canal, the legs crossed as if sitting on the ground Indian style.

Contact dermatitis Localized skin irritation or rash caused by contact with an allergen or other irritant.

Cytotec Trade name for misoprostol, a synthetic prostaglandin used to treat ulcers, induce abortion, enhance erectile function, and hasten labor. In August 2000 Searle (the manufacturer) wrote a warning letter to doctors that there had been cases of uterine rupture and death, among other complications, and asking it not be used on pregnant women.

Direct-entry midwife A midwife who is not also an R.N. but has trained at a school of midwifery, a university program distinct from nursing, as an apprentice or through self-study. Direct-entry midwives focus on births outside a hospital setting, such as at home or in a birthing center. In the United States, direct-entry midwives can obtain national certification as a CPM (certified professional midwife) administered from the North American Registry of Midwives. Most state governments also provide for licensure of qualified practitioners, while some states make no legal allowance for the practice.

Double-blind experiment A rigorous method of experimentation in which neither the experimenter nor the subject knows who is part of the control group. This protocol eliminates experimenter bias, which has been known to influence results.

Doula A birth attendant who offers non-medical assistance to a woman during pregnancy, labor, and after the baby is born. From the ancient Greek word for "slave woman," the term *doula* started to become popular in the 1970s in America when researchers found that women who had support during labor had better outcomes.

Down syndrome A chromosomal error in which an additional (third) copy of chromosome 21 is present in a person's DNA, resulting in a number of differences from typical development, particularly in height and cognition.

Effacement Cervical thinning (sometimes referred to as "ripening") that happens in preparation for labor and is estimated in percentages. The first stage of labor is complete when the cervix is fully dilated to ten centimeters and 100 percent effaced.

Endometriosis A painful condition in which uterine cells spread outside the uterus to other parts of the pelvis and react to the woman's hormonal cycle as the uterus does. Endometriosis can cause irregular periods and infertility.

Endorphin Hormone that acts as a neurotransmitter, producing pleasant feelings and inhibiting pain.

Epidural Pain relief for childbirth effected by injecting local anesthetic into the spinal cord, numbing everything below the site of injection. Though popular, epidurals have been associated with longer labors, a higher risk of maternal fever (leading to obstetric intervention and lower newborn Apgar scores), and lingering numbness.

Episiotomy Operation cutting open the perineum from the vaginal opening toward the anus, once thought to help avoid tearing of the vaginal opening during birth. Research has shown that outcomes are better without episiotomy than with it, but the practice continues among those ignorant of the medical evidence.

Excitatory cells Neurons (brain cells) that tend to propagate nerve signals by releasing glutamate. Inhibitory cells, by contrast, release GABA (gamma-amniobutyric acid), which inhibits the signal propagation.

Fetal Survey A detailed ultrasound assessing the size and physiology of a fetus, given typically between eighteen to twenty weeks gestation.

Flatulence The release of gas generated by bacteria in the large intestine and colon, commonly called farting.

Footling breech In footling breech, the feet push against the birth canal opening without the baby's bottom nearby.

Forceps Obstetrical forceps are large gripping tools with plierslike handles and spoon-shaped loops of metal for holding the fetus's head to pull it out of the mother's pelvis.

Foreskin Tube-like overlap of skin covering the glans (head) of the penis that protects the skin of the glans, which is a mucous membrane, contains erogenous tissue, and acts as a lubricated sheath for the penis to glide in and out of during intercourse.

Frank breech A frank breech is when the baby's bottom faces the birth canal with the legs sticking straight up in front so that the feet are close to the head, like a diver doing a jackknife.

Germ cells Undifferentiated cells in the brain or elsewhere in the body that have the capacity to grow into various types of specialized cells.

Gestational diabetes Temporary high levels of glucose (sugar) in the blood during pregnancy, a common condition without the dangers of type 1 or type 2 diabetes, readily controlled by diet and lifestyle changes.

Glucose tolerance test A test usually given between twenty-four and twenty-eight weeks of gestation to detect pregnancy-induced diabetes, involving quickly drinking a large dose of pure glucose sugar and then being monitored to evaluate how well the body processes it.

Gray matter Gray matter, which makes up most of the surface of the brain, consists of nerve cells and their supportive tissue, other specialized cells in the brain (glial cells), and capillaries. Gray matter includes regions of the brain involved in cognition, speech, muscle control, memory, emotions, and sensory perception.

Gynecologist Physician specialized in the female reproductive system. Most are also specialized in obstetrics, which concerns women and their children during and immediately following pregnancy.

Halitosis An unusually unpleasant odor in the breath.

Hemorrhoids Natural blood vessels in the anus, which can become a problem when they become and remain engorged with blood, a common occurrence in pregnancy. Hemorrhoids can also develop postpartum caused by a woman delivering vaginally flat on her back.

High-risk pregnancy Pregnancy is medically considered to have higher risks when the mother is particularly young or old, petite or obese, or with any of a long list of complicating factors. Diseases can complicate a pregnancy, whether congenital (such as maternal birth defects making birth difficult), acquired (such as cancer), or contagious (such as venereal diseases). Otherwise healthy women may be classified high-risk because of doctor or hospital protocol, prior Cesarean birth, or prenatal testing.

Hydraulic fracturing ("fracking") Pumping water and chemicals into natural cracks in underground bedrock, forcing them open to allow oil and natural gas far below to rise up for extraction. Oil and gas can be obtained from areas inaccessible to conventional drilling, increasing the supply of fossil fuels, at the cost of changing the rocky strata of the Earth's crust, and filling deep underground strata with polluted water.

Hypoglycemia Abnormally low level of blood sugar (glucose) in the blood.

Iatrogenic "Physician-induced," referring to any problem, injury, or reaction caused by the action of a physician or by a medical procedure. "Iatros" means *physician* in Greek, and "genic," means *generated*. Childbed fever, caused by doctors not washing their hands after dissecting corpses or treating ill patients, was an iatrogenic disease.

Implantation When a woman's egg, within a few days after being fertilized by a sperm, implants itself in the prepared lining of the uterus, where it is nourished and begins to develop part of itself into the placenta.

Inhibitory cells Brain cells (neurons) that work to restrain stimulation. The transmission of stimuli in the brain is guided by a balance between excitory cells, which promote it, and inhibitory cells, which restrain it.

Insulin A hormone regulating metabolism, insulin governs how the body deals with glucose sugar. When blood sugar levels rise, insulin is released by the pancreas, which causes the body to pull the glucose out of the blood and store it. If tissues become resistant to insulin, diabetes can develop.

Intrauterine Growth Restriction (IUGR) A term used to describe fetuses thought to weigh less than 90 percent of fetuses of the same gestational age. Also known as fetal growth restriction, this condition can be caused by malnourishment, high altitude, twins, maternal high blood pressure, infection, or a congenital disorder. Since it is difficult to accurately measure a fetus's size in utero and is normal for some women to gestate more slowly, IUGR is often misdiagnosed, leading to unnecessary interventions like preterm C-section.

Intussusception A sudden intestinal blockage where a part of the intestine pulls into itself, intussusception can result in bleeding, infection, shock, and dehydration. This serious condition sometimes requires emergency surgery and can be fatal.

Ionizing radiation Radiation of a kind and intensity that can electrically charge atoms, which turns the atom into a differently charged version called an ion. Ions react differently from the original form, so an ionized atom in your body won't work quite the same as the original. Ionizing radiation can break chemical bonds in molecules such as DNA. No amount of ionizing radiation is considered safe, though some exposure from the natural environment is unavoidable. All X-rays are ionizing radiation.

Jaundice A yellow skin tone caused by an excess of bilirubin in the bloodstream, affecting more than half of all newborns as the infant's metabolism catches up with the hemoglobin cycle in the blood. In the vast majority of cases jaundice is a normal condition that disappears in one to two weeks.

Kernicterus Damage to an infant's brain by excessive bilirubin levels in the blood, resulting from the breakdown of red blood cells. Usually caused by mother and fetus having different Rh blood types or by a genetic disorder.

Low-risk pregnancy A normal healthy pregnancy with no known complications (i.e., not high risk); over 90 percent of pregnancies in the United States fall into this category, which is typically used to describe the candidates for a vaginal birth unlikely to require medical intervention.

Meatal stenosis A narrowing at the tip of the penis of the opening of the hole that urine passes through (the urethra), that interferes with normal urination. Usually caused by swelling and irritation from a circumcision that results in scar tissue growing across the urethra. This condition can lead to painful urination, urinary tract infection, bleeding from the end of the penis, daytime incontinence, and bed-wetting. An operation to enlarge the urethra is often necessary.

Medicaid Government program providing limited payment for health care for some low-income Americans.

Midwife Birth attendant who specializes in physiological (natural) childbirth. Midwives assist women in delivering their children vaginally and with prenatal and postnatal care. The midwifery model of care is to follow the normal course of birth and facilitate successful vaginal delivery. Midwives monitor for complications requiring intervention, transferring care if necessary to obstetricians, who specialize in the illnesses and complications that can arise in pregnancy and birth. Compared to OBs, births with midwives have lower infant and mother mortality and fewer complications and interventions.

Minicolumn A vertical arrangement of neurons in the layers of the cerebral cortex. The brain develops a fine three-dimensional structure of cells, necessary to normal cognitive function. When these stacks of cells are too tightly or too loosely spaced, intellectual function is abnormal.

Naturopath Physician educated at a naturopathic medical school, following a holistic medical model emphasizing diet and botanical treatments as an alternative to drugs when possible.

Necrotizing Condition in which cells die abnormally and are not cleared away by the body, leading to dead tissue rotting while attached to living tissue. Often self-spreading, if not removed surgically, necrotizing can lead to gangrene, ending in amputation or death.

Neural tube defects Birth defects in which the neural tube, which forms the spinal cord and brain, fails to close completely in the development of the fetus, leading to complications of various severity.

Nurse anesthetist Registered nurse who specializes in giving anesthesia. These nurses do epidurals and spinals.

Otitis media Normally harmless infection between the eardrum, the Eustachian tube, and the inner ear, also known as an ear infection.

Otoscope Cone-shaped viewer, mounted on a handle, that is used by physicians to examine inside the ear; bane of children, second only to the tongue depressor.

Ovulate Part of menstrual cycle in which a mature ovum (egg) or ova (eggs) are released from the ovary, ready to travel down the fallopian tube to the uterus to be fertilized by sperm and implanted into the uterus, resulting in pregnancy.

Oxytocin Hormone that causes uterine contraction during and after birth, oxytocin is also released during breastfeeding. In the brain, it is active in producing feelings of empathy and bonding.

Pediatrician Physician specialized in the primary care treatment of children from birth through adolescence.

Perineum The area (skin and underlying tissue) between the genitals and anus. The woman's perineum stretches and thins to make way for the baby's head during childbirth.

Physician Health care provider who diagnoses and treats disease or injury. Though often thought synonymous with M.D., physicians also include D.O.s (osteopathic doctors), N.D.s (naturopathic doctor), D.C.s (doctor of chiropractic), and N.P.s (nurse practitioner).

Pitocin Synthetic (artificial) form of the hormone oxytocin used to induce labor in pregnant women or stimulate uterine contractions once labor has already started. It is synthesized from the pituitary glands of cattle and contains the preservative chloretone.

Placenta Two-part organ connecting the uterus and fetus, with part formed from each of them, the placenta allows gases, fluid, nutrients, and waste to be passed between mother and fetus.

Placenta previa Obstruction of the cervical opening by the placenta, which can cause a variety of problems, including hemorrhage.

Preeclampsia Syndrome combining high blood pressure and protein in the urine during pregnancy, which can lead to various hypertension problems, blood cell disfunction, and other complications, and in a small percentage of cases to full eclampsia with its associated seizures.

Prepuce Synonym for foreskin.

Public aid Any system of financial assistance for low-income needs, including health care, as in Medicaid or the Illinois Public Aid medical program.

Pulmonary embolism A blockage in the main artery of the lungs, either by a clot or by amniotic fluid; a major cause of maternal mortality. A danger to anyone who remains lying immobile for extended periods.

Sciatica Pain in the large nerves running down the backs of the legs. A frequent condition in the second and third trimesters of pregnancy as the growing baby puts pressure on the mother's sciatic nerve.

Scrub nurse Nurse who assists a surgeon and remains within the sterile area of the operating room.

Septicemia Blood poisoning from harmful bacteria that often occurs with severe infections. Chills, high fever, rapid breathing, rapid heartbeat, and discoloration of the skin can be signs of septicemia.

Serial Sequential Testing, also called serial sequential screening. A series of tests for Down Syndrome and other congenital problems, compounding an ultrasound of skull development with two successive blood tests. The nuchal fold ultrasound is the first test in this series, and must be performed before fifteen weeks.

Shoulder dystocia When a baby's head emerges from the birth canal but a shoulder gets stuck behind the mother's pubic bone. This is a dangerous place for delivery to stop, as the blood supply from the umbilical cord is likely to be cut off. Several techniques, as simple as pulling the mother's knees to her chest or rolling her onto all fours, resolve the problem in the majority of cases.

Smegma Secretion of skin oils and exfoliated cells under the foreskin that cleans, lubricates, and protects the glans (head) of the penis. There is little of it in childhood; it increases at puberty.

Sonogram Also called an ultrasound, imaging of a fetus (and the interior of the mother's abdomen) by making an echo-reflection image using very-high-frequency sound waves.

Sonography The process or practice of using ultrasound to image the interior of the body.

Spinal Anesthesia delivered directly into the spine to completely numb everything below for surgery. Often used with women who are having elective Cesarean sections.

Stem cells Cells that can turn into any of a variety of specialized cell types. These self-perpetuating cells are found in many places in the adult body, where they are used to replace dying cells. In the first several days after a human ovum is fertilized, before it implants in the uterus and begins to develop from a blastocyst into a differentiated embryo, *all* of the cells are stem cells that can become *any* type of human cell.

Superabsorbent polymer (SAP) Molecule that bonds chemically to water molecules and bonds to itself to form long molecular chains. In this way it can occupy three hundred times the SAP's own weight in water, and the water cannot be squeezed out because it has become part of the molecule.

Transverse Dimension of the body in which torso-twisting movements are performed; a plane parallel to the floor. Contrasted to coronal and sagittal planes of the body. A fetus is transverse when lying horizontally across the mother's uterus. A transverse Cesarean incision is made horizontally.

Ultrasound Imaging of a fetus (and the interior of the mother's abdomen) by making an echo-reflection image using very high-frequency sound waves. The intensity of the waves is not regulated, leading to possible risks to the fetus. According to the FDA, "There are no federal radiation safety performance standards for diagnostic ultrasound." Though user education and licensure exist, they are not required in most states.

Urethra Tube conducting urine from the bladder out to be eliminated. In women, it emerges in the vulva, above the vagina.

Uterus The womb; the organ in which implantation of the fertilized egg takes place and the fetus matures.

VBAC Pronounced "vee-back," vaginal birth after cesarean. Giving birth vaginally after a previous C-section has been found to be as safe as or safer than Cesarean birth, even for a woman who has had a previous C-section. However, because the uterine scar could separate and result in uterine rupture (often caused by pregnancy induction and aggressive labor management with Pitocin) and the fear of litigation, many hospitals do not allow women to attempt VBAC.

Ventricles The two larger chambers in the heart (left and right ventricles) that do the main work in pumping blood throughout the body.

Vernix A waxy secretion covering the fetus in utero and the newborn's skin at birth. Vernix protects and lubricates the delicate skin of the newborn, holding in heat and moisture, and perhaps acting as a barrier to harmful bacteria while it lasts.

X-rays High-frequency light used to image structures inside the body. X-rays are a form of ionizing radiation, which can break chemical bonds between atoms and change their electron charge. Scientists agree that no amount of ionizing radiation is safe and that there is no lower limit below which there are no health risks.

Appendix

RESOURCES

PREGNANCY SUPPORT

Birthing From Within (805-964-6611, http://www.birthingfromwithin.com/)—Birth empowerment organization started by midwife, writer, and VBAC mom, Pam England, that offers childbirth classes, events, and information for expectant parents and birth companions.

Centering Healthcare Institute (857-284-7570, https://www.centering healthcare.org/index.php)—A nonprofit founded by Sharon Rising that provides consultations, trainings, and materials to clinical practices to begin group care (Centering Pregnancy, Centering Parenting).

Childbirth Connection (212-777-5000, http://www.childbirthconnection.org/)—Nonprofit dedicated to helping women, their partners, and health professionals make scientific and evidence-based decisions about best maternity care practices. Their website is loaded with helpful, referenced articles about pregnancy, labor, and delivery.

Planned Parenthood Federation of America (212-541-7800, http://www.plannedparenthood.org/)—Nonprofit organization that runs health care centers throughout the United States to give women and their partners access to free or low-cost pap smears, birth control, pregnancy testing, and abortion.

CHILDBIRTH SUPPORT

Coalition for Improving Maternity Services (866-424-3635, http://www.motherfriendly.org/)—Coalition of individuals and national organizations dedicated to promoting evidence-based mother-friendly childbirth.

DONA International (888-788-DONA, http://www.dona.org/)—International nonprofit organization that trains, certifies, and promotes doulas (birth assistants) and helps expectant families find doula support.

Midwifery Alliance of North America (888-923-MANA, http://mana.org/)—Nonprofit organization promoting collaboration and support among midwives.

Waterbirth International (954-821-9125, http://www.waterbirth.org/)—Nonprofit organization dedicated to helping women have waterbirths.

BREASTFEEDING SUPPORT

La Leche League International (800-LA-LECHE, http://www.llli.org/)—Nonprofit organization that promotes breastfeeding through mother-to-mother support. Trained volunteers run monthly support groups for new moms in every state in America and dozens of countries around the world.

To buy breast milk: Only the Breast (http://www.onlythebreast.com/) is an online community of moms who want to buy, sell, and donate breast milk.

To find free breast milk near you: Human Milk For Human Babies (http://www.facebook.com/hm4hb) is a global network that has local chapters in Oregon, Washington, Oklahoma, South Carolina, and elsewhere. Each chapter has their own Facebook page where moms who have extra milk and moms who need milk can find each other.

To obtain breast milk from a milk bank: The Human Milk Banking Association of North America (817-810-9984, https://www.hmbana.org/) collects donated breast milk, processes it, and then ships it to babies with a doctor's prescription indicating medical need. Some insurance companies cover the cost of the milk.

POSTPARTUM SUPPORT

Doulas of North America (888-788-DONA, http://www.dona.org/mothers/faqs_postpartum. php)—Doulas are not only of great assistance during childbirth, they also can help families adjust to a new baby. They attend to the basic needs of the new mom and educate her and the family about appropriate care of an infant.

Postpartum Support International (800-944-4PPD, http://www.postpartum.net/)—Nonprofit to support awareness, prevention, and treatment of postpartum depression and other mental health issues that arise after giving birth.

DIAPERING AND POTTYING SUPPORT

Diaper-Free Baby (http://www.diaperfreebaby.org/)—Volunteer-led nonprofit that runs local support groups to educate and help parents who want to practice elimination communication.

Real Diaper Association (http://www.realdiaperassociation.org/)—Nonprofit trade association for the cloth diaper industry that educates parents about cloth diapering, promoting it as a cheaper, safer, and more environmental diapering choice.

INFORMATION ABOUT CIRCUMCISION

American Academy of Pediatrics, the trade organization of American pediatricians, issued "Technical Report: Male Circumcision" (August 2012) from its task force that concludes that the benefits of male infant circumcision outweigh the risks but not enough to recommend routine circumcision for all newborn males, that the choice should be up to the parents, and that insurance companies should pay for the procedure. The full report is available online: http://pediatrics.aappublications.org/content/130/3/e756.full.

Circumcision Information and Resource Pages (CIRP) contains an exhaustive reference library of research, history, and statistics related to circumcision: http://www.cirp.org/.

Circumcision Resource Center (http://www.circumcision.org/)—Nonprofit educational organization that raises awareness about male circumcision with the goal of discouraging parents from doing the procedure.

Intact America (http://www.intactamerica.org/)—Nonprofit child advocacy organization overseen by a board of health professionals that educates parents about the harms of circumcision.

The Royal Australasian College of Physicians—Professional organization of doctors in Australia and New Zealand. They have written a twenty-eight-page referenced position statement on male circumcision, "Circumcision of Infant Males" (2010), detailing the anatomy of the foreskin, the medical debate, and the reason it is not recommended or routinely done in Australia or New Zealand: http://www.kids.vic.gov.au/downloads/male_circumcision.pdf.

INFORMATION ABOUT VACCINES

CDC (800-232-4636, http://www.cdc.gov/vaccines/)—The American government's Centers for Disease Control and Prevention post exhaustive information on their website about childhood vaccines, including current immunization schedules, breaking news, revised recommendations, vaccine ingredients, and adverse effects. Detailed information slated toward health care providers can be found at: http://www.cdc.gov/vaccines/hcp.htm.

National Vaccine Information Center (703-938-0342, http://www.nvic.org/)—Nonprofit educational vaccine safety watchdog group that is dedicated to preventing vaccine-induced injuries. Funds research into vaccine safety, lobbies the government to better identify which children may be prone to an adverse vaccine reaction, and provides exhaustive and up-to-date information on vaccine ingredients, current guidelines, legal issues, and informed consent.

The Vaccine Adverse Event Reporting System VAERS (http://vaers.hhs.gov/index)—National vaccine safety surveillance program cosponsored by the Centers for Disease Control and Prevention (CDC) and the Food and Drug Administration (FDA). Parents can access reports submitted to VAERS (http://vaers.hhs.gov/data/index) and report vaccine reactions online, by fax, or by mail (http://vaers.hhs.gov/esub/index).

RECOMMENDED READING

Suzanne Arms, *Immaculate Deception II: Myth, Magic, & Birth*

Naomi Baumslag, M.D., and Dia L. Michels, *Milk, Money, and Madness: The Culture and Politics of Breastfeeding*

Grantly Dick-Read, *Childbirth Without Fear: The Principles and Practices of Natural Childbirth*

Ina May Gaskin, *Ina May's Guide to Childbirth*

David Gollaher, *Circumcision: A History of the World's Most Controversial Surgery*

Christine Gross-Loh, *The Diaper-Free Baby: The Natural Toilet Training Alternative*

Lise Eliot, Ph.D., *What's Going on in There?: How the Brain and Mind Develop in the First Five Years of Life*

Susan Markel, M.D., *What Your Pediatrician Doesn't Know Can Hurt Your Child*

Robert S. Mendelsohn, M.D., *How to Raise a Healthy Child . . . In Spite of Your Doctor*

Gabrielle Palmer, *The Politics of Breastfeeding: Why Breasts Are Bad for Business*

Robert W. Sears, M.D., *The Vaccine Book: Making the Right Decision for Your Child*

Alecia Swasy, *Soap Opera: The Inside Story of Procter & Gamble*

Marsden Wagner, M.D., *Born in the USA: How a Broken Maternity System Must Be Fixed to Put Women and Children First*

Diane Wiessinger et al., *The Womanly Art of Breastfeeding*

NORWEGIAN CHILDHOOD VACCINE SCHEDULE, BIRTH TO AGE 5

No routine vaccines given before 3 months of life*
At 3 months, 5 months, and 12 months:
> Diphtheria
>
> Tetanus
>
> Acellular pertussis
>
> Hib
>
> Polio
>
> Pneumococcal conjugate

At 15 months:
> Measles, mumps, and rubella vaccine

*Hepatitis B given to at-risk groups only

AMERICAN CHILDHOOD VACCINE SCHEDULE, BIRTH TO AGE 5

At birth, 1–2 months, 6–18 months:
> Hepatitis B

At 2, 4, 6, 15–18 months:
> Diphtheria
>
> Tetanus
>
> Acellular pertussis

At 12–23 months:
> Hepatitis A

At 2, 4, 6, 12–15 months:
> Hib
>
> Pneumococcal conjugate

At 6–23 months and every year thereafter:
> Influenza

At 2, 4, 6–18 months, 4–6 years:
> Polio

At 2, 4, 6 months (or 2, 4 months depending on vaccine):
> Rotavirus

At 12–15 months, 4–6 years:
> Measles
>
> Mumps
>
> Rubella

At 12–18 months, 4–6 years:
> Varicella (chickenpox)

Source: WHO vaccine-preventable diseases: monitoring system 2012 global summary

Photo Credits

Illustration 1: Newborn baby born via C-section, photo courtesy of Keren Fenton

Illustration 2: Belly shot, photo by Jennifer Margulis

Illustration 3: Young Woman at Bus Stop, photo by Jennifer Margulis

Illustration 4: Chicago's Lying-In Hospital, photo courtesy of the University of Chicago Medical Center

Illustration 5: Woman Prepped for a C-section, photo by Ginny Adkins

Illustration 6: Baby Oliver Just After Birth, photo by Jennifer Margulis

Illustration 7: Screen shot of Circumstraint advertisement, photo from http://www.quickmedical.com/olympicmedical/circumstraint/immobolizer.html

Illustration 8: Sign, "Go Ahead & Breastfeed, We Like Both Babies and Boobs!" photo by Jennifer Margulis

Illustration 9: T. Berry Brazelton in Pampers Advertisement, photo by Jennifer Margulis

Illustration 10: Infant with a Band-Aid, photo by Jennifer Margulis

Illustration 11: Mom, Baby, and Pediatrician at a Well-Baby Visit, photo by Jennifer Margulis

Illustration 12: Anna's First Birthday, Korean Style, photo courtesy of Christine Gross-Loh

Illustration 13: Three Children Walking Away, photo by Jennifer Margulis

Notes

Introduction

x *They had been married for almost two years:* Maria McCullough, "Teachers Joined in Birth, Death," *Philadelphia Inquirer,* May 10, 2007.

x *dying as a result of childbirth is five times greater:* Save the Children, *Nutrition in the First 1,000 Days: State of the World's Mothers 2012* (Save the Children: Westport, Conn.: 2012), 53, accessed at http://www.savethechildren.org/atf/cf/{9def2ebe-10ae-432c-9bd0-df91d2eba74a}/STATEOFTHEWORLDSMOTHERSREPORT2012.PDF.

x *Eight American children per 1,000:* Ibid.

x *A child in the United States is more than twice:* Ibid., 53, 55.

xi *contaminated infant formula:* Alan Scher Zagier, "More Retailers Pull Formula After Baby's Death," Children's Health, MSNBC.com, December 23, 2011, accessed at http://www.msnbc.msn.com/id/45762595/ns/health-childrens_health/t/more-retailers-pull-formula-after-babys-death/#.T1Um7phR4mE. Though the CDC found evidence of cronobacter in an opened container of infant formula, and prepared infant formula samples provided by the Missouri Department of Health and Senior Services, according to the *Chicago Sun-Times,* the CDC and the FDA were unable to find evidence of the bacteria in sealed infant formula with the same lot number.

xi *complications due to an out-of-hospital circumcision:* Thomas Zambito, "Baby Dies in Circumcision," *New York Daily News,* March 3, 2012, http://www.nydailynews.com/new-york/infant-death-maimonides-hospital-linked-circumcision-article-1.1032432. Though the circumcision took place out of the hospital, the baby was later treated at Maimonides.

xi *severe reaction to the birth dose of the hepatitis B shot:* Accessed at http://www.ageofautism.com/2009/02/managing-editors-note-below-is-the-story-of-ian-gromowski-a-boy-who-lived-47-days-after-his-hepatitis-b-vaccination-thank.html.

xi *safer to be born into forty-eight countries:* CIA, *World Factbook,* "Country Comparison: Infant Mortality Rate," 2012 estimated numbers. Available on line at https://www.cia.gov/library/publications/the-world-factbook/rankorder/2091rank.html. According to the CIA: "This entry gives the number of deaths of infants under one year old in a given year per 1,000 live births in the same year; included is the total death rate, and deaths by sex, male and female. This rate is often used as an indicator of the level of health in a country."

xi *4.3 million babies born in America:* CIA *World Fact Book* estimated 2011 statistics (based on a population of 313,847,465 and a live birth rate of 13.68 births/1,000 population), accessed at https://www.cia.gov/library/publications/the-world-factbook/geos/us.html.

xi *more than 25,000 will die in their first year:* Calculating infant mortality rate of 6 per 1,000 multiplied by 4.3 million births.

xi *maternal mortality rate in the United States:* Accessed at http://www.guardian.co.uk/news/datablog/2010/apr/12/maternal-mortality-rates-millennium-development-goals.

xi *via C-section in April 2007:* McCullough, "Teachers Joined in Birth, Death."

xi *most of these deaths go unnoted:* Amnesty International, *Deadly Delivery: The Maternal Health Care Crisis in the USA* (London: Amnesty International Secretariat, 2010), accessed at http://www.amnestyusa.org/sites/default/files/pdfs/deadlydelivery.pdf.

xi *evidence that autopsy rates in hospitals are declining:* Donna L. Hoyert, Ph.D., "The Changing Profile of Autopsied Deaths in the United States, 1972–2007," NCHS Data Brief, no. 67 (August 2011): 2–3, accessed at http://www.cdc.gov/nchs/data/databriefs/db67.htm. Pregnancy-related deaths are more often autopsied than deaths from disease. Only about 5 percent of *all* hospital deaths are autopsied. Habiba Nosheen, "Fewer Autopsies Mean Crucial Information Goes to the Grave," *All Things Considered,* February 5, 2012, accessed at http://www.npr.org/2012/02/05/146355717/fewer-autopsies-mean-crucial-info-goes-to-the-grave.

xi *a review of the death is almost always conducted:* Marsden Wagner, M.D., M.S., *Born in the USA: How a Broken Maternity System Must Be Fixed to Put Women and Children First* (Berkeley: University of California Press, 2006), 23.

xii *Only twenty-four states require:* Ina May Gaskin, C.P.M., M.A., "Maternal Death in the United States: A Problem Solved or a Problem Ignored?" *Journal of Perinatal Education* 17, no. 2 (Spring 2008): 9–13, accessed at http://www.ncbi.nlm.nih.gov/pmc/articles/PMC2409165/.

xii *including Ohio:* According to the State of Ohio's Hospital Compare website: "In November 2006, House Bill 197 was passed requiring Ohio to form a council (the Hospital Measures Advisory Council) appointed by the House and Senate to make recommendations to the Director of Health on hospital performance measures and a publically available website. The Hospital Measures Advisory Council was created pursuant to Ohio Revised Code section 3727.31 and each member of the Council appointed a representative to the Data Expert Group pursuant to Ohio Revised Code section 3727.32. The Infection Control Group was appointed by the Advisory Council and is a group of health care consumers, nurses, and experts in infection control convened to provide information about infection issues to the Council as needed for the Council to perform its duties. The Council also created two other specialty groups: Pediatric Workgroup and Perinatal Workgroup. Both group's memberships are in the area of their respective expertise and provided guidance to the Hospital Measures Advisory Council in recommending pediatric and perinatal measures." "FAQs," Ohio Hospital Compare, Ohio Department of Health, http://www.ohiohospitalcompare.ohio.gov/documents/FAQs.pdf.

xii *New York:* New York State's Maternity Information Act (MIA) was enacted into law in 1989. The MIA requires hospitals to make public via a brochure their annual rates of "cesarean sections, primary and repeat; women with previous cesarean sections who have had a subsequent successful vaginal birth; deliveries in birthing rooms; deliveries by certified nurse-midwives; fetal monitoring listed on the basis of auscultation, external and internal; births utilizing forceps, listed on the basis of low forceps and mid forceps delivery; births utilizing breech vaginal delivery; vaginal births utilizing analgesia; vaginal births utilizing anesthesia including general, spinals, epidural, and paracervical; births utilizing induction of labor; births utilizing augmentation of labor; births utilizing episiotomies; and mothers breast feeding upon discharge." (Alliance for the Improvement of Maternity Services [AIMS], "Legislation Affecting Maternity Care," 2000, http://www.aimsusa.org/laws.htm.) However, the Office of the New York City Public Advocate has consistently found that most New York State hospitals are not in compliance with this law. See Betsy Gotbaum, *Giving Birth in the Dark: City Hospitals Still Failing to Provide Legally Mandated Maternity Information* (New York: Office of the New York City Public Advocate, 2006), accessed at http://publicadvocategotbaum.com/policy/documents/GivingBirthInTheDark12.06.pdf.

xii *California:* Accessed at http://escholarship.org/uc/item/20h9h32c#page-1.

xii *"no maternal mortality review process at all":* Amnesty International, *Deadly Delivery: The Maternal Health Care Crisis in the USA.*

xii *Her husband told the BBC:* "Maternal Mortality Across the World: US," BBC News, October 26, 2009, Jim Scythes speaks after wife's death in childbirth, accessed at http://news.bbc.co.uk/2/hi/americas/8325685.stm.

xii *"an act of God":* Rita Rubin, "Answers Prove Elusive as C-Section Rate Rises," *USA Today,*

January 8, 2008, accessed at http://www.usatoday.com/news/health/2008-01-07-csections_N. htm. David Birnbach, an obstetrical anesthesia specialist who directs the Center for Patient Safety at the University of Miami is quoted in the article: "Unfortunately, it can be due to misadventure or an error, but more often than not, it's an act of God."

xii *older women having first babies:* Rob Stein, "Number of Twins Soar as Older Moms Turn to Fertility Treatments," NPR, January 4, 2012.

xii *increase in twin and premature births:* Denise Grady, "Premature Births Are Fueling Higher Rates of Infant Mortality in the U.S., Report Says," *New York Times*, November 3, 2009, accessed at http://www.nytimes.com/2009/11/04/health/04infant.html?_r=1.

xii *obese when they get pregnant:* Anemona Hartocollis, "Growing Obesity Increases Perils of Childbearing," *New York Times*, June 5, 2010, accessed at http://www.nytimes. com/2010/06/06/health/06obese.html.

xii *making labor more dangerous:* U.S. Department of Agriculture and U.S. Department of Health and Human Services, *Dietary Guidelines for Americans 2010*, 7th ed. (Washington, D.C.: U.S. Government Printing Office, December 2010), 10, accessed at http://health.gov/ dietaryguidelines/dga2010/DietaryGuidelines2010.pdf. See also: Kathleen M. Rasmussen and Ann L. Yaktine, eds., Institute of Medicine and National Research Council, *Weight Gain During Pregnancy: Reexamining the Guidelines* (Washington, D.C.: National Academies Press, 2009), 55–56.

xiii *"the least evidence-based discipline":* Stefan Topolski, M.D., in an interview with the author, June 21, 2012.

xiv *ultrasound scans have not been shown:* J. P. Newnham, "Effects of Frequent Ultrasound During Pregnancy: A Randomised Controlled Trial," *Lancet* 342, no. 8876 (October 9, 1993): 887–891, accessed at http://www.ncbi.nlm.nih.gov/pubmed/8105165.

xiv *"routine ultrasound during pregnancy":* Marsden Wagner, "Ultrasound: More Harm Than Good?" *Midwifery Today,* no. 50 (Summer 1999), accessed at http://www.midwiferytoday. com/articles/ultrasoundwagner.asp.

xv *birth dose of hepatitis B:* See American Academy of Pediatrics, "Joint Statement of the American Academy of Pediatrics (AAP) and the United States Public Health Services (USPHS)," *Pediatrics* 104, no. 3 (September 1999): 568–569. See also Department of Health and Human Services, Centers for Disease Control and Prevention, "Notice to Readers: Thimerosal in Vaccines: A Joint Statement of the American Academy of Pediatrics and the Public Health Service," *Morbidity and Mortality Weekly Report* 48, no. 26 (July 9,1999): 563– 565.

Chapter One: Gestation Matters: The Problem with Prenatal Care

2 *When she and her husband, Matt:* Matthew Logelin, *Two Kisses for Maddy: A Memoir of Loss and Love* (New York: Grand Central Publishing, 2011), 16–17.

2 *Greggory DeVore, M.D.:* Ibid., 19. According to Matt, Liz's husband, DeVore is known to many as "Dr. Doom" for his pessimistic way of focusing on worst-case scenarios.

2 *the cord was wrapped around her neck:* Ibid., 20.

2 *Terrified first-time parents:* Ibid., 22.

2 *At 3 pounds 13.5 ounces:* Ibid., 45.

3 *Liz most likely died:* Ibid., 67.

3 *one study showed that embolism:* G. J. Kovacevich et al., "The Prevalence of Thromboembolic Events Among Women with Extended Bed Rest Prescribed as Part of the Treatment for Premature Labor or Preterm Premature Rupture of Membranes," *American Journal of Obstetrics & Gynecology* 182, no. 5 (May 2000): 1089–1092.

3 *says one mother of four from Pennsylvanis:* Mom of four from Pennsylvania, in interview with the author, September 14, 2012.

3 *"moderately accurate"*: E. F. Magann et al., "The Accuracy of Ultrasound Evaluation of Amniotic Fluid Volume in Singleton Pregnancies: The Effect of Operator Experience and Ultrasound Interpretive Technique," *Journal of Clinical Ultrasound* 25, no. 5 (June 1997): 249–253.

3 *more pregnancy interventions:* A. Alchalabi et al., "Induction of Labor and Perinatal Outcome: The Impact of Amniotic Fluid Index," *European Journal of Obstetrics, Gynecology, and Reproductive Biology* 129, no. 2 (December 2006): 124–127.

3 *leads to more pregnancy interventions:* L. Leeman and D. Almond, "Isolated Oligohydramnios at Term: Is Induction Indicated?" *Journal of Family Practice* 54, no. 1 (January 2005): 25–32.

3 *"lead to overdiagnosis of problems"*: Sarah J. Buckley, M.D., *Gentle Birth, Gentle Mothering* (Berkeley, Calif.: Celestial Arts, 2009), 67. Beth Israel Deaconess Medical Center warns its patients that "ultrasound scans can . . . result in overdiagnosis and cause unnecessary worry to expectant parents." "Ultrasound Screening," Beth Israel Deaconess Medical Center, 2012, accessed at http://www.bidmc.org/CentersandDepartments/Departments/ObstetricsGynecology/PatientEducationResources/YourPregnancy/UltrasoundScreening.aspx.

3 *"Routine use [of anmiotic fluid index]"*: J. M. Morris et al., "The Usefulness of Ultrasound Assessment of Amniotic Fluid in Predicting Adverse Outcome in Prolonged Pregnancy: A Prospective Blinded Observational Study," *BJOG: An International Journal of Obstetrics & Gynaecology* 110, no. 11 (November 2003): 989–994. Accessed at http://onlinelibrary.wiley.com/doi/10.1111/j.1471-0528.2003.02417.x/full.

3 *"lead to increased obstetric intervention"*: Another study by researchers from the Department of Obstetrics and Gynecology, Naval Medical Center-Portsmouth in Portsmouth, Virginia, drew a similar conclusion: One popular method used to determine amniotic fluid levels, "excessively characterizes a greater number of pregnancies as having oligohydramnios [a deficiency of amniotic fluid] leading to more interventions without improvement in perinatal outcome." E. F. Magann, "The Evidence for Abandoning the Amniotic Fluid Index in Favor of the Single Deepest Pocket," *American Journal of Perinatology* 24, no. 9 (October 2007): 549–555, accessed at http://www.ncbi.nlm.nih.gov/pubmed/17909990.

3 *not enough available evidence:* The most recent systematic review of the available literature on bed rest found that the available evidence did not support or refute the idea that bed rest can help prevent preterm labor. C. Sosa et al., "Bed Rest in Singleton Pregnancies for Preventing Preterm Birth," *Cochrane Database of Systematic Reviews* 1 (2004). For a short discussion of the other studies that have found no benefit to bed rest or not enough evidence to make a determination either way, see F. G. Cunningham et al., *Williams Obstetrics,* 23rd ed. (New York: McGraw Hill, 2010), 823.

3 *dramatically increases the likelihood of getting blood clots:* G. J. Kovacevich et al., "The Prevalence of Thromboembolic Events Among Women with Extended Bed Rest Prescribed as Part of the Treatment for Premature Labor or Preterm Premature Rupture of Membranes," *American Journal of Obstetric Gynecology* 182, no. 5 (May 2000): 1089–1092.

3 *significant bone loss in pregnant women:* J. H. Promislow et al., "Bed Rest and Other Determinants of Bone Loss During Pregnancy," *American Journal of Obstetrics and Gynecology* 191, no. 4 (October 2004): 1077–1083.

4 *Reynir Tómas Geirsson, M.D.:* Icelanders use patronymics rather than family surnames and go by their first names. See "Author's Note."

4 *"But enforced strict bed rest has never been proven of use"*: Dr. Reynir Tómas Geirsson, M.D., chair, Department of Obstetrics and Gynecology at the Landspítali University Hospital, Reykjavik, Iceland, in an email communication with the author, August 4, 2011.

4 *"The twenty-first century"*: This and subsequent quotations: Michael Klaper, M.D., in an interview with the author, March 20, 2012.

4 *"Between the false negatives and the false positives"*: Anonymous father, in an interview with the author, March 14, 2012.

6 *"increase the anxiety"*: Brian Price, M.D., associate director, Harvard Vanguard Medical Associates, in discussion with the author, December 15, 2010.

6 *about 7 percent of pregnancies become complicated*: American Diabetes Association, "Gestational Diabetes Mellitus: Definition, Detection, and Diagnosis," *Diabetes Care*, last revised 2000, accessed at http://care.diabetesjournals.org/content/26/suppl_1/s103.full.

6 *It is usually a mild condition*: Buckley, *Gentle Birth, Gentle Mothering*, 46.

6 *when paired with artificial food additives*: D. McCann, A. Barrett, A. Cooper, et al., "Food Additives and Hyperactive Behaviour in 3-Year-Old and 8/9-Year-Old Children in the Community: A Randomised, Double-Blinded, Placebo-Controlled Trial," *Lancet* 370, no. 9598 (2007): 1560–1567.

6 *a weak acid with a tart taste*: "More About Glucose Tolerance Test," Biofile Diagnostics, last updated February 27, 2012, accessed at http://www.biofilediagnostics.com/glucose-tolerance-test/read-more.

6 *she vomited it up in the waiting room*: Angela Decker, parent, in an interview with the author, March 16, 2012.

7 *what glucose response is elevated enough*: Buckley, *Gentle Birth, Gentle Mothering*, 47.

7 *every woman be screened for gestational diabetes*: "Screening and Diagnosis of Gestational Diabetes Mellitus," Committee Opinion No. 504. American College of Obstetricians and Gynecologists, *Obstetrics & Gynecology* 118, no. 3 (September 2011): 751–753, accessed at http://www.acog.org/Resources_And_Publications/Committee_Opinions/Committee_on_Obstetric_Practice/Screening_and_Diagnosis_of_Gestational_Diabetes_Mellitus.

7 *"evidence is insufficient"*: U.S. Department of Health & Human Services, "Developing and Promoting the Use of Evidence" in *Agency for Healthcare Research and Quality Annual Highlights 2008* (Rockville, Md.: AHRQ, 2008), accessed at http://www.ahrq.gov/about/highlt08b.htm; and *Screening for Gestational Diabetes Mellitus*, Topic Page, U.S. Preventive Services Task Force, May 2008, accessed at http://www.uspreventiveservicestaskforce.org/uspstf/uspsgdm.htm.

7 *"low-risk status requires no glucose testing"*: American Diabetes Association, "Gestational Diabetes Mellitus: Definition, Detection, and Diagnosis," *Diabetes Care* 26, no. supp. 1 (January 2003): s103–s105, accessed at http://care.diabetesjournals.org/content/26/suppl_1/s103.full.

7 *"Baby wasn't ready to come out"*: Kristen Boyle, parent, in an interview with the author, March 16, 2012.

8 *how to care for their health during pregnancy*: American College of Nurse Midwives, "Our Moment of Truth Survey," August 2012. Accessed at http://ourmomentoftruth.midwife.org/ACNM/files/ccLibraryFiles/Filename/000000002595/ACNM%20Our%20Moment%20of%20Truth%20Survey%20Findings%20Overview%2009%2022_MG.pdf.

9 *"What do you eat?"*: This and subsequent quotations: Michael Klaper, M.D., in an interview with the author, March 20, 2012.

10 *She lay as still as she could on the couch*: Jenna Nichols, parent, in an interview with the author, March 7, 2012.

10 *every time she took her conventional prenatal vitamin*: Sarah Jane Nelson Millan, parent, in an email communication with the author, March 6, 2012.

10 *she threw up twenty minutes later*: Katherine Womack, parent, in an interview with the author, March 7, 2012. After trying several brands, Katherine found that a prenatal vitamin made from whole foods did not give her mouth sores. Katherine and her doctor realized together that the high iron content was making her sick, and he prescribed an expensive prenatal without the iron, which cleared up the problem.

10 *painful constipation*: "Pregnancy Week by Week: Prenatal Vitamins, Why They Matter,

How to Choose," Mayo Clinic, April 2012, accessed at http://www.mayoclinic.com/health/prenatal-vitamins/PR00160/NSECTIONGROUP=2.

10 *horrible stomach pains:* "Problems with Prenatal Vitamins," *Berkeley Parents Network*, page updated, December 7, 2004, accessed on March 17, 2012, at http://parents.berkeley.edu/advice/pregnancy/vitamins.html.

10 *dizziness:* Dizziness and nausea were two side effects I experienced myself. Internet chat and advice reveals that I am not alone, see http://community.babycenter.com/post/a24874817/are_you_dizzy_beware_of_one_a_day_prenatal_vitamins; and http://www.livestrong.com/article/412222-is-it-normal-to-be-dizzy-when-taking-prenatal-vitamins/

11 *a large study of pregnant women and infants in China:* Robert J. Berry, Adolfo Correa, et al., "Prevention of Neural-Tube Defects with Folic Acid in China," *New England Journal of Medicine* 341, no. 20 (November 11, 1999): 1485–1490, accessed at http://www.nejm.org/doi/full/10.1056/NEJM199911113412001.

11 *"nutritional insurance policy":* "Prenatal Vitamins: A Nutritional Insurance Policy," BabyCenter, last updated July 2010, accessed at http://www.babycenter.com/0_prenatal-vitamins-a-nutritional-insurance-policy_287.bc.

11 *some doctors now believe:* Lester Voutsos, M.D., section chief of obstetrics, Providence Hospital, Novi, Michigan, in an interview with the author, March 7, 2012.

11 *Sundown "Naturals" contain:* Many vitamin ingredient lists can be found at Drugstore.com; "Sundown Naturals Prenatal, Vitamin & Mineral Formula, Tablets 100 ea," accessed at http://www.drugstore.com/sundown-naturals-prenatal-vitamin-and-mineral-formula-tablets/qxp311300.

12 *Titanium dioxide:* Thomas C. Long and Bellina Veronesi, "Nanosize Titanium Dioxide Stimulates Reactive Oxygen Species in Brain Microglia and Damages Neurons *in Vitro*," *Environmental Health Perspectives* 115, no. 11 (November 2007): 1631–1637, accessed at http://dx.doi.org/10.1289/ehp.10216.

12 *cell injury, mutation:* World Health Organization, International Agency for Research on Cancer, "Carbon Black, Titanium Dioxide, and Talc," *IARC Monographs on the Evaluation of Carcinogenic Risks to Humans* 93 (Lyon, France: International Agency for Research on Cancer, 2010), accessed at http://monographs.iarc.fr/ENG/Monographs/vol93/mono93.pdf.

12 *respiratory tract cancer in rodent experiments:* Eun-Jung Park et al., "Induction of Chronic Inflammation in Mice Treated with Titanium Dioxide Nanoparticles by Intratracheal Instillation," *Toxicology* 260, nos. 1–3 (June 16, 2009): 37–46, accessed at http://dx.doi.org/10.1016/j.tox.2009.03.005.

12 *carcinogenic in some forms to humans:* Eric Y. T. Chen et al., "Mucin Secretion Induced by Titanium Dioxide Nanoparticles," *PLoS ONE* 6, no. 1 (January 19, 2011): accessed at http://www.plosone.org/article/info%3Adoi%2F10.1371%2Fjournal.pone.0016198.

12 *harm marine animals:* Robert J. Miller et al., "TiO2 Nanoparticles Are Phototoxic to Marine Phytoplankton," *PLoS ONE* 7, no. 1 (January 20, 2012): e30321, accessed at http://doi:10.1371/journal.pone.0030321; M. Madhupratap et al., "Toxicity of Effluent from a Titanium Dioxide Factory on Some Marine Animals," reprinted from *Indian Journal of Marine Sciences* 8, no. 1 (March 1979): 41–42, accessed at http://tinyurl.com/7uu6gps.

12 *linked to autoimmune disorders:* Vera D. M. Stejskal, "Human Hapten-Specific Lymphocytes: Biomarkers of Allergy in Man," *Drug Information Journal* 31 (1997): 1379–1382, accessed at http://www.melisa.org/pdf/dij063.pdf.

12 *Red 40 and Yellow 6 have also been found:* Sarah Kobylewski and Michael F. Jacobson, *Food Dyes: A Rainbow of Risks* (Washington, D.C.: Center for Science in the Public Interest, June 2010), accessed at http://cspinet.org/new/pdf/food-dyes-rainbow-of-risks.pdf.

12 *Stuart Prenatal Mutlivitamin/Multimineral supplement tablets:* "Stuart Prenatal Multivitamin/Multimineral Supplement, Tablets 100 ea," Drugstore.com, accessed on March 17, 2012, at http://www.drugstore.com/products/prod.asp?pid=58617&catid=183042.

12 four times *as much as Sundown Naturals:* Sundown Naturals cost $6.90 for 100 tablets if you buy them from Drugstore.com: "Sundown Naturals Prenatal, Vitamin & Mineral Formula, Tablets 100 ea," accessed at http://www.drugstore.com/sundown-naturals-prenatal-vitamin-and-mineral-formula-tablets/qxp311300.

12 *sodium aluminosilicate:* "Material Safety Data Sheet: Sodium Aluminosilicate MSDS," ScienceLab.com, last updated November 1, 2010 at 12:00 p.m., http://www.sciencelab.com/msds.php?msdsId=9924957.

12 *cornstarch and sugar, to name just a few:* Another ingredient, Carnauba (*Copernicia Cerifera*) Wax, which is also used in automobile wax and shoe polish, is made by bleaching the naturally occurring wax found on a plant native to South America, so it is not necessarily an unnatural product.

12 *gummy vitamin was contaminated with high amounts of lead:* "Consumers Warned of Pitfalls with Some Multivitamins and Vitamin Waters: Testing by ConsumerLab.com Uncovers Problems with Many Brands," news release, ConsumerLab.com, May 21, 2004, accessed at http://www.consumerlab.com/news/Mutivitamin_Vitamin_Waters_Tests_Supplements/5_21_2004/.

12 *did not contain the amount of nutrients listed:* This June 15, 2011, report did not find problems with three prenatal brands, but did find gross inaccuracies in vitamins for children, and that the price of the vitamins had no relationship to the accuracy of the labeling. (Linda Carroll, "Many Multivitamins Don't Have Nutrients Claimed in Label," Diet and Nutrition on MSNBC.com, updated June 20, 2011, accessed at http://www.msnbc.msn.com/id/43429680/ns/health-diet_and_nutrition/t/many-multivitamins-dont-have-nutrients-claimed-label/#.T1_Bf5hR4mE.) When ConsumerLab.com compared Rite Aid Prenatal Tablets with Folic Acid, which costs 4 cents per day, they found it provided the same vitamin and minerals as Stuart Prenatal, which cost 30 cents per day, over seven times as much. ConsumerLab.com, Product Review, "Multivitamin and Multimineral Supplements Review," June 28, 2011, accessed at https://www.consumerlab.com/reviews/review_multivitamin_compare/multivitamins/ (available to members only).

13 *The manufacturer does not even have to:* "Food: Overview of Dietary Supplements," FDA, last updated October 14, 2009, accessed at http://www.fda.gov/food/dietarysupplements/consumerinformation/ucm110417.htm.

13 *"I had her switch each day":* Tod Cooperman, M.D., president, ConsumerLab.com, in an interview with the author, March 14, 2012.

13 *$26.7 billion on supplements in 2009:* "What's Behind Our Dietary Supplements Coverage," ConsumerReports.org, last updated January 2011, accessed at http://www.consumerreports.org/health/natural-health/dietary-supplements-coverage/overview/index.htm.

13 *30 cents a pill:* http://children.costhelper.com/prenatal-vitamins.html.

14 *"These include Vitamine E":* As quoted by Jennifer Margulis in "Wheat? Whole Wheat? What?" *Pregnancy Magazine,* April 2004, 86.

14 *"Unlike vitamines and minerals, phytochemicals":* Ibid.

15 *"They have more fiber":* Ibid.

15 *A 2012 meta-analysis of available research on white rice:* E. A. Hu et al., "White Rice Consumption and Risk of Type 2 Diabetes: Meta-analysis and Systematic Review," *British Medical Journal* 344, no. 7851 (April 7, 2012): e1454, accessed at http://www.bmj.com/content/344/bmj.e1454.

15 *"They do a high-volume practice":* This and subsequent quotations: Paul Qualtere-Burcher, M.D., obstetrician, in an interview with the author, March 8, 2012.

16 *Relative Value Units:* Relative Value Units (RVUs) are a way physicians groups and hospitals calculate compensation for staff by using a formula tied to various physician services.

16 *At Qualtere-Burcher's last job:* He was employed by PeaceHealth Medical Group, a nonprofit Catholic community health organization that owns eight hospitals and forty-two clinics in Alaska, Washington, and Oregon.

17 *"They're looking for the billable opportunity"*: Edward Linn, M.D., in an interview with the author, August 18, 2011. A follow-up interview was conducted on March 18, 2012.

18 *"When the outcomes aren't great you need to change the system"*: Sharon Rising, founder and CEO, Centering Healthcare Institute, in an interview with the author, March 18, 2012.

18 *"Dreger, a pants-wearing omnivore"*: Alice Dreger, "The Most Scientific Birth Is Often the Least Technological Birth," *Atlantic*, March 20, 2012, accessed at http://atmo4.theatlantic. com/health/archive/2012/03/the-most-scientific-birth-is-often-the-least-technological-birth/254420/.

18 *"was committed to being much more modern"*: Ibid.

19 *"The medical model of obstetrics is reactive"*: Stuart Fischbein, M.D., obstetrician, in an interview with the author, November 15, 2011.

19 *"If they have a patient who gets into trouble"*: This and subsequent quotations: Paul Qualtere-Burcher, M.D., obstetrician, in an interview with the author, March 8, 2012.

20 *"We have never allowed 'free' pharmaceutical samples"*: Brian Price, M.D., obstetrician, email communication with the author, September 26, 2012.

21 *American College of Obstetricians and Gynecologists gross receipts:* The exact number was 80,522,676.

21 *Average salary of a high-risk obstetrician:* 2011 MGMA National Survey Data.

21 *Average salary of a hospital midwife:* MGMA Physician Compensation and Production Survey: 2012 Report Based on 2011 Data.

21 *Total costs of prenatal visits with a doctor:* The actual number is $3,942.49. The total cost of prenatal visits with a doctor varies widely, depending on the practice, location, and level of care. This number is based on an average of thirteen prenatal appointments (women have typically from eleven to fifteen) multiplied by $180 per visit, plus two ultrasound scans, one at less than fourteen weeks ($842.08), and one five-month anatomy scan ($761.41), which is typical for southern Oregon, where I live.

21 *Total costs of prenatal visits with a homebirth midwife:* Homebirth midwives in our area charge between $50 to $150 per prenatal visit and usually see clients on a schedule similar to a doctor's (thirteen visits multiplied by $100 per visit equals $1,300). The cost of ultrasound scans, which are not always part of homebirth care in low-risk pregnancies in our area, would be extra.

21 *Cost per minute to have pregnancy supervised by a doctor:* Calculated based on a doctor spending twenty minutes on average with a pregnant patient.

21 *Cost per minute to have pregnancy supervised by a homebirth midwife:* Homebirth midwives spend an average of one hour with their patients.

21 *Nine-month supply of brand-name prenatals:* One bottle of Trimedisyn 800 mg, which is a one-month supply, costs $129.95, though it is offered at $69.95 as a trial price. Accessed on September 26, 2012, http://www.trimedisyn.com/.

21 *Nine-month supply of generic prenatals:* One bottle of CVS women's prenatal with DHA, which is a one-month supply, costs $14.99 (though if you buy it from the web, it was discounted to $11.24) on September 26, 2012. Accessed at http://www.cvs.com/shop/product-detail/CVS-Womens-Prenatal--DHA-Vitamins-&-Minerals?skuId=460461.

21 *Jennifer Penick:* As told to the author on September 14, 2012.

Chapter 2: Sonic Boom: The Downside of Ultrasound

25 *couldn't prescribe pain medication:* Karen Bridges, parent, in an interview with the author, April 9, 2012.

25 *First used for obstetrics by a Scottish doctor:* Historians differ on the exact date sonograms were introduced. F. G. Cunningham et al., *Williams Obstetrics*, 23rd ed. (New York: McGraw Hill, 2010), 349, gives 1958 as the date. Margaret B. McNay and John E. E. Fleming, "Forty

Years of Obstetric Ultrasound 1957–1997: From A-scope to Three Dimensions," *Ultrasound in Medicine & Biology* 25, no. 1 (1999): 50, says they were introduced in 1957.

25 *ultrasounds had become a routine part:* By the mid-1960s, obstetric ultrasound was being used in many hospitals and doctors had begun buying scanning equipment for private practices. For an extended discussion of this, see McNay and Fleming, "Forty Years of Obstetric Ultrasound 1957–1997: From A-scope to Three Dimensions," *Ultrasound in Medicine & Biology* 25, no. 1 (1999): 3–56.

25 *67 percent of pregnant women:* National Center for Health Statistics, 2002. As quoted in F. G. Cunningham et al., *Williams Obstetrics*, 22nd ed. (New York: McGraw Hill, 2005), 390.

25 *three ultrasounds per woman:* What Mothers Say: The Canadian Maternity Experiences Survey. Ottawa: Public Health Agency of Canada, 2009. Available online at http://www.phac-aspc.gc.ca/rhs-ssg/pdf/survey-eng.pdf, 13.

25 *high-risk pregnancies:* According to current American obstetrical practices, a high-risk pregnancy includes women carrying multiples and any mother over age thirty-five.

25 *twenty-five ultrasounds per pregnacy:* In response to the question "How Many Ultrasounds Will You Have While Pregnant?" at The Stir (blog), one mom wrote that because she was carrying twins, was considered high risk, and was punched in the stomach by her ex (which caused her to miscarry one of the twins), she had a total of twenty-five ultrasounds in the thirty-seven weeks she gestated (see http://thestir.cafemom.com/pregnancy/1686/How_Many_Ultrasounds_Will_You). While that was an unusual situation, discussions on pregnancy chat groups reveal that many women expect between four and eight ultrasounds per pregnancy.

26 *"We recommend an eighteen-week ultrasound":* Stephanie Koontz, M.D., obstetrician, in discussion with the author, December 15, 2010.

27 *"the skill of the technician reading the scan":* Felicia Cohen, M.D., obstetrician, in an interview with the author, August 24, 2011. When I checked this quote with Dr. Cohen for accuracy, she asked me to add the following: "Ultrasound technology has great value as a diagnostic tool, especially earlier in pregnancy, when it can detect a lot of potential complications that a physical exam alone would miss. And even late in pregnancy, it can help us decide whether an elective induction or Cesarean section is indicated, especially for complicated obstetrical patients. But as a tool for estimating fetal weight in a full-term patient? We know it's not especially accurate for that, and I counsel my patients that the actual fetal weight could be a pound or so lower or higher" (email communication with the author, September 25, 2012).

27 *gender identification before fourteen weeks:* B. J. Whitlow et al., "First trimester diagnosis of gender," *Ultrasound in Obstetrics and Gynecology* 13 (1999): 301–304.

28 *"Ultrasound can't promise us a healthy baby":* This and subsequent quotations: Colleen Forbes, midwife, in an interview with the author, August 12, 2011.

28 *"My husband and I liked the tests":* Rachelle Eisenstat, parent, in an interview with the author, November 3, 2011.

28 *this stress can have a lasting effect:* E. J. H. Mulder et al., "Prenatal Maternal Stress: Effects on Pregnancy and the (Unborn) Child," *Early Human Development* 70 (December 2002): 3–14.

29 *"I think it's a psychological lie for women":* Louana George, midwife, in an interview with the author, October 26, 2011.

29 *carries a risk of miscarriage:* J. W. Seeds, "Diagnostic Mid Trimester Amniocentesis: How Safe?" *American Journal of Obstetrics and Gynecology* 191, no. 2 (August 2004): 607–615.

29 *carries a miscarriage rate of between 1 in 100:* Mayo Clinic Staff, "Down Syndrome: Tests and Diagnosis," April 7, 2011, accessed at http://www.mayoclinic.com/health/down-syndrome/DS00182/DSECTION=tests-and-diagnosis.

29 *and 3 in 100:* According to the CDC, the risk is from between 1 in 100 to 1 in 200; see: CDC, "Chorionic Villus Sampling and Amniocentesis: Recommendations for Prenatal Counseling," in *Morbidity and Mortality Weekly Report* (*MMWR*) 44, no. R-99 (July 21, 1995): 1–12. A more recent analysis of the risk of miscarriage from CVS conducted at just one clinic

found the risk of miscarriage over a twenty-year period to be 3.12 percent overall; see: A. B. Caughey, M.D., Ph.D., et al., "Chorionic Villus Sampling Compared with Amniocentesis and the Difference in the Rate of Pregnancy Loss," *Obstetrics and Gynecology* 108, no. 3, part 1 (September 2006): 612–616.

29 *small risk of uterine infection:* G. G. Rhoads et al., "The Safety and Efficacy of Chorionic Villus Sampling for Early Prenatal Diagnosis of Cytogenetic Abnormalities," *New England Journal of Medicine* 320 (1989): 609–617.

29 *having a baby with a limb missing:* F. J. Hsieh et al., "Limb Defects After Chorionic Villus Sampling," *Obstetrics and Gynecology* 85, no. 1 (January 1995): 84–88.

29 *"I feel like all the testing":* Stephanie La Croix Hinkaty, parent, in an interview with the author, November 7, 2011.

30 *a false positive rate of 5 percent:* Cunningham, *Williams Obstetrics,* 23rd ed., 293.

30 *"It depends on how you're reimbursed":* Edward Linn, chair of obstetrics and gynecology, Chicago Cook County Health and Hospitals System, in an interview with the author, August 18, 2011.

31 *"[T]his practice-based trial demonstrates":* Bernard G. Ewigman and the RADIUS Study Group et al., "Effect of Prenatal Ultrasound Screening on Perinatal Outcome," *New England Journal of Medicine* 329, no. 12 (September 16, 1993): 821–827, accessed at http://www.nejm.org/doi/full/10.1056/NEJM199309163291201#t=abstract.

31 *more likely to experience intrauterine growth restriction:* J. P. Newnham, "Effects of Frequent Ultrasound During Pregnancy: A Randomised Controlled Trial," *The Lancet* 342, no. 8876 (October 9, 1993): 887–891, accessed at http://www.ncbi.nlm.nih.gov/pubmed/8105165.

31 *Ironically, intrauterine growth restriction is:* For a more extended discussion of this study, see Marsden Wagner, "Ultrasound: More Harm Than Good?" *Midwifery Today,* no. 50 (Summer 1999), accessed at http://www.midwiferytoday.com/articles/ultrasoundwagner.asp. When the lead author of the 1993 *Lancet* study followed up to test the children's intelligence at eight years of age, he and his team did not find evidence of long-term neurological damage. However, "Reassurances provided by our results do not lessen our need to undertake further studies of potential bio-effects of prenatal ultrasound scans," the authors write. ". . . In view of the widespread and liberal use of this technology we are responsible for ensuring the safety of its use. Uncertainty remains about several potential issues . . ." See John P. Newnham et al., "Effects of Repeated Prenatal Ultrasound Examinations on Childhood Outcome Up to 8 Years of Age: Follow-up of a Randomised Controlled Trial," *Lancet* 364 (December 2004): 2038–2044, http://www.slredultrasound.com/Filesandpictures/Risk3.pdf.

31 *did not reveal lasting neurological damage:* Newnham, "Effects of Repeated Prenatal Ultrasound Examinations on Childhood Outcome Up to 8 Years of Age: Follow-up of a Randomised Controlled Trial."

31 *experience long-term developmental delays:* Y. Leitner et al., "Six-Year Follow-up of Children with Intrauterine Growth Retardation: Long-Term, Prospective Study," *Journal of Child Neurology* 15, no. 12 (December 2000): 781–786.

31 *prenatal exposure to ultrasound waves changed:* Pasko Rakic et al., "Prenatal Exposure to Ultrasound Waves Impacts Neuronal Migration in Mice," *PNAS* 103, no. 34 (August 2006): 12903–12910.

32 *"We should be using the same care with ultrasound as with X-rays":* "Ultrasound Effects on Fetal Brains Questioned," *RSNA News* 16, no. 11 (November 2006): 8, accessed at http://www.rsna.org/uploadedFiles/RSNA/Content/News/nov2006.pdf.

32 *Parkinson's disease:* Parkinson's disease occurs when the nerve cells in the brain that make dopamine, which is used to control muscle movement, are destroyed. Without dopamine, the nerve cells in the substantia nigra can't send messages properly, leading to abnormal motor (tremors, rigidity) and nonmotor (mood, sleep disturbances) features, which worsen over time. See Stanley Fahn and Serge Przedborski, "Parkinson Disease," in *Merritt's Neurology,*

12th ed., edited by Lewis P. Rowland and Timothy A. Pedley (Philadelphia: Lippincott, Williams & Wilkins, 2010), 751–769.

32 *Alzheimer's:* According to the Alzheimer's Foundation of America, "Alzheimer's disease is a progressive, degenerative disorder that attacks the brain's nerve cells, or neurons, resulting in loss of memory, thinking and language skills, and behavioral changes. These neurons, which produce the brain chemical, or neurotransmitter, acetylcholine, break connections with other nerve cells and ultimately die. For example, short-term memory fails when Alzheimer's disease first destroys nerve cells in the hippocampus, and language skills and judgment decline when neurons die in the cerebral cortex."

    M. F. Casanova et al., "Clinicopathological Correlates of Behavioral and Psychological Symptoms of Dementia," *Acta Neuropathologica* 122, no. 2 (August 2011): 117–135, accessed at http://www.ncbi.nlm.nih.gov/pubmed/21455688.

32 *all the neurons in the line:* These findings are described in Mountcastle's two seminal papers: V. B. Mountcastle et al., "Response Properties of Neurons of Cat's Somatic Sensory Cortex to Peripheral Stimuli," *Journal of Neurophysiology* 20, no. 4 (July 1957): 374–407; and V. B. Mountcastle, "Modality and Topographic Properties of Single Neurons of Cat's Somatic Sensory Cortex," *Journal of Neurophysiology* 20, no. 4 (July 1957): 408–434.

32 *"minicolumns":* "Mini" because they are microscopic (they span a tiny amount of tissue too small to see with the naked eye, 25–60 microns) and "columns" because the neurons seemed stacked upon each other.

32 *higher cognitive functions:* M. F. Casanova and C. Tillquist, "Encephalization, Emergent Properties, and Psychiatry: A Minicolumnar Perspective," *The Neuroscientist* 14, no. 1 (February 2008): 101–118.

33 *abnormal in the brains of autistic children:* Peter Mundy and Courtney Burnette, "Joint Attention and Neurodevelopmental Models of Autism," *Handbook of Autism and Pervasive Developmental Disorders,* 3rd ed., edited by Fred R. Volkmar et al. (Hoboken, N.J.: John Wiley and Sons, 2005), 650–681.

33 *brains of autistic patients:* M. F. Casanova, "Minicolumnar Pathology in Autism," *Neurology* 58, no. 3 (February 12, 2002): 428–432, accessed at http://www.neurology.org/content/58/3/428.

33 *"You know that a shower curtain":* Manuel Casanova, M.D., neuroscientist, in an interview with the author, October 27, 2011.

33 *known to deform cell membranes:* In the ear, a sound wave makes the tympanic membrane vibrate, which activates mechanisms to allow you to hear. Ultrasound waves work the same way. Casanova says that the energy of sound can put pressure on, and even penetrate, the cell membrane. The cell membrane itself is a liquid formed of fats and therefore easier to penetrate than a solid. When sound puts pressure on the water surrounding the cell, that water can do two things depending on the force of the sound: (1) it can form gas bubbles from the water, which subsequently spin and implode, thereby disrupting the cell membranes of nearby cells; and (2) the water can place mechanical pressure on the cell membrane itself. Both the implosion of bubbles (otherwise known as cavitation) and the force of the water pressure can disrupt the cell membrane, making transient holes. This brief break in the boundaries of the cell can let molecules both in and out, which acutely alter how the cell behaves and which have the potential to alter its behavior long term as well.

33 *use of ultrasound to treat bone fractures:* California Department of Health Care Services, "Osteogenesis Stimulator Devices to Accelerate the Healing of Selected Bone Fractures," *Criteria Manual,* chap. 13.2, R-19-99E, accessed at http://www.dhcs.ca.gov/services/medical/Documents/ManCriteria_35_OstStimDev.htm.

33 *increases cell division:* N. Doan et al., "In Vitro Effects of Therapeutic Ultrasound on Cell Proliferation, Protein Synthesis, and Cytokine Production by Human Fibroblasts, Osteoblasts, and Monocytes," *Journal of Oral and Maxillofacial Surgery* 57, no. 4 (April 1999): 409–419.

33 *Prolonged or inappropriate ultrasound exposure:* E. L. Williams and M. F. Casanova, "Potential

Teratogenic Effects of Ultrasound on Corticogenesis: Implications for Autism," *Medical Hypotheses* 75, no. 1 (July 2010): 53–58.

34 *High-risk women who receive multiple ultrasound:* Pregnancies deemed high-risk due to diabetes, hypertension, and obesity have all shown to be at higher risk of autism: see http://www.webmd.com/baby/news/20110511/diabetes-hypertension-obesity-linked-to-autism. Women of advanced maternal age—also considered high-risk—have a greater tendency to have autistic children: see Janie F. Shelton, "Independent and Dependent Contributions of Advanced Maternal and Paternal Ages to Autism Risk," *Autism Research* 3, no. 1 (February 2010): 30–39, accessed at http://onlinelibrary.wiley.com/doi/10.1002/aur.116/abstract. And women carrying multiples, who receive an average of six scans per pregnancy, are also at higher risk: see C. Betancur et al., "Increased Rate of Twins Among Affected Sibling Pairs with Autism," *American Journal of Human Genetics* 70, no. 5 (May 2002): 1381–1383, accessed at http://www.ncbi.nlm.nih.gov/pmc/articles/PMC447617/?tool=pubmed.

34 *Autism is much more common among educated:* Maureen S. Durkin et al., "Socioeconomic Inequality in the Prevalence of Autism Spectrum Disorder: Evidence from a U.S. Cross-Sectional Study," *PLoS ONE* 5, no. 7 (2010): accessed at http://www.plosone.org/article/info:doi/10.1371/journal.pone.0011551.

Tia Ghose, "Autism in Kids More Prevalent Among Wealthier Parents, Study Finds," *Journal Sentinel*, July 23, 2010, accessed at http://www.jsonline.com/blogs/news/99128024.html.

L. A. Croen, "Descriptive Epidemiology of Autism in a California Population: Who Is at Risk?" *Journal of Autism and Developmental Disorders* 32, no. 3 (June 2002): 217–224.

Simon Baron-Cohen, "Does Autism Occur More Often in Families of Physicists, Engineers, and Mathematicians?" *Autism* 2, no. 3 (September 1998): 296–301.

34 *In Somalia, autism is virtually unheard of:* Minnesota Department of Health, "Autism and the Somali Community—Report of the Study Fact Sheet," 2008, accessed at http://www.health.state.mn.us/ommh/projects/autism/reportfs090331.cfm.

David Kirby, "Minneapolis and the Somali Autism Riddle," November 14, 2008, accessed at http://www.huffingtonpost.com/david-kirby/minneapolis-and-the-somal_b_143967.html.

M. Barnevik-Olsson, "Prevalence of Autism in Children Born to Somali Parents Living in Sweden: A Brief Report," *Developmental Medicine and Child Neurology* 50, no. 8 (August 2008): 598–601.

34 *Amish, are at lower risk for autism:* Dan Olmsted, "The Age of Autism: The Amish Anomaly," United Press International, April–May 2005, accessed at http://www.putchildrenfirst.org/media/e.4.pdf.

34 *the entire population of Finland:* See Population Register Centre at http://vrk.fi/default.aspx?site=4.

34 *diagnosed with attention disorders:* CDC, "Attention-Deficit/Hyperactivity Disorder: Data and Statistics in the United States," accessed at http://www.cdc.gov/ncbddd/adhd/data.html.

34 *one in every eighty-eight children in America:* CDC, "New Data on Autism Spectrum Disorders," accessed at http://www.cdc.gov/Features/CountingAutism/.

34 *Norway:* J. Isaksen et al., "Observed Prevalence of Autism Spectrum Disorders in Two Norwegian Counties," *European Journal of Paediatric Neurology* (February 18, 2012).

34 *industrialized nations that are seeing:* F. E. Yazbak, "Autism Seems to Be Increasing Worldwide, if Not in London," *British Medical Journal* 328, no. 7433 (January 24, 2004): 226–227.

C. M. Zaroff and S. Y. Uhm, "Prevalence of Autism Spectrum Disorders and Influence of Country Measurement and Ethnicity," *Social Psychiatry and Psychiatric Epidemiology* 47, no. 3 (March 2012): 395–398.

34 *and Japan:* Kishi R et al., "Japanese Women's Experiences from Pregnancy Through Early Postpartum Period," *Health Care for Women International* 32, no. 1 (2011): 57–71.

35 *No single genetic or environmental factor:* Caroline Rodgers, "Questions About Prenatal Ultrasound and the Alarming Increase in Autism," *Midwifery Today*, no. 80 (Winter 2006): accessed at http://www.midwiferytoday.com/articles/ultrasoundrodgers.asp.

35 *But all these countries do have one thing in common:* Ibid.

35 *"I have spent most of my working life in medical research":* Varyanna, "Autism Links to Ultrasound and Other Obstetrical Procedures," Banned from Baby Showers (blog), April 30, 2009, accessed at http://banned-from-baby-showers.blogspot.com/2009/04/autism-links-to-ultrasound-and-other.html.

35 *sound waves with eight times the intensity:* "In revising its regulations in 1993, the FDA altered its approach to ultrasound safety. The new regulations combine an overall limit of I-SPTA of 720 mW/cm 2 for all equipment with a system of output displays to allow users to employ effective and judicious levels of ultrasound appropriate to the examination undertaken. The new regulations allow an eightfold increase in ultrasound intensity to be used in fetal examinations." Quote from Colin Deane, "Safety of Diagnostic Ultrasound in Fetal Scanning," *Doppler in Obstetrics,* updated 2002, accessed at http://www.centrus.com.br/DiplomaFMF/SeriesFMF/doppler/capitulos-html/chapter_02.htm.

35 *the majority of technicians using ultrasound machines:* E. Sheiner, I. Shoham-Vardi, and J. S. Abramowicz, "What Do Clinical Users Know Regarding Safety of Ultrasound During Pregnancy?" *Journal of Ultrasound in Medicine* 26, no. 3 (March 2007): 319–325.

36 *"[A]ntenatal work without":* L. N. Reece, "The Estimation of Foetal Maturity by a New Method of X-ray Cephalometry: Its Bearing on Clinical Midwifery," *Proceedings of the Royal Society of Medicine,* January 18, 1935, accessed at http://www.ncbi.nlm.nih.gov/pmc/articles/PMC2205881/pdf/procrsmed00681-0021.pdf.

36 *early obstetrical textbooks denied that exposure:* Marsden Wagner, "Ultrasound: More Harm Than Good?" *Midwifery Today,* no. 50 (Summer 1999): accessed at http://www.midwiferytoday.com/articles/ultrasoundwagner.asp.

36 *miscarriage, mental retardation, and birth defects:* Cunningham, *Williams Obstetrics,* 23rd ed., 915.

36 *leukemia and other kinds of childhood cancer:* John D. Boice Jr. and Robert W. Miller, "Childhood and Adult Cancer After Intrauterine Exposure to Ionizing Radiation," *Teratology* 59, no. 4 (April 1999): 227–233. See also T. Sorahan et al., "Childhood Cancer and Paternal Exposure to Ionizing Radiation: A Second Report from the Oxford Survey of Childhood Cancers," *American Journal of Industrial Medicine* 28, no. 1 (July 1995): 71–78.

37 *"We were super excited when":* This and subsequent quotations: Lisa Nguyen, parent, in an interview with the author, April 9, 2012.

37 *"We never limit the number of guests you can bring":* Before the Stork 4D, http://www.beforethestork4d.com/.

37 *Fetal Fotos has branches:* Fetal Fotos, http://www.fetalfotosusa.com/location.aspx?i=34.

38 *"the person performing the scan may not be adequately trained":* Sheiner, Shoham-Vardi, Abramowicz, "What Do Clinical Users Know Regarding Safety of Ultrasound During Pregnancy?"

38 *the fetus is often being exposed to sound waves:* U.S. Food and Drug Administration, "Avoid Fetal 'Keepsake' Images, Heartbeat Monitors," Consumer Update, March 24, 2008, accessed on September 28, 2011, at http://www.fda.gov/ForConsumers/ConsumerUpdates/ucm095508.htm.

38 *The long-term effects of tissue heating:* Ibid.

38 *"The baby looked like it was in pain":* Danielle Driscoll, parent, in an interview with the author, October 26, 2011.

39 *ultrasound has not been proven to be effective:* ACOG Practice Bulletin Number 101, "Ultrasonography in Pregnancy," *Obstetrics & Gynecology* 113, no. 2, part 1 (February 2009): 451–461.

39 *"Sonography should be performed":* Cunningham, *Williams Obstetrics,* 23rd ed., 349.

39 *In 2004 when Lia Joy Rundle:* Lia Joy Rundle, parent, in an interview with the author, June 4, 2012.

39 *"After the fact I was so upset"*: Wendy Scharp, parent, in an interview with the author, October 31, 2011.

41 *"Every other person I know"*: Jennifer Cario, parent, in an interview with the author, June 5, 2012.

42 *Salary of a radiologist:* MGMA Physician Compensation and Production Survey: 2012 Report Based on 2011 Data.

42 *Voluson 730 Pro GE ultrasound system:* As quoted by a salesman, email communication, September 28, 2012, accessed at http://kpiultrasound.com/Ultrasound-Systems/Voluson-730-Pro/flypage_images.tpl.html.

42 *Prenatal office visit without ultrasound:* The cost of an ultrasound varies widely, depending on the provider and the geographic region. These are prices quoted from a provider in the Pacific Northwest and thought to be fairly standard.

42 *Prenatal visit with first-trimester ultrasound:* These prices were advertised at My Sunshine Baby in Charlotte, North Carolina, on September 26, 2012. Accessed at http://mysunshinebaby.com/packages.htm.

42 *Cost of 4D ultrasound scan with video clip:* These prices were advertised at My Sunshine Baby in Charlotte, North Carolina, on September 26, 2012. Accessed at http://mysunshinebaby.com/packages.htm.

43 *Louana George:* As told to the author on October 26, 2011.

44 *actively dividing cells:* Stanley B. Barnett et al., "The Sensitivity of Biological Tissue to Ultrasound," *Ultrasound in Medicine and Biology* 23, no. 6 (1997): 805–812.

Chapter 3 Emerging Expenses: The Real Cost of Childbirth

46 *For British journalist Molly Castle:* Cedric Belfrage, *They All Hold Swords*, a memoir, quoted in Jessica Mitford, *The American Way of Birth* (New York: Dutton, 1992), 54–55.

46 *from as early as 1906:* Christopher Cumo, *Science and Technology in 20th-Century American Life* (Westport, Conn.: Greenwood Publishing, 2007), 49.

46 *until the 1960s:* Judith Walzer Leavitt, "Birthing and Anesthesia: The Debate Over Twilight Sleep," *Signs: Journal of Women in Culture and Society* (University of Chicago) 6, no. 1 (1980): 163.

46 *American women were promised that Twilight Sleep:* "Twilight Sleep Is the Subject of a New Investigation: American Woman, After Study of Thousands of Cases, Reports Favorably," *New York Times*, January 31, 1915.

46 *"one of the most dangerous of all poisons":* Ibid.

46 *made other patients "wild" instead of calm:* Ibid.

46 *Women in Twilight Sleep would become out of control:* Judith Walzer Leavitt, "Birthing and Anesthesia: The Debate Over Twilight Sleep," 161.

47 *leather straps to secure women to the bed:* G. D. Schultz, "Cruelty in the Maternity Wards," *Ladies' Home Journal*, May 1958.

47 *Some used loops of lamb's wool:* Michelle Harrison, M.D., *A Woman in Residence: A Dedicated Doctor's Personal Story of Her Struggle with the Fierce Challenges of a Major American Hospital* (New York: Fawcett, 1993).

47 *scopolamine functioned as an amnesiac:* Asbury Somerville, M.D., "Hyoscine (Scopolamine) Amnesia in Labour," *Canadian Medical Association Journal* 24, no. 6 (June 1931): 818–820, accessed at http://www.ncbi.nlm.nih.gov/pmc/articles/PMC382502/?page=1.

47 *like Boehringer & Son:* James Clifton Edgar, *The Practice of Obstetrics: Designed for the Use of Students and Practitioners of Medicine* (Philadelphia: P. Blakiston's Son & Co., 1916), 838.

47 *made an "educational film":* Jacqueline H. Wolf, *Deliver Me from Pain: Anesthesia and Birth in America* (Baltimore, Md.: Johns Hopkins University Press, 2009), 56.

47 *synthesizing it themselves:* In their 1915 book, Dr. Bertha Van Hoosen and Elisabeth Ross Shaw exhort: "To the chemist of the future we must look not for the commercial scopolamine, but for the special preparation that shall excel for purity and strength." Dr. Bertha Van Hoosen

and Elisabeth Ross Shaw, *Scopolamine-Morphine Anaesthesia* (Chicago: House of Manz, 1915), 28.

47 *leased land for private sanatoria:* One such sanitorium was located on Riverside Drive in New York City. See Thomas Lathrop Stedman, ed., "News of the Week: Object to 'Twilight Sleep Home,' " *Medical Record* 89, no. 2 (January 8, 1916): 70.

47 *"I see almost every day comments on this":* Dr. R. L. Thomas quoted in W. N. Mundy, M.D., "Twilight Sleep," *Eclectic Medical Journal* 75 (January–December 1915): 422.

47 *The experience was so flawless:* Wolf, *Deliver Me from Pain,* 55.

48 *as late as 1974:* A certified nurse midwife practicing in Atlanta saw Twilight Sleep drugs being administered to every patient when she was doing her training in 1974. As she described to investigative journalist Jessica Mitford, "I can recall we would have hordes of laboring women—the doctors would knock them out . . . with scopolamine, an amnesia drug, heavy-duty narcotics, and sedatives. The women would be thrashing about in bed and yelling—but totally unaware of any of this. You had to put the rails up to keep them safe . . . those women were left alone in there for hours. They were drugged up and knocked out. And the babies were often born unconscious themselves. You'd have to give them drugs to reverse the narcotics the mother had, and they'd stay sleepy for days." Mitford, *The American Way of Birth,* 56–57.

48 *By the beginning of the twentieth century:* Judith Walzer Leavitt, *Brought to Bed: Childbearing in America, 1750–1950* (New York: Oxford University Press, 1988), 61.

48 *In 1929 the first Indiana limestone:* John Easton, senior science writer, University of Chicago Medicine, interview with the author, August 18, 2011.

48 *build the University of Chicago's Lying-In Hospital:* Lying-in is synonymous with childbirth, but it was first used to define a postpartum woman, who was supposed to lie in (that is, rest and not go out) after she gave birth. Jan Nusche, "Lying in," *Canadian Medical Association Journal* 167, no. 6 (September 17, 2002), accessed at http://www.cmaj.ca/content/167/6/675.full.

48 *"In the early days of Lying-In":* John Easton, senior science writer, University of Chicago Medicine, interview with the author, August 18, 2011.

48 *"destructive," "pathogenic," and "pathologic":* Joseph DeLee, "Progress Towards Ideal Obstetrics," *American Journal of Obstetrics and Diseases of Women and Children* 73, no. 1 (January 1916): 407–415.

49 *"Mother Nature's methods":* J. B. DeLee, Mother's Day Address, May 12, 1940. J. B. DeLee, M.D., *Papers,* Northwestern Memorial Hospital Archives, quoted in Wolf, *Deliver Me from Pain,* 242.

49 *"My mother-in-law was":* Mary Fauls, doula, CenteringPregnancy director and obstetric patient liaison, John H. Stroger Jr. Hospital, Chicago, interview with the author, August 18, 2011.

49 *Other women report being hit in the face:* Schultz, "Cruelty in the Maternity Wards."

50 *The United States spends more money:* Organisation for Economic Co-operation and Development, Frequently Requested Data, "Health Expenditure: Total Expenditure on Health, % Gross Domestic Product," *OECD Health Data 2011* (November 2011), accessed at http://www.oecd.org/document/16/0,3746,en_2649_33929_2085200_1_1_1_1,00.html.

50 *over $98 billion in 2008:* Lauren M. Wier and Roxanne M. Andrews, "The National Hospital Bill: The Most Expensive Conditions by Payer, 2008," Healthcare Cost and Utilization Project, Statistical Brief no. 107, March 2011, accessed at http://www.hcup-us.ahrq.gov/reports/statbriefs/sb107.pdf.

50 *a "crisis":* Amnesty International, Deadly Delivery: The Maternal Health Care Crisis in the USA (London: Amnesty International Secretariat, 2010), accessed at http://www.amnesty-usa.org/sites/default/files/pdfs/deadlydelivery.pdf.

50 *greater lifetime risk of dying of pregnancy-related complications:* Ibid., 1.

50 *the maternal mortality rate in the United States:* World Health Organization, UNICEF, UNFPA, and The World Bank, *Trends in Maternal Mortality: 1990 to 2008: Estimates Developed by WHO, UNICEF, UNFPA, and The World Bank* (Geneva, Switzerland: World Health Organization, 2010), accessed at http://www.who.int/reproductivehealth/publications/monitoring/9789241500265/en/index.html.

50 *More than two women die every day:* Amnesty International, *Deadly Delivery*, 1.

50 *Diane Rizk McCabe:* Paul Grondah, "Joyful Day Turns to Grief as Mother Bleeds to Death After Birth," *Houston Chronicle*, July 30, 2009, accessed at http://www.chron.com/news/article/Joyful-day-turns-to-grief-as-mother-bleeds-to-1730203.php#page-1.

51 *Karen Vasques:* Allison Goldsberry, "Medford Woman Dies During C-Section Birth," InsideMedford.com, October 22, 2008, accessed at http://insidemedford.com/2008/10/22/medford-woman-dies-during-c-section/.

51 *Jennifer Tait:* Obituary of Jennifer Tait, *Grand Rapids Press*, March 23, 2011, accessed at http://obits.mlive.com/obituaries/grandrapids/obituary.aspx?n=jennifer-tait&pid=149573315.

51 *Candice Boyle:* "West Branch Mourns Death of Teacher," *Vindicator*, February 26, 2011, accessed at http://www.vindy.com/news/2011/feb/26/west-branch-mourns-death-of-teacher/.

51 *may actually be higher:* Amnesty International, *Deadly Delivery*, 87.

51 *35 deaths per 100,000 births:* Stacy Fine, "Interview with Ina May Gaskin," *Ecomall*, August 2006, accessed at http://www.ecomall.com/greenshopping/inamay.htm.

51 *maternal death rate for women in New York City:* Pregnancy-Associated Mortality, New York City 2001–2005, accessed at http://www.nyc.gov/html/doh/downloads/pdf/ms/ms-report-online.pdf.

51 *national goal:* Anemona Hartocollis, "High Rate for Deaths of Pregnant Women in New York State," *New York Times*, June 18, 2010, accessed at http://www.nytimes.com/2010/06/19/nyregion/19obese.html.

51 *more than* five *times higher:* Pat MacEnulty, "Oh Baby: Ina May Gaskin on the Medicalization of Birth," *Sun*, no. 433 (January 2012), accessed at http://www.thesunmagazine.org/issues/433/oh_baby.

51 *"near misses":* Amnesty International, *Deadly Delivery*, 1.

51 *Abbie Dorn:* Maria L. La Ganga, "Severely Disabled, Is She Still a Mom? Battle Nears Over Visitation Rights of a Woman Injured in Childbirth," *Los Angeles Times*, April 11, 2010, accessed at http://articles.latimes.com/2010/apr/11/local/la-me-abbie11-2010apr11.

52 *could not walk unassisted for weeks:* Nicole Dennis, parent, in an interview with the author, January 17, 2012.

52 *near misses increased by 25 percent:* Amnesty International, *Deadly Delivery*, 1.

52 *34,000 women every year:* E. Kuklina et al., "Severe Obstetric Morbidity in the United States, 1998–2005," *Obstetrics and Gynecology* 113 no. 2, part 1 (February 2009): 293–299.

53 *"What position did I birth my first in?":* Karen, December 7, 2010, comment on "Latin American Countries Campaign for 'Vertical Birth,'" Mother's Advocate (blog), December 6, 2010, accessed at http://mothersadvocate.wordpress.com/2010/12/06/latin-american-countries-campaign-for-vertical-birth/.

54 *Since monitoring became routine:* "Intrapartum Fetal Heart Rate Monitoring," *ACOG Practice Bulletin: Clinical Management Guidelines for Obstetrician–Gynecologists*, no. 70 (December 2005): 7, accessed at http://rumcobgyn.org/ACOGFHR.pdf.

54 *The false positive rate:* Ibid., 3.

         Z. Alfirevic et al., "Continuous Cardiotocography (CTG) as a Form of Electronic Fetal Monitoring (EFM) for Fetal Assessment During Labour," *Cochrane Library*, no. 3 (July 19, 2006). A further review, published by *Cochrane Library* (an independent international non-profit partnership that assesses scientific data to promote evidence-led health practices), of 12 trials involving more than 37,000 women that compared studies of electronic fetal monitoring with intermittent listening to the baby's heartbeat confirmed ACOG's earlier findings. The reviewers discovered that though continuous monitoring was found to be associated with a reduction in already rare neonatal seizures, electronic fetal monitoring made no difference in the number of babies who died during or shortly after birth, or in the incidences of cerebral palsy. But fetal monitoring did make a significant difference in *how* a baby was born: continuous monitoring was associated with a significant increase in Caesarean section and instrumental vaginal births.

55 *continuous support during labor:* E. D. Hodnett et al., *Continuous Support for Women During Childbirth* (Hoboken, N.J.: John Wiley & Sons, 2007), accessed at http://apps.who.int/rhl/reviews/langs/CD003766.pdf.

55 *"We often see people":* Stuart Fischbein, M.D., obstetrician, in an interview with the author, November 15, 2011.

56 *"She made it seem like I was an idiot":* Kristy Boone, mother, in an interview with the author, January 27, 2012.

56 *"Labor and delivery nurses all over the country":* " 'Pit to Distress': A Disturbing Reality," Nursing Birth (blog), July 8, 2009, accessed at http://nursingbirth.com/2009/07/08/"pit-to-distress"-a-disturbing-reality/. In the interest of readability, I fixed the syntax and grammar on this quote without marking the places I had changed. Parts of the story left out are indicated by ellipses.

57 *Too much Pitocin:* The danger of Pitocin, even in normal doses, is spelled out in detail in the "Precaution" section of the package inserts: "Maternal deaths due to hypertensive episodes, subarachnoid hemorrhage, rupture of the uterus, and fetal deaths due to various causes have been reported associated with the use of parenteral oxytocic drugs for induction of labor or for augmentation in the first and second stages of labor." Overdosage: "Overdosage with oxytocin depends essentially on uterine hyperactivity whether or not due to hypersensitivity to this agent. Hyperstimulation with strong (hypertonic) or prolonged (tetanic) contractions, or a resting tone of 15 to 20 mm H_2O or more between contractions can lead to tumultuous labor, uterine rupture, cervical and vaginal lacerations, postpartum hemorrhage, uteroplacental hypoperfusion, and variable deceleration of fetal heart, fetal hypoxia, hypercapnia, perinatal hepatic necrosis or death. Water intoxication with convulsions, which is caused by the inherent antidiuretic effect of oxytocin, is a serious complication that may occur if large doses (40 to 50 milliunits/minute) are infused for long periods. Management consists of immediate discontinuation of oxytocin and symptomatic and supportive therapy." Accessed at http://www.jhppharma.com/products/PI/07112011/Pitocin-42023-116-02-Package-Insert-2011.pdf JHP Pharmaceuticals, 2011.

57 *"Even if the oxytocin order":* Michelle L. Murray and Gayle M. Huelsmann, *Labor and Delivery Nursing: A Guide to Evidence-Based Practice* (New York: Springer, 2008), 182.

58 *investor-owned, for-profit institutions:* American Hospital Association, "Fast Facts on U.S. Hospitals," January 3, 2012, accessed at http://www.aha.org/aha/resource-center/Statistics-and-Studies/fast-facts.html.

58 *They bill private and public insurance:* AHA Resource Center anonymous source, in an interview with the author, April 10, 2012.

58 *Only 213 hospitals:* American Hospital Association, "Fast Facts on U.S. Hospitals."

58 *These hospitals' operating costs:* AHA Resource Center anonymous source, in an interview with the author, April 10, 2012.

58 *"The rushed atmosphere":* Marsha Walker, registered nurse and women's health advocate, in an interview with the author, May 24, 2011.

58 *"In and out":* Richard Anderson, health care financial consultant, in an interview with the author, February 5, 2012.

59 *Women delivering at for-profit hospitals:* Nathanael Johnson, "For-profit hospitals performing more C-sections," California Watch, September 11, 2010, accessed at http://californiawatch.org/health-and-welfare/profit-hospitals-performing-more-c-sections-4069.

59 *$36,625:* Anna Wilde Mathews, "Tallying the Cost to Bring Baby Home," *Wall Street Journal,* May 7, 2009, accessed at http://online.wsj.com/article/SB124165279035493687.html.

59 *On her itemized bill:* Ibid.

59 *$530.29 just for:* Ibid.

60 *and may even pay:* Dr. Maggie Kozel, pediatrician, in an interview with the author, April 6, 2012. See also Alex Lickerman, M.D., "A Proposal to Contain Health Care Costs: Combating Health Care Overutilization with the Careful Placement of Incentives," *Psychology Today,* January 30, 2011.

60 *"crate rate":* "Crate rate" was the term used by Edward Linn, M.D., chair of obstetrics and gynecology, Chicago's Cook County Health and Hospitals System, in an interview with the author, August 18, 2011.

60 *you have to deliver as many women as you can:* Ibid.

60 *C-section rate was too low:* Anonymous obstetrician, in a interview with the author, October 15, 2012.

61 *"the most important person":* MacEnulty, "Oh Baby: Ina May Gaskin on the Medicalization of Birth," 6.

61 *teaching and advocating safe childbirth:* The Right Livelihood Award, Ina May Gaskin, 2011, accessed at http://www.rightlivelihood.org/inamay_gaskin.html.

61 *have safely delivered their babies at the Farm:* MacEnulty, "Oh Baby," 6.

61 *C-section rate of 1.7 percent:* Ibid.

61 *never lost a mother:* Ina May Gaskin, *Ina May Gaskin's Guide to Childbirth* (New York: Bantam Books, 2003), 322.

62 *"Those who are used to":* MacEnulty, "Oh Baby," 6.

62 *Early that evening, after Sara:* Sara Schley, parent, in an interview with the author, February 6, 2012. See also Sara Schley, "Anatomy of a Miracle," Mothering.com, accessed at http://mothering.com/pregnancy-birth/anatomy-of-a-miracle.

64 *unwavering support of her loved ones:* Kristen was a certified nurse midwife who had attended more than a thousand births, Lynnie was a doula who had been to seventy-five, and Alisa was a women's empowerment leader who had also witnessed more than fifty births.

64 *restriction of food and drink:* M. Singata, J. Tranmer, G. M. Gyte, "Restricting Oral Fluid and Food Intake During Labor," accessed at http://www.ncbi.nlm.nih.gov/pubmed/20091553.

65 *"You're taught the model":* Stuart Fischbein, M.D., in an interview with the author, November 15, 2011.

65 *"Birth is an inherently dangerous process":* Kurt Wiese, obstetrician, in an interview with the author, April 2, 2012.

65 *"Childbirth is not safe":* Mary Elizabeth Soper, obstetrician, in an interview with the author, April 13, 2012.

67 *"The resident came back to see me":* Laura Swaminathan, parent, in an interview with the author, January 19, 2012.

67 *Norway:* The estimated infant mortality rate in Norway in 2012 is 3.5 deaths per 1,000 live births. In America it is 5.98 deaths per 1,000 live births for the same year. In comparison, in Niger, a destitute country in West Africa where record keeping is not always accurate, the infant death rate in 2012 is estimated at 109.98 per 1,000 live births. (See CIA, *World Factbook,* "Country Comparison: Infant Mortality Rate," accessed at https://www.cia.gov/library/publications/the-world-factbook/rankorder/2091rank.html?countryName=Norway&countryCode=no&regionCode=eur&rank=209#no.)

67 more than three times *as likely:* The maternal death rate in Norway in 2008 was 7 per 100,000 live births. In the United States it was 24 per 100,000 live births in the same year. (See CIA, *World Factbook,* "Country Comparison: Maternal Mortality Rate," https://www.cia.gov/library/publications/the-world-factbook/rankorder/2223rank.html?countryName=Norway&countryCode=no&regionCode=eur&rank=159#no.)

67 *attended by medically trained midwives:* Marit Heiberg, president, Norwegian Association of Midwives, in an interview with the author, September 15, 2011.

67 *one big difference between America and Norway:* Anne Flem Jacobsen, M.D., head of obstetrics, Ullevål University Hospital, in an interview with the author, September 14, 2011.

67 *one third of births take place at home:* Philip Steer, "How Safe Is Home Birth?" *BJOG: An International Journal of Obstetrics & Gynaecology* 115, no. 5 (April 2008): i–ii. A follow-up study in the Netherlands concluded that homebirth is as safe as hospital birth and other factors account for the Dutch perimortality rates: A. de Jonge, B. Y. van der Goes, A. C. J. Ravelli, et al., "Perinatal Mortality and Morbidity in a Nationwide Cohort of 529,688 Low-Risk Planned Home and Hospital Births," *BJOG: An International Journal of Obstetrics and Gynaecology* 116, no. 9 (August 2009). Published online April 15, 2009, accessed at http://onlinelibrary.wiley.com/doi/10.1111/j.1471-0528.2009.02175.x/full.

67 *highly trained midwives, who work in collaboration:* Marsden Wagner, M.D., *Born in the USA: How a Broken Maternity System Must Be Fixed to Put Women and Children First* (Berkeley: University of California Press, 2006), 243.

67 almost five times *more likely:* The maternal death rate in Iceland in 2008 was 5 per 100,000 live births, according to the CIA. In the United States it was 24 per 100,000 live births in the same year. (See CIA, *World Factbook,* "Country Comparison: Maternal Mortality Rate.") However, according to the meticulous record keeping done by the Icelandic government, there were no maternal deaths in Iceland in 2008. (Directorate of Health Annual Report, 2008, accessed at http://www.landspitali.is/lisalib/getfile.aspx?itemid=24179 in chapter xvii, page 25.) Regardless of which statistics are correct, it is much safer to give birth in Iceland than in America, and this has been the case for at least the past twenty years.

68 *a doctor's salary is set by the state:* Geir Gunnlaugsson, surgeon general, Iceland, in an interview with the author, September 7, 2011.

68 *Lawsuits . . . are not nearly as common:* Ibid.

68 *a matter of public record:* Ibid.

68 *14.6 percent:* The most-up-to-date health statistics in Iceland, made available by the Directorate of Health, can be found at http://landlaeknir.is/Heilbrigdistolfraedi/Faedingar.

68 *32.8 percent:* B. E. Hamilton, J. A. Martin, S. J. Ventura, "Births: Preliminary Data for 2010," *National Vital Statistics Reports* 60, no. 2 (2011): 1–25. Available at www.cdc.gov/nchs/data/nvsr/nvsr60/nvsr60_02.pdf.

68 *infant mortality . . . a fraction of ours:* The maternal death rate in Iceland in 2008 was 5 per 100,000 live births. In the United States it was 24 per 100,000 live births in the same year. (See CIA, *World Factbook,* "Country Comparison: Maternal Mortality Rate.") The infant mortality rate in Iceland in 2012 is 3.18 per 1,000 live births and in the United States 5.98 per 1,000 live births. (See CIA, *World Factbook,* "Country Comparison: Infant Mortality Rate.")

68 *nine months paid leave:* "The Icelandic Act on Maternity/Paternity and Parental Leave underwent significant changes in the year 2000. The leave was extended from six months to nine, parents who were active in the labor market were paid 80 percent of their average salaries during the leave and the payments were to come from a specific fund, financed through an insurance levy. The leave was furthermore distributed so that fathers were given three months' leave, mothers three months and the parents were given three months to share as they wished. The Act has been well received by society and around 90 percent of fathers take advantage of their right, using on average 97 days while mothers use an average of 180 days. It is therefore likely that more fathers than ever are active in the caring for young children." Ingólfur V. Gíslason, *Parental Leave in Iceland Bringing the Fathers In: Developments in the Wake of New Legislation in 2000* (Akureyri: Ásprent, 2007), 3, accessed at http://www.jafnretti.is/D10/_Files/parentalleave.pdf.

68 *At the largest hospital in the country:* Helga Sigurðardóttir, head midwife, post- and prepartum, Landspítali, in an interview with the author, September 12, 2011.

69 *virtually the same:* In the year 2009, 5,015 infants were born in Iceland, of whom 21 were stillborn, 4 died in the first week after birth, and 5 died from day 8 to day 365. In the year 2010, 4,903 infants were born in Iceland, of whom 17 were stillborn, 5 died in the first week after birth, and 5 died from day 8 to day 365. This is summarized in papers from Landspítali http://www.landspitali.is/gagnasafn?branch=4810.40 (year 2003–2010) and the Directorate of Health http://landlaeknir.is/Heilbrigdistolfraedi/Faedingar (for the year 2010).

69 *trained in acupuncture:* Zita West, *Acupuncture in Pregnancy and Childbirth* (Philadelphia: Elsevier Health Sciences, 2001).

70 *"We all promote normal vaginal delivery":* Hildur Harðardóttir, M.D., head of obstetrics, Landspítali, in an interview with the author, September 8, 2011.

70 *Hildur, however, does not champion unmedicated birth:* Hildur Harðardóttir, M.D., head of obstetrics, Landspítali, email communication with the author, November 15, 2011.

70 *a midwife in Selfoss:* Selfoss is a small town located east of Reykjavik.

70 *"Doctors-in-training train with midwives":* Dagný Zoega, midwife, in discussion with the author, September 6, 2011.

70 *"after her birth ended disappointingly in a C-section":* Anonymous parent, in discussion with the author, September 5, 2011.

71 *birthed her second baby at home:* Emma Swift, midwife, in an interview with the author, August 15, 2011.

71 *meet face-to-face with the midwife or obstetrician:* Guðrún Eggertsdóttir, head midwife, labor ward, Landspítali, in an interview with the author, September 12, 2011.

71 *"We are human. We make mistakes":* Guðrún Eggertsdóttir, head midwife, labor ward, Landspítali, in an interview with the author, September 12, 2011.

71 *Average charge for C-section:* Ryan Ramos, M.S., M.A., et al., *Complications of Pregnancy and Childbirth in Orange County* (Santa Ana, Calif.: Orange County Health Care Agency, August 2011), accessed at http://www.ochealthinfo.com/docs/admin/ComplicationsPregnancyChildbirth_OC.pdf.

71 *Average charge for vaginal birth in a hospital:* Ibid.

71 *Average cost of homebirth in Southern California:* Homebirth midwives charge between $2,000–$7,000 for all prenatal care and the delivery, depending on the state. The average is about $3,000. In our area, a homebirth midwife would charge between $500 and $1,000 to attend the birth and provide postpartum care.

71 *Average time to deliver a baby vaginally in hospital:* http://www.mountsinai.org/patient-care/health-library/treatments-and-procedures/labor-and-delivery-vaginal-birth.

71 *Average time to deliver a baby via C-section:* http://womenshealth.gov/pregnancy/childbirth-beyond/labor-birth.cfm.

71 *Average charge for:* Traven Health Analytics. "The Cost of Having a Baby in the United States." Ann Arbor, THA, January 2013.

72 Lauren Shaddox: As told to the author, May 26, 2011.

Chapter 4 Cutting Costs: The Business of Cesarean Birth

79 *"We exteriorize the uterus":* Anonymous obstetrician, in an interview with the author, January 18, 2012.

80 *more than 1.4 million women in America:* The C-section rate in the United States in 2010 was 32.8 percent (of approximately 4.3 million births).

80 *5 percent:* F. Menacker and B. E. Hamilton, "Recent Trends in Cesarean Delivery in the United States," *NCHS Data Brief* 35 (March 2010): 1–8, accessed at http://www.cdc.gov/nchs/data/databriefs/db35.pdf.

80 *starting to steadily rise:* S. C. Zahniser, J. S. Kendrick, A. L. Franks, and A. F. Saftlas, "Trends in Obstetric Operative Procedures, 1980 to 1987," *American Journal of Public Health* 82,

no. 10 (October 1992): 1340–1344, accessed at http://www.ncbi.nlm.nih.gov/pmc/articles/PMC1695853/pdf/amjph00547-0030.pdf.

80 *"Cesarean epidemic"*: Diana Korte and Roberta Scaer, *A Good Birth, A Safe Birth* (New York: Bantam Books, 1984), 138–167.

80 *32.9 percent*: "Births: Final Data for 2009," *National Vital Statistics Reports* 60, no. 1 (November 3, 2011), accessed at http://www.cdc.gov/nchs/data/nvsr/nvsr60/nvsr60_01.pdf.

80 *more than 50 percent*: John T. Queenan, M.D., "How to Stop the Relentless Rise in Cesarean Deliveries," Editorial, *Obstetrics & Gynecology* 118, no. 2, part 1 (August 2011): 199–200.

81 *"a relic of the past in bulldogs and women"*: I. Nygaard, M.D., "Vaginal Birth: A Relic of the Past in Bulldogs and Women?" *Obstetrics & Gynecology* 118, no. 4 (October 2011): 774–776.

81 *baby may be at higher risk from a vaginal birth*: It was once believed that any herpes infection counterindicated vaginal childbirth because it could lead to problems for the neonate, including blindness, but while some obstetricians may insist on C-section if there are active herpes lesions on the vulva, the homebirth midwives we used for our second child's birth told us they simply cover the sores with beeswax and have never had a problem. According to *Midwifery Today*: "Neonatal herpes is a remarkably rare event," says Zane Brown, M.D., an expert on neonatal herpes and a member of the Department of Obstetrics and Gynecology at the University of Washington. "Compared to all the other possible risks in a pregnancy, the risk of neonatal herpes is extremely small. Transmission rates are lowest for women who acquire herpes before pregnancy—one study (Randolph, *JAMA*, 1993) placed the risk at about 0.04 percent for such women who have no signs or symptoms of an outbreak at delivery." (*Midwifery Today* 3, no. 30, July 25, 2001, accessed at http://www.midwiferytoday.com/enews/enews0330.asp.) A California study of hospital discharges over a period of ten years found no increase in neonatal herpes infections despite an increase in vaginal births, see http://www.ncbi.nlm.nih.gov/pubmed/10353880?ordinalpos=9&itool=Entrez System2.PEntrez.Pubmed.Pubmed_ResultsPanel.Pubmed_DefaultReportPanel.Pubmed_RVDocSum. Another study, this one of sixteen adult patients done at the Dubai Specialized Medical Center in 2004, found that topical honey applications were more effective in reducing the duration and intensity of herpes lesions than acyclovir cream, see http://www.ncbi.nlm.nih.gov/pubmed/15278008.

81 *10 to 15 percent*: World Health Organization, "Appropriate Technology for Birth," *Lancet* 2 (1985): 436–437.

81 *Seventy-nine percent of the 121 women*: New York City Maternal Mortality Review Project Team, *Pregnancy Associated Mortality: New York City 2001–2005*, Bureau of Maternal, Infant, and Reproductive Health, New York, 19–23, accessed at http://www.nyc.gov/html/doh/downloads/pdf/ms/ms-report-online.pdf.

81 *36 women in every 100,000*: M. A. Harper et al., "Pregnancy-Related Death and Health Care Services," *Obstetrics & Gynecology* 102, no. 2 (August 2003): 273–278.

82 *three times more likely*: M. H. Hall and S. Bewley, "Maternal Mortality and Mode of Delivery," *Lancet* 354, no. 9180 (August 28, 1999): 776.

82 *"At worst, C-sections can kill"*: Patji Alnaes-Katjavivi, obstetrician, Oslo University Hospital, in an interview with the author, September 14, 2011.

82 *serious side effects*: H. Goer, "Step 6: Does Not Routinely Employ Practices, Procedures Unsupported by Scientific Evidence," *Journal of Perinatal Education* 16, no. 1 (Winter 2007): 32S–64S.

82 *accidental cuts to internal organs*: R. M. Silver et al., "Maternal Morbidity Associated with Multiple Repeat Cesarean Deliveries," *Obstetrics & Gynecology* 107, no. 6 (June 2006): 1226–1232.

82 *emergency hysterectomy*: J. Kacmar, "Route of Delivery as a Risk Factor for Emergent Peripartum Hysterectomy: A Case-Control Study," *Obstetrics & Gynecology* 102, no. 1 (July 2003): 141–145; M. Knight et al., "Cesarean Delivery and Peripartum Hysterectomy," *Obstetrics & Gynecology* 111, no. 1 (January 2008): 97–105.

A. C. Rossi, R. H. Lee, and R. H. Chmait, "Emergency Postpartum Hysterectomy for Uncontrolled Postpartum Bleeding: A Systematic Review," *Obstetrics & Gynecology* 115, no. 3 (March 2010): 637–644.

82 *complications from anesthesia:* S. M. Koroukian, "Relative Risk of Postpartum Complications in the Ohio Medicaid Population: Vaginal versus Cesarean Delivery," *Medical Care Research and Review* 61, no. 2 (2004): 203–224.

82 *chronic pain:* E. Declercq, "Mothers' Reports of Postpartum Pain Associated with Vaginal and Cesarean Deliveries: Results of a National Survey," *Birth* 35, no. 1 (March 2008): 16–24.

P. Latthe, L. Mignini, R. Gray, et al., "Factors Predisposing Women to Chronic Pelvic Pain: Systematic Review," *British Medical Journal* 332, no. 7544 (2006): 749–755.

82 *endometriosis:* M. Eogan and P. McKenna, "Endometriosis in Caesarean Section Scars," *Irish Medical Journal* 95, no. 8 (September 2002): 247; K. B. Gajjar, "Caesarean Scar Endometriosis Presenting as an Acute Abdomen: A Case Report and Review of Literature," *Archives of Gynecology and Obstetrics* 277, no. 2 (February 2008): 167–169; A. Gaunt et al., "Caesarean Scar Endometrioma," *Lancet* 364, no. 9431 (July 24, 2004): 368; M. Gunes, "Incisional Endometriosis After Cesarean Section, Episiotomy and Other Gynecologic Procedures," *Journal of Obstetrics and Gynaecology Research* 31, no. 5 (October 2005): 471–475; A. Kafkasli, "Endometriosis in the Uterine Wall Cesarean Section Scar," *Gynecologic and Obstetric Investigation* 42, no. 3 (1996): 211–213; P. Kaloo, "Caesarean Section Scar Endometriosis: Two Cases of Recurrent Disease and a Literature Review," *Australian and New Zealand Journal of Obstetrics and Gynaecology* 42, no. 2 (May 2002): 218–220; J. Leng, "Carcinosarcoma Arising from Atypical Endometriosis in a Cesarean Section Scar," *International Journal of Gynecological Cancer* 16, no. 1 (January–February 2006): 432–435; S. Luisi, "Surgical Scar Endometriosis After Cesarean Section: A Case Report," *Gynecological Endocrinology* 22, no. 5 (May 2006): 284–285; S. Minaglia, "Incisional Endometriomas After Cesarean Section: A Case Series," *Journal of Reproductive Medicine* 52, no. 7 (July 2007): 630–634. O. Olufowobi, "Scar Endometrioma: A Cause for Concern in the Light of the Rising Caesarean Section Rate," *Journal of Obstetrics & Gynaecology* 23, no. 1 (January 2003): 86; G. K. Patterson and G. B. Winburn, "Abdominal Wall Endometriomas: Report of Eight Cases," *American Surgery* 65, no. 1 (January 1999), 36–39; V. Phupong and S. Triratanachat, "Cesarean Section Scar Endometriosis: A Case Report and Review of the Literature," *Journal of the Medical Association of Thailand* 85, no. 6 (June 2002): 733–738; S. L. Sholapurkar et al., "Life-Threatening Uterine Haemorrhage Six Weeks After Caesarean Section Due to Uterine Scar Endometriosis: Case Report and Review of Literature," *Australian and New Zealand Journal of Obstetrics and Gynaecology* 45, no. 3 (June 2005): 256–258; L. Wicherek et al., "The Obstetrical History in Patients with Pfannenstiel Scar Endometriomas—An Analysis of 81 Patients," *Gynecologic and Obstetric Investigation* 63, no. 2 (2007): 107–113; G. C. Wolf and K. B. Singh, "Cesarean Scar Endometriosis: A Review," *Obstetrical & Gynecological Survey* 44, no. 2 (February 1989): 89–95; Z. Zhu, "Clinical Characteristic Analysis of 32 Patients with Abdominal Incision Endometriosis," *Journal of Obstetrics and Gynaecology* 28, no. 7 (October 2008): 742–745.

82 *twice the risk:* J. Villar et al., "Maternal and Neonatal Individual Risks and Benefits Associated with Caesarean Delivery: Multicentre Prospective Study," *British Medical Journal* 335, no. 7628 (November 17, 2007): 1025.

82 *"good" bacteria:* Jennifer Ackerman, "How Bacteria in Our Body Protect Our Health," *Scientific American*, May 15, 2012.

82 *colonized by sometimes lethal hospital bacteria:* Maria G. Dominguez-Bello et al., "Delivery Mode Shapes the Acquisition and Structure of the Initial Microbiota Across Multiple Body Habitats in Newborns," *PNAS* 107, no. 26 (June 29, 2010): 11971–11975.

82 *digestive tracts . . . were disturbed:* M. M. Grönlund, "Fecal Microflora in Healthy Infants Born by Different Methods of Delivery: Permanent Changes in Intestinal Flora After Cesarean Delivery," *Journal of Pediatric Gastroenterology and Nutrition* 28, no. 1 (January 1999): 19–25.

83 *breathing problems:* J. Madar et al., "Surfacant-Deficient Respiratory Distress After Elective Delivery at 'Term,' " *Acta Paediatrica* 88, no. 11 (November 1999): 1244–1248, accessed at http://onlinelibrary.wiley.com/doi/10.1111/j.1651-2227.1999.tb01025.x/abstract.

D. J. Annibale et al., "Comparative Neonatal Morbidity of Abdominal and Vaginal Deliveries After Uncomplicated Pregnancies," *Archives of Pediatrics & Adolescent Medicine* 149, no. 8 (August 1995): 862–867.

Nicholas S. Fogelson et al., "Neonatal Impact of Elective Repeat Cesarean Delivery at Term: A Comment on Patient Choice Cesarean Delivery," *American Journal of Obstetrics and Gynecology* 192 (January 2005): 1433–1436.

83 *difficulty breastfeeding:* K. G. Dewey, "Risk Factors for Suboptimal Infant Breastfeeding Behavior, Delayed Onset of Lactation, and Excess Neonatal Weight Loss," *Pediatrics* 112, no. 2, part 1 (September 2003): 607–619.

83 *infection:* Dao M. Nguyen, "Risk Factors for Neonatal Methicillin-Resistant *Staphylococcus aureus* Infection in a Well-Infant Nursery," *Infection Control and Hospital Epidemiology* 28, no. 4 (April 2007): 406–411, accessed at http://www.jstor.org/stable/10.1086/513122#rf3.

83 *severe childhood asthma:* M. C. Tollanes et al., "Cesarean Section and Risk of Severe Childhood Asthma: A Population-Based Cohort Study," *Journal of Pediatrics* 153, no. 1 (July 2008): 112–116, accessed at http://www.jpeds.com/article/S0022-3476(08)00070-X/abstract.

83 *twice as likely to be obese by age three:* Susanna Y. Huh et al., "Delivery by Caesarean Section and Risk of Obesity in Preschool-Age Children: A Prospective Cohort Study," *Archives of Disease in Childhood* (May 23, 2012), accessed at http://adc.bmj.com/content/early/2012/05/09/archdischild-2011-301141.abstract.

83 *nicked or otherwise harmed:* J. M. Alexander, "Fetal Injury Associated with Cesarean Delivery," *Obstetrics & Gynecology* 108, no. 4 (October 2006): 885–890.

83 *mistakenly amputated the baby's finger:* Hisham Aburezq, M.D., et al., "Iatrogenic Fetal Injury," *Obstetrics & Gynecology* 106, no. 5, part 2 (November 2005): 1172–1174.

83 *"I was totally robbed":* Karen Bridges, parent, in an interview with the author, November 14, 2011.

84 *necrotizing fasciitis:* A. R. Goepfert, "Necrotizing Fasciitis After Cesarean Delivery," *Obstetrics & Gynecology* 89, no. 3 (March 1997): 409–412.

84 *6 percent of women who have C-sections:* F. G. Cunningham et al., *Williams Obstetrics,* 23rd ed. (New York: McGraw-Hill, 2010), 665.

84 *microorganisms . . . in the hospital:* S. L. Emmons, "Development of Wound Infections Among Women Undergoing Cesarean Section," *Obstetrics and Gynecology* 72, no. 4 (October 1988): 559–564; J. Owen and W. W. Andrews, "Wound Complications After Cesarean Sections," *Clinical Obstetrics and Gynecology* 37, no. 4 (December 1994): 842–855.

84 *"I never saw my doctor":* Denise Schipani, parent, in an interview with the author, November 17, 2011.

85 *"I wanted to have my baby":* Poppy Street-Heywood, parent, in an interview with the author, January 11, 2011.

85 *The average time for labor:* Ina May Gaskin, *Ina May's Guide to Childbirth* (New York: Bantam Dell, 2003), 146.

86 *induced labor is much more likely to end in Cesarean birth:* "Currently all Ventura County hospitals, except Los Robles Hospital in Thousand Oaks, have VBAC bans in place. Cottage Hospital, the only hospital offering labor and delivery in Santa Barbara, also has a de facto ban in place, in that the environment does not encourage or even allow OBs to offer this option," Birth Action Coalition, news release, July 27, 2010.

Stuart Fischbein, M.D., obstetrician, in an interview with the author, November 15, 2011.

86 *"emergency" C-section without question:* As reported by Nathanael Johnson, "For-Profit Hospitals Performing More C-Sections," *California Watch,* September 11, 2010, accessed at http://californiawatch.org/health-and-welfare/profit-hospitals-performing-more-c-sections-4069.

86 *before shift change at the hospital:* Ontario Maternity Care Expert Panel, Appendix K, K-1 Consumer Complaint to Wendy Katherine, Project Manager, OMCEP, accessed at http:// www.cmo.on.ca/downloads/OMCEP_App_K_1.pdf.

86 *rises sharply before long weekends:* Naomi Wolf, *Misconceptions: Truth, Lies, and the Unexpected on the Journey to Motherhood* (New York: Anchor Books, 2003), 178.

O. Goldstick, "The Circadian Rhythm of 'Urgent' Operative Deliveries," *Israel Medical Association Journal* 5, no. 8 (August 2003): 564–566.

"Births by Day of the Year," Peltier Technical Services, 2012, accessed at http://pelti-ertech.com/Excel/Commentary/BirthsByDayOfYear.html#ixzz18d0vAidL.

86 *timing of births in America has shifted dramatically:* Marsden Wagner, *Born in the USA* (Berkeley: University of California Press, 2006), 39; Joshua S. Gans and Andrew Leigh, "What Explains the Fall in Weekend Births?" (September 26, 2008), accessed at http://www.mbs. edu/home/jgans/papers/weekend%20shifting-08-09-26%20(ms%20only).pdf.

86 *born Monday through Friday:* Joyce A. Martin, "Births: Final Data for 2005," *National Vital Statistics Reports* 56, no. 6 (December 5, 2007), accessed at http://www.cdc.gov/nchs/data/nvsr/nvsr56/nvsr56_06.pdf.

87 *"If I do a breech":* Stuart Fischbein, M.D., in an interview with the author, November 15, 2011.

87 *"suggests that other non-medical factors":* Ryan Ramos, M.S., M.A., et al., *Complications of Pregnancy and Childbirth in Orange County* (Santa Ana, Calif.: Orange County Health Care Agency, August 2011), 20, accessed at http://www.ochealthinfo.com/docs/admin/ComplicationsPregnancyChildbirth_OC.pdf.

87 *17 percent more likely:* Nathanael Johnson, "For-Profit Hospitals Performing More C-Sections," *California Watch,* September 11, 2010, accessed at http://californiawatch.org/health-and-welfare/profit-hospitals-performing-more-c-sections-4069.

87 *20 percent more likely:* S. C. Zahniser, J. S. Kendrick, A. L. Franks, and A. F. Saftlas, "Trends in Obstetric Procedures, 1980 to 1987," *American Journal of Public Health* 82, no. 10 (October 1992): 1342, accessed at http://www.ncbi.nlm.nih.gov/pmc/articles/PMC1695853/pdf/amjph00547-0030.pdf.

87 *$20,228 . . . $11,114:* Ramos, *Complications of Pregnancy and Childbirth in Orange County,* 18.

88 *"whose health plans can afford":* Wolf, *Misconceptions,* 177.

88 *"Reimbursement rates for C-sections":* Mark C. Hornbrook, medical economist, in an email communication with the author, December 28, 2010.

88 *"It doesn't have anything to do with hospital finances":* Anonymous former CEO in an interview with the author, September 1, 2011.

88 *sued more frequently:* Carol K. Kane, *Policy Research Perspectives: Medical Liability Claim Frequency: A 2007–2008 Snapshot of Physicians* (Chicago: American Medical Association, 2010), 7, accessed at http://www.ama-assn.org/ama1/pub/upload/mm/363/prp-201001-claim-freq.pdf.

88 *more than 50 percent . . . have been sued:* Ibid., 8.

89 *$84,000 a year for medical malpractice:* Jaime Holguin, "High Cost of Malpractice Insurance," *CBS Evening News with Scott Pelley,* December 5, 2007, accessed at http://www.cbsnews.com/2100-500262_162-610102.html.

89 *$92,000 . . . $201,000 a year:* U.S. General Accounting Office, *Report to Congressional Requesters, Medical Malpractice Insurance: Multiple Factors Have Contributed to Increased Premium Rates* (Washington, D.C.: U.S. General Accounting Office, 2003), 14, accessed at http://www.gao.gov/new.items/d03702.pdf.

89 *"Most of the large malpractice cases":* Kristina Goodnough, "Researcher Studies Rates of Cesarean Sections, Malpractice Suits," *Advance* (University of Connecticut), September 2, 2008, accessed at http://www.advance.uconn.edu/2008/080902/08090203.htm.

89 *bowel was severed during a C-section:* "Suit Claims Doctor, Elkin Hospital at Fault in Death," *Mount Airy News,* accessed at http://www.mtairynews.com/view/full_story/14710002/

article-Suit-claims-doctor--Elkin-hospital-at-fault-in-death?instance=secondary_news_ left_column.

89 *died from sepsis infection:* "Medical Malpractice Lawyer Files Suit for Family in Obstetrical Error Case," Illinois Medical Malpractice (blog), July 17, 2011, accessed at http:// medicalmalpractice.levinperconti.com/2011/07/medical_malpractice_lawyer_fil.html.

89 *Jana Pokorny:* Corrinne Hess, "Froedtert Sued for Malpractice," *Business Journal,* November 25, 2009, accessed at http://www.bizjournals.com/milwaukee/blog/health_care/2009/11/ froedtert_sued_for_malpractice.html.

89 *$7.62 million:* Steven M. Levin and John J. Perconti, "Successful Cases: Medical Malpractice," accessed at http://www.levinperconti.com/lawyer-attorney-1090251.html.

89 *died from bleeding:* John Flynn Rooney, "HMO, MD to Pay $7 Million Over Med-mal," *Chicago Daily Law Bulletin* 147, no. 67 (April 5, 2001).

89 *New York had the tenth highest C-section rate:* Mary Beth Pfeiffer, "C-Section Rates Tick Upward as Doctors Fear Being Sued," *Poughkeepsie Journal,* May 8, 2010, accessed at http:// www.poughkeepsiejournal.com/article/20100509/NEWS01/5090346/C-section-rates-tick-upward-doctors-fear-being-sued.

90 *One thirty-two-year-old woman:* Ibid.

90 *"I feel really bad for them":* Felicia Cohen, M.D., in an interview with the author, August 24, 2011.

90 *"Doctors don't admit they make mistakes":* Marsden Wagner, *Born in the USA,* 153.

91 *detailed report:* "CMACE Release: Saving Mothers' Lives Report: Reviewing Maternal Deaths 2006–2008," Royal College of Obstetricians and Gynaecologists, January 3, 2011, accessed at http://www.rcog.org.uk/news/cmace-release-saving-mothers'-lives-report---reviewing-maternal-deaths-2006-2008.

91 *"If something goes wrong":* Emma Swift, birth professional, in an interview with the author, August 15, 2011.

91 *"A fundamental principle of medical practice":* Marsden Wagner, *Born in the USA,* 155.

91 *In 2003, when Patricia Roe:* This story, and the subsequent quotations, Patricia Roe, parent, in an interview with the author, March 4, 2010. A version of this story was first published in *Mothering* magazine (September/October 2010).

92 *C-section rate in America:* Joyce A. Martin, M.P.H., et al., "Births: Final Data for 2010," *National Vital Statistics Reports* 61, no. 1 (August 2012): 2.

92 *C-section rate in Norway:* Luz Gibbons et al., "The Global Numbers and Costs of Additionally Needed and Unnecessary Caesarean Sections Performed Per Year: Overuse as a Barrier to Universal Coverage," *World Health Report* (2010), Background Paper, 30 (Geneva: World Health Organization, 2010).

92 *C-section rate in Iceland:* The most-up-to-date health statistics in Iceland, made available by the Directorate of Health, can be found at http://landlaeknir.is/Heilbrigdistolfraedi/Faedingar.

92 *Maternal mortality rate in America:* See CIA, *World Factbook,* "Country Comparison: Maternal Mortality Rate."

92 *Maternal mortality rate in Norway:* : Ibid.

92 *Maternal mortality rate in Iceland:* Ibid.

92 *Number of midwives to doctors attending births in Scandinavia:* Marit Heiberg, president, Norwegian Association of Midwives, in an interview with the author, September 15, 2011. While the exact number will vary by country, geographical region, and health care facility, this ratio was also confirmed by the ratio of doctors to midwives in Iceland and Norway at the hospital I visited.

92 *Number of midwives to doctors attending births in America:* According to the CDC, only 8 percent of births in America are attended by midwives. Joyce A. Martin, M.P.H., "Births: Final Data for 2008," *National Vital Statistics Reports* 59, no. 1 (December 8, 2010): 9, accessed at http://www.cdc.gov/nchs/data/nvsr/nvsr59/nvsr59_01.pdf.

93 *98 percent of childbed fever:* M. Best and D. Neuhauser, "Ignaz Semmelweis and the Birth

of Infection Control," *Quality & Safe Health Care* 13 (2004): 233–234, accessed at http://qualitysafety.bmj.com/content/13/3/233.full.

93 *elaborate ruses to hide their knowledge:* Peter M. Dunn, "Perinatal Lessons from the Past: The Chamberlen Family (1560–1728) and Obstetric Forceps," *Archives of Disease in Childhood Fetal Neonatal Edition* 81 (1999): 232–235, accessed at http://www.ncbi.nlm.nih.gov/pmc/articles/PMC1721004/pdf/v081p0F232.pdf.

94 *episiotomy* increases *the risk:* Cunningham, *Williams Obstetrics,* 23rd ed., 401.

Chapter 5 Perinatal Prices: Profit-Mongering After the Baby Is Born

97 *"It was hard because you just want":* This and subsequent quotations: Cyndi Sellers, parent, in an interview with the author, February 1, 2012.

98 *outcome for infant health and for mother-baby bonding:* B. E. Morgan et al., "Should Neonates Sleep Alone?" *Biological Psychiatry* 70, no. 9 (November 1, 2011): 817–825; R. Dalbye, E. Calais, and M. Berg, "Mothers' Experiences of Skin-to-Skin Care of Healthy Full-Term Newborns—A Phenomenology Study," *Sexual & Reproductive Healthcare* 2, no. 3 (2011): 107–111, accessed at http://www.ncbi.nlm.nih.gov/pubmed/21742289; S. Vincent, "Skin-to-Skin Contact. Part 2: The Evidence," UNICEF UK Baby Friendly Initiative, *Practical Midwifery* 14, no. 6 (June 2011): 44–46, accessed at http://www.ncbi.nlm.nih.gov/pubmed/21739738; M. Velandia, A. S. Matthisen, K. Uvnäs-Moberg, and E. Nissen, "Onset of Vocal Interaction Between Parents and Newborns in Skin-to-Skin Contact Immediately After Elective Cesarean Section," *Birth* 37, no. 3 (September 2010): 192–201, accessed at http://www.ncbi.nlm.nih.gov/pubmed/20887535; E. R. Moore, G. C. Anderson, and N. Berman, "Early Skin-to-Skin Contact for Their Mothers and Healthy Newborn Infants," *Cochrane Review,* accessed at http://apps.who.int/rhl/reviews/CD003519.pdf.

   N. J. Bergman et al., "Randomized Controlled Trial of Skin-to-Skin Contact from Birth Versus Conventional Incubator for Physiological Stabilization in 1200- to 2199-Gram Newborns," *Acta Paediatrica* 93, no. 6 (June 2004): 779–785.

   E. S. Rey and H. G. Martínez, "Maneio Racional del Niño Premature," in *Curso de Medicina Fetal* (Bogotá: Universidad Nacional, 1983).

   K. Christensson and C. Siles, et al., "Temperature, Metabolic Adaptation and Crying, in Healthy, Full-Term Newborns Cared for Skin-to-Skin or in a Cot," *Acta Paediatrica* 81 (1992): 488–493; E. Nissen et al., "Elevation of Oxytocin Levels Early Post Partum in Women," *Acta Obstet Gynecol Scand* (Sweden) 74, no. 7 (August 1995): 530–533.

98 *enhance the feeling that can only be described as love:* See, for example: K. M. Kendrick et al., "Changes in the Sensory Processing of Olfactory Signals Induced by Birth in Sheep," *Science* (England) 256, no. 5058 (May 1992): 833–836; P. Popok and J. Vetulani, "Opposite Action of Oxytocin and Its Peptide Antagonists on Social Memory in Rats," *Neuropeptides* 18, no. 1 (January 1991): 23–27.

98 *Early uninterrupted skin-to-skin contact:* G. Puig and Y. Sguassero, "Early Skin-to-Skin Contact for Mothers and Their Healthy Newborn Infants: RHL Commentary," *WHO Reproductive Health Library* (Geneva: World Health Organization, 2007), accessed at http://apps.who.int/rhl/newborn/gpcom/en/index.html.

   G. C. Anderson et al., "Early Skin-to-Skin Contact for Mothers and Their Healthy Newborn Infants (Review)," *Cochrane Database of Systematic Reviews,* no. 3 (2007), accessed at http://apps.who.int/rhl/reviews/CD003519.pdf.

98 *separated newborns were anxious:* Barak E. Morgan, Alan R. Horn, and Nils J. Bergman, "Should Neonates Sleep Alone?" *Biological Psychiatry* 70, no. 9 (November 2011): 817–825.

98 *"Though they were sleeping":* This and subsequent quotations: Nils J. Bergman, M.D., M.P.H., Ph.D., independent researcher, in an interview with the author, February 6, 2012.

98 *to carry them skin-to-skin in pouches:* Skin-to-Skin Contact, "Kangaroo Mother Care: The Public Health Imperative," accessed at http://skintoskincontact.com/manama-story.aspx.

98 damage *on the developing newborn brain:* "Maternal Separation Stresses the Baby, Research Finds," *ScienceDaily,* November 2, 2011, accessed at http://www.sciencedaily.com/releases/2011/11/111102124955.htm.

99 *"Inexperienced doctors, especially":* Linda Hopkins, M.D., obstetrician, in an interview with the author, February 5, 2012.

99 *"Based on no scientific evidence":* John H. Kennell, "Commentary: Randomized Controlled Trial of Skin-to-Skin Contact from Birth versus Conventional Incubator for Physiological Stabilization in 1200 g to 2199 g Newborns," *Acta Paediatrica* 95, no. 1 (January 2006): 15–16.

99 *"They move mom":* Anonymous birth professional, in an interview with the author, February 10, 2012.

100 *between about 20 and 24 inches long:* G. Ente and P. H. Penzer, "The Umbilical Cord: Normal Parameters," *Journal of the Royal Society of Health* 111, no. 4 (August 1991): 138–140, accessed at http://www.ncbi.nlm.nih.gov/pubmed/1941874.

100 *diameter of .3 to .8 inches:* F. G. Cunningham, *Williams Obstetrics,* 23rd ed. (New York: McGraw-Hill, 2010), 62.

100 *But then he started to change his mind:* Nicholas Fogelson, "Delayed Cord Clamping Should Be Standard Practice in Obstetrics," Academic OB/GYN (blog), December 3, 2009, accessed at http://academicobgyn.com/2009/12/03/delayed-cord-clamping-should-be-standard-practice-in-obstetrics/.

101 *Up to 40 percent of his blood:* Susan Markel, M.D., "Peace of Mind . . . From Birth Onward," keynote address, Lamaze Conference, Fort Worth, Texas, September 17, 2011. This study shows that premature cord clamping can result in reducing the number of blood cells an infant receives at birth by 50 percent: J. S. Mercer, "Current Best Evidence: A Review of the Literature on Umbilical Cord Clamping, *Journal of Midwifery & Women's Health* 46, no. 6 (November–December 2001): 402–414, accessed at http://www.ncbi.nlm.nih.gov/pubmed/11783688.

101 *higher iron stores:* C. M. Chaparro et al., "Effect of Timing of Umbilical Cord Clamping on Iron Status in Mexican Infants: A Randomised Controlled Trial," *Lancet* 367, no. 9527 (June 17, 2006): 1997–2004.

101 *less likely to hemorrhage:* J. S. Mercer, "Delayed Cord Clamping in Very Preterm Infants Reduces the Incidence of Intraventricular Hemorrhage and Late-Onset Sepsis: A Randomized, Controlled Trial," *Pediatrics* 117, no. 4 (April 2006): 1235–1242.

101 *delaying cord clamping for at least two minutes:* E. K. Hutton and E. S. Hassan, "Late vs Early Clamping of the Umbilical Cord in Full-Term Neonates: Systematic Review and Meta-analysis of Controlled Trials," *Journal of the American Medical Association* 297, no. 11 (March 21, 2007): 1241–1252.

101 *reimbursed at a fixed rate:* For a discussion of this see chapter 3, "Emerging Expenses."

101 *charged more than a thousand dollars:* "Fee Comparison Chart," AlphaCord, accessed February 7, 2012, at http://www.alphacord.com/fee_comparison.htm?gclid=CJ3QyZ7WjK4 CFQ8yhwodP0Yagw.

101 *"Obtaining cord blood for future":* T. Levy and I. Blickstein, "Timing of Cord Clamping Revisited," *Journal of Perinatal Medicine* 34, no. 4 (2006): 293–297.

101 *"I wonder at times why":* Fogelson, "Delayed Cord Clamping Should Be Standard Practice in Obstetrics."

102 *"The message is that birth is dirty":* MaryBeth Foard-Nance, doula and mother, in an interview with the author, February 12, 2012.

102 *soothe the baby:* H. Varendi et al., "Soothing Effect of Amniotic Fluid Smell in Newborn Infants," *Early Human Development* 51, no. 1 (April 17, 1998): 47–55.

102 *help him connect with his mom:* B. Schaal et al., "Olfactory Function in the Human Fetus:

Evidence from Selective Neonatal Responsiveness to the Odor of Amniotic Fluid," *Behavioral Neuroscience* 112, no. 6 (December 1998): 1438–1449.

102 *the smell of amniotic fluid:* B. Schaal and L. Marlier, "Maternal and Paternal Perception of Individual Odor Signatures in Human Amniotic Fluid—Potential Role in Early Bonding?" *Biology of the Neonate* 74, no. 4 (October 1998): 266–273.

102 *prefer the smell of their own amniotic fluid:* Schaal, "Olfactory Function in the Human Fetus," 1438–1449.

102 *vernix on a baby's body:* M. Tollin et al., "Vernix Caseosa as a Multi-component Defence System Based on Polypeptides, Lipids, and Their Interactions," *Cellular and Molecular Life Sciences* 62, nos. 19–20 (October 2005): 2390–2399.

102 *Bathing a newborn causes:* Susan Markel, M.D., with Linda F. Palmer, *What Your Pediatrician Doesn't Know Can Hurt Your Child: A More Natural Approach to Parenting* (Dallas, Tex.: BenBella Books, 2010), 11.

102 *"the #1 choice of hospitals":* Johnson's Head-to-Toe Baby Wash, accessed at http://www.johnsonsbaby.com/johnsons-head-to-toe-body-wash.

102 *"milder than baby soap":* Ibid.

103 *suggests consumers read ingredient lists:* Michael Pollan, *Food Rules: An Eater's Manual* (New York: Penguin Books, 2009).

103 *Quaternium-15:* National Institute of Environmental Health Sciences, "New Substances Added to HHS Report on Carcinogens," news release, June 10, 2011, accessed at www.niehs.nih.gov/news/newsroom/releases/2011/june10/.

103 *most common cause of contact dermatitis:* E. M. Warshaw et al., "Contact Dermatitis of the Hands: Cross-sectional Analyses of North American Contact Dermatitis Group Data, 1994–2004," *Journal of the American Academy of Dermatology* 57, no. 2 (August 2007): 301–314.

103 *1,4-dioxane:* National Toxicology Program, *Report on Carcinogens*, 12th ed. (Research Triangle Park, N.C.: U.S. Department of Health and Human Services, Public Health Service, National Toxicology Program, 2011), 176, accessed at http://ntp.niehs.nih.gov/ntp/roc/twelfth/profiles/Dioxane.pdf.

103 *banned in Europe:* Carolyn Butler, "Soaps, Makeup and Other Items Contain Deadly Ingredients, Say Consumer Advocates," *Washington Post,* January 30, 2012, accessed at http://www.washingtonpost.com/national/health-science/soaps-makeup-and-other-items-contain-deadly-ingredients-say-consumer-advocates/2012/01/24/gIQAeJ56cQ_story.html.

103 *carcinogenic by-product:* "Contaminants in Bath Products," Campaign for Safe Cosmetics, accessed at http://safecosmetics.org/article.php?id=221.

103 *not included on ingredient lists:* Heather Sarantis, M.S., with Stacy Malkan and Lisa Archer, *No More Toxic Tub: Getting Contaminants Out of Children's Bath & Personal Care Products* (Campaign for Safe Cosmetics, 2009), 4, accessed at http://safecosmetics.org/downloads/NoMoreToxicTub_Mar09Report.pdf.

103 *"Babies are much more vulnerable":* Kathleen Green, "Read Baby-product Labels Closely," *Dallas Morning News,* January 5, 2010, accessed at http://www.dallasnews.com/health/headlines/20100105-Read-baby-product-labels-closely-5379.ece.

103 *PEGs:* "PEG Compounds and Their Contaminants," David Suzuki Foundation, accessed on April 11, 2012, at http://www.davidsuzuki.org/issues/health/science/toxics/chemicals-in-your-cosmetics---peg-compounds-and-their-contaminants/.
    "Contaminants in Bath Products," Campaign for Safe Cosmetics.

103 *skin is your largest organ:* National Geographic, Health and Human Body, "Skin: The Body's Protective Cover," accessed at http://science.nationalgeographic.com/science/health-and-human-body/human-body/skin-article/.

103 *wholly absorbed into your body:* Thomas J. Franz, M.D., "Percutaneous Absorption. On the Relevance of In Vitro Data," *Journal of Investigative Dermatology* 64 (1975): 190–195, accessed at http://www.nature.com/jid/journal/v64/n3/pdf/5617309a.pdf.

103 *"When you are slathering stuff"*: Rex Rombach, organic perfumer, in an interview with the author, January 20, 2011.

104 *Between 1973 and 1999:* Cancer Prevention Coalition, *The Stop Cancer Before It Starts Campaign: How to Win the Losing War Against Cancer* (Chicago: Cancer Prevention Coalition, 2003), 5, accessed at http://www.preventcancer.com/press/pdfs/Stop_Cancer_Book.pdf.

104 *10,700 children under age fifteen:* National Cancer Institute, "Childhood Cancers," fact sheet accessed at http://www.cancer.gov/cancertopics/factsheet/Sites-Types/childhood.

104 *Childhood cancer is second leading cause of death:* A. Jemal, R. Siegel, E. Ward, et al., "Cancer Statistics, 2009," *CA: A Cancer Journal for Clinicians* 59, no. 4 (2009): 225–249.

104 *1,500 lives every year:* National Cancer Institute, "Childhood Cancers."

104 *claims there is substantial scientific evidence:* Samuel S. Epstein, *National Cancer Institute and American Cancer Society: Criminal Indifference to Cancer Prevention and Conflicts of Interest* (Xlibris, 2011), 12.

104 *baby products with nonformaldehyde preservatives:* See "Baby's Tub Is Still Toxic," Campaign for Safe Cosmetics, November 1, 2011, accessed at http://safecosmetics.org/article.php?id=887.

104 *announced they would phase out* some: Letter to Lisa Archer, director, Campaign for Safe Cosmetics, November 16, 2011, from Susan Nettensheim, vice president, Production Stewardship and Toxicology, accessed on April 11, 2012, at http://www.johnsonsbaby.com/a-statement-on-ingredients-in-the-news.

104 *Johnson & Johnson the most trusted brand in America:* Jennifer Rooney, "Brand Power to the People: J&J Takes Lead in Forbes Ranking," October 5, 2011, accessed at http://www.forbes.com/sites/jenniferrooney/2011/10/05/brand-power-to-the-people-jj-takes-lead-in-forbes-ranking/.

105 *"You can't just trust brands"*: Stacy Malkan, communications director, in an interview with the author, April 11, 2012.

105 *vitamin K deficiency bleed:* American Academy of Pediatrics, Committee on Fetus and Newborn, "Controversies Concerning Vitamin K and the Newborn," *Pediatrics* 112, no. 1 (July 1, 2003): 191–192, accessed at http://aappolicy.aappublications.org/cgi/content/full/pediatrics;112/1/191.

105 *human breast milk . . . naturally rich in vitamin K:* Christopher Duggan, John B. Watkins, and W. Allan Walker, *Nutrition in Pediatrics 4: Basic Science, Clinical Applications* (Hamilton, Ont.: BC Decker, 2008), 347. Foods rich in vitamin K include green vegetables like kale, broccoli, chard, and Brussels sprouts; fruits like avocado, grapes, kiwi; and animal products.

105 *immediate cord clamping:* J. S. Mercer, B. R. Vohr, M. M. McGrath, et al., "Delayed Cord Clamping in Very Preterm Infants Reduces the Incidence of Intraventricular Hemorrhage and Late-Onset Sepsis: A Randomized Controlled Trial," *Pediatrics* 117, no. 4 (April 2006): 1235–1242.

105 *harder for newborns to develop healthy intestinal bacteria:* K. Orrhage and C. E. Nord, "Factors Controlling the Bacterial Colonization of the Intestine in Breastfed Infants," *Acta Paediatrica* suppl. 88, no. 430 (August 1999): 47–57.

105 *colonized by bacteria found in the hospital:* J. Penders, C. Thijs, C. Vink, F. F. Stelma, et al., "Factors Influencing the Composition of the Intestinal Microbiota in Early Infancy," *Pediatrics* 118, no. 2 (2006): 511–521.

105 *long-lasting negative impact on the human digestive tract:* G. Biasucci, B. Benenati, L. Morelli, et al., "Cesarean Delivery May Affect the Early Biodiversity of Intestinal Bacteria," *Journal of Nutrition* 138, no. 9 (September 2008): 1796S–1800S; E. Decker, G. Engelmann, A. Findeisen, et al., "Cesarean Delivery Is Associated with Celiac Disease but Not Inflammatory Bowel Disease in Children," *Pediatrics* 125, no. 6 (June 2010): e1433–e1444, published online May 17, 2010; Minna-Maija Grölund, Olli-Pekka Lehtonen, Erkki Eerola, and Pentti Kero, "Fecal

Microflora in Healthy Infants Born by Different Methods of Delivery: Permanent Changes in Intestinal Flora After Cesarean Delivery," *Journal of Pediatric Gastroenterology & Nutrition* 28, no. 1 (January 1999): 19–25.

106 *the shot contains:* Merck & Co., Inc., 2002, INJECTION AquaMEPHYTON® (PHYTONADIONE) Aqueous Colloidal Solution of Vitamin K1, accessed at http://web. archive.org/web/20070213093306/http://www.fda.gov/medwatch/SAFETY/2003/03Feb_PI/ AquaMEPHYTON_PI.pdf.

106 *"Take your baby to the hospital":* This and subsequent quotations: Rachel Zaslow, parent, in an interview with the author, January 31, 2012.

106 *jaundice after birth:* Sarah J. Buckley, M.D., *Gentle Birth, Gentle Mothering: A Doctor's Guide to Natural Childbirth and Gentle Early Parenting Choices* (Berkeley, Calif.: Celestial Arts, 2009), 168.

107 *elevated bilirubin levels:* Shahab M. Shekeeb et al., "Evaluation of Oxidant and Antioxidant Status in Term Neonates: A Plausible Protective Role of Bilirubin," *Mollecular Cellular Biochemistry* 317, nos. 1–2 (October 2008): 51–59.

   Thomas W. Sedlak, M.D., Ph.D., and Solomon H. Snyder, M.D., "Bilirubin Benefits: Cellular Protection by a Biliverdin Reductase Antioxidant Cycle," *Pediatrics* 113, no. 6 (June 1, 2004): 1776–1782.

   A study from 1937 found that bilirubin may help protect against invasive bacteria: Najib-Farah, "Defensive Role of Bilirubinaemia in Pneumococcal Infection," *Lancet* 229, no. 5922 (February 27, 1937): 505–506.

107 *kernicterus:* "Kernicterus: Bilirubin Encephalopathy," PubMed Health, *A.D.A.M. Medical Encyclopedia,* National Center for Biotechnology Information, U.S. National Library of Medicine, May 9, 2011, accessed at http://www.ncbi.nlm.nih.gov/pubmedhealth/ PMH0004562/.

107 *what would be considered normal:* M. Jeffrey Maisels and Kathleen Gifford, "Normal Serum Bilirubin Levels in the Newborn and the Effect of Breast-Feeding," *Pediatrics* 78, no. 5 (November 1, 1986): 837–843.

107 *The one-room NICU:* Rachel Zaslow, parent, in an interview with the author, January 31, 2012.

108 *place a jaundiced baby in a sunny window:* David Perlstein, M.D., "Newborn Jaundice: Newborn Jaundice Self-Care at Home," accessed at http://www.emedicinehealth.com/ newborn_jaundice/page5_em.htm.

109 *between $3,000 and $10,000:* Jonathan Muraskas, M.D., and Kayhan Parsi, J.D., Ph.D., "The Cost of Saving the Tiniest Lives: NICUs versus Prevention," *Virtual Mentor* 10, no. 10 (October 2008): 655–658, accessed at http://virtualmentor.ama-assn.org/2008/10/pfor1-0810.html.

109 *15 percent of infants:* Cunningham, *Williams Obstetrics,* 23rd edition, 625.

109 *1 to 2 percent of newborns:* Ibid.

109 *highest premature birth rates in the world:* C. P. Howson, M. V. Kinney, and J. E. Lawneds, *Born Too Soon: The Global Action Report on Preterm Birth* (Geneva, Switzerland: World Health Organization, 2012), 2, accessed at http://whqlibdoc.who.int/ publications/2012/9789241503433_eng.pdf.

109 *risen 30 percent since 1981:* Ibid., 21.

109 *half a million babies:* Donald G. McNeil Jr., "U.S. Lags in Global Measure of Premature Births, *New York Times,* May 2, 2012, accessed at http://www.nytimes.com/2012/05/03/health/ us-lags-in-global-measure-of-preterm-births.html.

110 *40 percent of premature births:* Howson, Kinney, and Lawneds, *Born Too Soon,* 21.

110 *higher infant mortality rate than forty-eight other countries:* CIA *World Factbook,* "The United States infant mortality ranking," accessed at https://www.cia.gov/library/publications/the-world-factbook/rankorder/2091rank.html.

110 *rise in NICUs in America:* E. M. Howell et al., "Deregionalization of Neonatal Intensive Care in Urban Areas," *American Journal of Public Health* 92, no. 1 (January 2002): 119–124.

Ciaran S. Phibbs, Ph.D., Laurence C. Baker, Ph.D., Aaron B. Caughey, M.D., Ph.D., et al., "Level and Volume of Neonatal Intensive Care and Mortality in Very-Low-Birth-Weight Infants," *New England Journal of Medicine* 356 (2007): 2165–2175.

110 *Rates of reimbursement:* Michael Kornhauser, M.D., and Roy Schneiderman, M.D., "How Plans Can Improve Outcomes and Cut Costs for Preterm Infant Care: Ten Percent of Newborns Are Admitted to a Neonatal Intensive Care Unit. NICU Costs Are High but Controllable," *Managed Care,* January 2010, accessed on April 11, 2012, at http://www.managedcaremag. com/archives/1001/1001.preterm.html.

110 *costs Medicaid $45,000:* Emily Ramshaw, "In Search of Cuts, Health Officials Question NICU Overuse," *New York Times,* March 19, 2011, accessed at http://www.nytimes.com/2011/03/20/ us/20ttnicus.html.

110 *"The NICU is a moneymaker":* Stuart Fischbein, M.D., in an interview with the author, November 15, 2011.

111 *"When we look at the data":* Ramshaw, "In Search of Cuts, Health Officials Question NICU Overuse."

111 *without the need for NICUs:* Howson, Kinney, and Lawneds, *Born Too Soon,* 2.

111 *"promote and support successful breastfeeding":* American Academy of Pediatrics Subcommittee on Hyperbilirubinemia, "Management of Hyperbilirubinemia in the Newborn Infant 35 or More Weeks of Gestation," *Pediatrics* 114, no. 1 (July 2004): 297–316, accessed at http://aappolicy.aappublications.org/cgi/content/full/pediatrics;114/1/297.

111 *"minimize the risk of unintended harm":* Ibid.

111 *"Parents are often not aware":* Markel, "Peace of Mind . . . From Birth Onward."

112 *the threat of a call to the Department of Social Services:* After Jodi Ferris had her baby in the ambulance on the way to Penn State Hershey Medical Center, she was belittled by hospital staff, and her healthy daughter was taken into protective custody because Jodi asked to be tested for hepatitis B before being administered the shot. http://www.thehealthyhomeeconomist. com/mother-who-questions-vax-at-hospital-has-newborn-taken-away/.

112 *"Babies are manhandled":* MaryBeth Foard-Nance, doula, in an interview with the author, February 10, 2012.

112 *Sarah Vaile's itemized hospital bill:* Sarah Vaile, parent, in an interview with the author, February 4, 2012.

113 *about $.085 a tablet:* "Tylenol Pain Reliever/Fever Reducer 325 Mg Regular Strength Tablets," CVS Pharmacy, accessed at http://www.cvs.com/CVSApp/catalog/shop_product_detail.jsp? skuId=410579&productId=410579.

113 *protocol dictates that hospitalized patients:* Linda Hopkins, M.D., in discussion with the author, February 5, 2012.

113 *It was a rainy day in late January 2012:* Angelina Mendenhall, parent, in an interview with the author, February 7, 2012.

113 *had briefed family members beforehand:* Augustine Colebrook, owner and midwife, Trillium Water Birth Center, in an interview with the author, February 1, 2011.

114 *"holding the space":* Ibid.

114 *until it has entirely stopped pulsing:* At Trillium Water Birth Center, the cord is often not cut until after the placenta is delivered.

114 *"Moms have nothing but eyes":* Augustine Colebrook, owner and midwife, Trillium Water Birth Center, in an interview with the author, February 1, 2011.

115 *Babies born:* CDC, National Prematurity Awareness Month, November 7, 2012.

115 *Average cost of one-day stay in the hospital:* International Federation of Health Plans 2011 Comparative Price Report, 8, accessed at http://www.ifhp/documents/2011;FHPPriceReports Graphs_version3.PDF.

115 *Average cost of one-day stay in the NICU:* Emily Ramshaw, "Maternity Wards, NICUs Face Budget Scrutiny," *Texas Tribune,* March 21, 2011.

115 *Average charges:* Traven Health Analytics. "The Cost of Having a Baby in the United States." Ann Arbor, THA, January 2013.

115 *Revenue lost by Seton Hospitals:* Ramshaw, "In Search of Cuts, Health Officials Question NICU overuse."

Chapter 6 Foreskins for Sale: The Business of Circumcision

120 *"The tissue is very stretchy":* Beth Hardiman, obstetrician, Mount Auburn Hospital, in an interview with the author, March 13, 2012.

121 *different experience watching a circumcision in March 2012:* Anonymous nursing student, SUNY Rockland Community College, in an interview with the author, March 21, 2012.

121 *fewer than 20 percent of men are circumcised: Male Circumcision: Global Trends and Determinants of Prevalence, Safety and Acceptability* (Geneva, Switzerland: World Health Organization, 2007), 9, accessed at http://www.malecircumcision.org/media/documents/MC_Global_Trends_Determinants.pdf.

121 *fewer than 2 percent of men:* Jake H. Waskett, "Global Circumcision Rates," Circumcision Independent Reference and Commentary Service, http://www.circs.org/index.php/Reviews/Rates/Global#n22.

121 *65 percent of all American boys:* "NCHS Health E-Stat: Trends in Circumcision Among Newborns," Centers for Disease Control and Prevention, February 3, 2010, accessed at http://www.cdc.gov/nchs/data/hestat/circumcisions/circumcisions.htm.

121 *Hispanics: Male Circumcision,* 11.

121 *African-Americans:* Centers for Disease Control and Prevention, "Circumcision by Region and Race 1979–2008," http://www.cdc.gov/nchs/data/nhds/9circumcision/2007circ9_regionracetrend.pdf.

121 *geographic location:* Ibid.

121 *circumcision became prevalent:* B. Barker-Benfield, "Sexual Surgery in Late-Nineteenth-Century America," *International Journal of Health Services* 5, no. 2 (1975): 279–288; Robert Darby, *A Surgical Temptation: The Demonization of the Foreskin and the Rise of Circumcision in Britain* (Chicago: University of Chicago Press, 2005), 7; F. Hodges, "The Antimasturbation Crusade in Antebellum American Medicine," *Journal of Sexual Medicine* 2, no. 5 (September 2005): 722–731.

121 *5 percent of men in England are circumcised:* Adam Liptak, "Circumcision Opponents Use the Legal System and Legislatures," *New York Times,* January 23, 2003, accessed at http://www.nytimes.com/2003/01/23/national/23CIRC.html?pagewanted=all.

122 *"[M]ost healthcare professionals":* "Circumcision," NHS Choices, last reviewed January 1, 2012, accessed at http://www.nhs.uk/Conditions/Circumcision/Pages/Introduction.aspx.

122 *In Canada, 31.9 percent:* Public Health Agency of Canada, *What Mothers Say: The Canadian Maternity Experiences Survey* (Ottawa, Ont.: Public Health Agency of Canada, 2009), 224, accessed at http://www.phac-aspc.gc.ca/rhs-ssg/pdf/survey-eng.pdf.

122 *New Zealand and Australia:* The Royal Australasian College of Physicians, Paediatrics & Child Health Division, *Circumcision of Infant Males* (Sydney, N.S.W., Australia: The Royal Australasian College of Physicians, 2010), 5.

122 *54.7 percent, of American newborn boys:* "Trends in In-Hospital Newborn Male Circumcision—United States, 1999–2010," *Morbidity and Mortality Weekly Report* 60, no. 34 (September 2, 2011): 1167–1168, accessed at http://www.cdc.gov/mmwr/preview/mmwrhtml/mm6034a4.htm?s_cid=mm6034a4_w.

122 *"We believed it was":* This and subsequent quotations: Aseem Shukla, pediatric urologist, University of Minnesota Amplatz Children's Hospital, in an interview with the author, July 5,

2011. Since our initial interview, Dr. Shukla has changed jobs. He is now director of minimally invasive surgery in the Division of Urology at the Children's Hospital in Philadelphia, as well as associate professor of surgery at the University of Pennsylvania Perelman School of Medicine.

122 *protect against urinary tract infections:* S. Mukherjee et al., "What Is the Effect of Circumcision on Risk of Urinary Tract Infection in Boys with Posterior Urethral Valves?" *Journal of Pediatric Surgery* 44, no. 2 (February 2009): 417–421; N. Shaikh et al., "Does This Child Have a Urinary Tract Infection?" *Journal of the American Medical Association* 298, no. 24 (December 26, 2007): 2895–2904.

122 *urinary tract infections among infants:* D. Singh-Grewal et al., "Circumcision for the Prevention of Urinary Tract Infection in Boys: A Systematic Review of Randomised Trials and Observational Studies," *Archives of Disease in Childhood* 90 (2005): 853–858, accessed at http://adc.bmj.com/content/90/8/853.full#ref-32.

122 *circumcised men can get penile cancer:* C. J. Cold et al., "Carcinoma in Situ of the Penis in a 76-Year-Old Circumcised Man," *Journal of Family Practice* 44, no. 4 (April 1997): 407–410.

122 *incidents of penile cancer vary greatly:* AAFP, "Circumcision: Position Paper on Neonatal Circumcision," August 2007, accessed at http://www.aafp.org/online/en/home/clinical/clinicalrecs/children/circumcision.html.

123 *penile cancer decreased in prevalence:* M. Frisch, "Falling Incidence of Penis Cancer in an Uncircumcised Population (Denmark 1943–90)" *British Medical Journal* 311 (December 2, 1995): 1471.

123 *circumcision reduces the risk of men getting HIV:* American Academy of Pediatrics, Task Force on Circumcision, "Technical Report: Male Circumcision," *Pediatrics* 130, no. 3 (Sept. 1, 2012): e756–e785. Accessed at http://pediatrics.aappublications.org/content/130/3/e756.full.

123 *circumcising adult men in African countries:* B. Auvert, D. Taljaard, E. Lagarde, et al., "A Randomized, Controlled Intervention Trial of Male Circumcision for Reduction of HIV Infection Risk: The ANRS 1265 Trial," *PLoS Med.* 2, no. 11 (November 2005): e298. Erratum in: *PLoS Med.* 3, no. 5 (May 2006): e298. R. C. Bailey et al., "Male Circumcision for HIV Prevention in Young Men in Kisumu, Kenya: A Randomised Controlled Trial," *Lancet* 369, no. 9562 (February 2007): 643–656; R. Gray et al., "Male Circumcision for HIV Prevention in Men in Rakai, Uganda: A Randomised Trial," *Lancet* 369, no. 9562 (February 2007): 657–666.

123 *"insufficient evidence":* G. A. Millett et al., "Circumcision Status and Risk of HIV and Sexually Transmitted Infections Among Men Who Have Sex with Men," *Journal of the American Medical Association* 300, no. 14 (October 2008): 1674–1684, accessed at http://jama.jamanetwork.com/article.aspx?volume=300&issue=14&page=1674.

123 *resume sexual relations without protection:* Michael Carter, "Quarter of Men Resume Sex Before Wounds from Circumcision Fully Healed in Zambian Study," *NAM*, January 31, 2012, http://www.aidsmap.com/Quarter-of-men-resume-sex-before-wounds-from-circumcision-fully-healed-in-Zambian-study/page/2227154/; P. C. Hewett et al., "Sex with Stitches: Assessing the Resumption of Sexual Activity during the Postcircumcision Wound-Healing Period," *AIDS* 26, no. 6 (March 27, 2012): 749–756.

123 *the equivalent of a surgical vaccine:* "South African Doctor Warns Against Using Circumcision to Fight HIV," *M&C*, "Health News," November 23, 2011, accessed at http://news.monstersandcritics.com/health/news/article_1677037.php/South-African-doctor-warns-against-using-circumcision-to-fight-HIV.

123 *"The proponents of this speculation choose":* James L. Snyder, "The Problem of Circumcision in America," *The Truth Seeker* (July/August 1989): 40, accessed at http://www.noharmm.org/problem.htm. See also this video of James L. Snyder, M.D., F.A.C.S., past president of the Virginia Urological Society, discussing the controversy around infant circumcision. Conversation recorded in San Francisco, California, July 29, 2010, http://www.youtube.com/watch?v=XrcMYq0ASB8.

123 *decision should be left up to parents:* Task Force on Circumcision, "Technical Report: Male Circumcision," *Pediatrics* 130 no. 3, September 1, 2012, accessed at http://pediatrics. aappublications.org/content/130/3/e756.full. Earlier statements from both AAP and ACOG also did not universally recommend the operation: ACOG Committee Opinion, "Circumcision," no. 260, October 2001 (reaffirmed 2011), accessed at http://www.acog.org/Resources_And_ Publications/Committee_Opinions/Committee_on_Obstetric_Practice/Circumcision. See also "American Academy of Pediatrics Circumcision Policy Statement," *Pediatrics* 103, no. 3 (March 1, 1999): 686–693, accessed at http://pediatrics.aappublications.org/content/103/3/686.full.

124 *ban on the procedure in Holland:* "The official viewpoint of KNMG and other related medical/ scientific organisations is that non-therapeutic circumcision of male minors is a violation of children's rights to autonomy and physical integrity," the Royal Dutch Medical Association explains on its website. "Contrary to popular belief, circumcision can cause complications— bleeding, infection, urethral stricture and panic attacks are particularly common. KNMG is therefore urging a strong policy of deterrence. KNMG is calling upon doctors to actively and insistently inform parents who are considering the procedure of the absence of medical benefits and the danger of complications." "Non-therapeutic Circumcision of Male Minors (2010)," accessed at http://knmg.artsennet.nl/Publicaties/KNMGpublicatie/Nontherapeutic-circumcision-of-male-minors-2010.htm.

124 *"All human beings should":* Holm Putzke, "Let Boys Decide When They're 16," Room for Debate, *New York Times*, July 10, 2012, accessed at http://www.nytimes.com/ roomfordebate/2012/07/10/an-age-of-consent-for-circumcision.

124 *considering banning the operation:* "Norwegian gov't coalition seeking to outlaw circumcision," *JTA*, June 25, 2012, accessed at http://www.jta.org/news/article/2012/06/25/3099116/ norwegian-political-party-seeks-to-outlaw-circumcision.

124 *less than 1 percent is Muslim:* Cathy Lynn Grossman, "Number of U.S. Muslims to Double," *USA Today*, January 27, 2011, accessed at http://www.usatoday.com/news/religion/2011-01-27-1Amuslim27_ST_N.htm. See also Pew Research Center, *The Future of the Global Muslim Population: Projections of 2010–2030* (Washington, D.C.: Pew Research Center's Forum on Religion & Public Life, 2011), 15, accessed at http://www.pewforum.org/uploadedFiles/ Topics/Religious_Affiliation/Muslim/FutureGlobalMuslimPopulation-WebPDF-Feb10.pdf.

124 *"We all awoke":* This and subsequent quotations: Deston Nokes, parent, in an interview with the author, March 1, 2012.

124 *"Statistically, if you look":* Jade Eagles, parent, in discussion with the author, June 29, 2011.

125 *"You put one drop of wine":* Adam Deutsch, internist, cardiologist, and parent, in an interview with the author, June 29, 2011.

125 *newborn's brain chemistry is altered:* Curtis L. Lowry, M.D., et al., "Neurodevelopmental Changes of Fetal Pain," *Seminars in Perinatology* 31, no. 5 (October 2007): 275–282, accessed at http://anes-som.ucsd.edu/VP%20Articles/Topic%20C.%20Anand.pdf.

125 *less able to cope with the pain:* A. Taddio, J. Katz, A. L. Ilersich, and G. Koren, "Effect of Neonatal Circumcision on Pain Response During Subsequent Routine Vaccination," *Lancet* 349, no. 9052 (1997): 599–603.

125 *"They exhibit all the":* Sylvia Fine, M.D., in an interview with the author, December 15, 2010.

126 *many parents think:* Beth Hardiman, M.D., in an interview with the author, December 13, 2010.

126 *"The baby goes home from the hospital":* Georganne Chapin, executive director, Intact America, in an interview with the author, August 28, 2009.

126 *the surgery itself is botched:* G. R. Gluckman et al., "Newborn Penile Glans Amputation During Circumcision and Successful Reattachment," *Journal of Urology* 153, no. 3, part 1 (March 1995): 778–779, accessed at http://www.ncbi.nlm.nih.gov/pubmed/7861536.

126 *In January 2003 a one-week-old:* "U.S. Judge Awards Payout for Botched Circumcision," Agence France-Presse, July 18, 2011, http://tinyurl.com/7q6rfef.

126 *In 2010 a Florida couple:* Ty Tagami, "Atlanta Lawyer Wins $11 Million Lawsuit for Family in

Botched Circumcision," *Atlanta Journal-Constitution*, July 19, 2010, accessed at http://www. ajc.com/news/nation-world/atlanta-lawyer-wins-11-573890.html.

126 *139 reports of problems from circumcision clamps:* Molly Hennessy-Fiske, "Injuries Linked to Circumcision Clamps," *Los Angeles Times*, September 26, 2011, accessed at http://articles. latimes.com/2011/sep/26/health/la-he-circumcision-20110926.

126 *blood loss:* Stanford School of Medicine, "Complications of Circumcision," 2012, accessed at http://newborns.stanford.edu/CircComplications.html.

126 *"We have about two":* Anonymous physician in an interview with the author, December 1, 2010.

126 *a tiny baby:* The average newborn weighs about 7.5 pounds or 3.4 kilograms. He will have approximately 76.5 ml/kilogram = 260.1 ml = 8.795 fluid ounces. For more on calculating human blood volume, see http://www.med.umich.edu/irbmed/guidance/blood_draw.htm.

127 *"I didn't realize a baby":* Suzanne Fournier, "Lack of Post-Surgery Info Angers Grieving Parents," *Province*, February 13, 2004, accessed at http://www.cirp.org/news/theprovince02-13-04/.

127 *ritual Jewish circumcision is performed:* Susan Markel, M.D., with Linda Palmer, *What Your Pediatrician Doesn't Know Can Hurt Your Child* (Dallas, Tex.: BenBella Books, 2010), 26–27.

127 *heart failure was unconnected to the surgery:* Martin Beckford, "Police Investigate Baby's Death After Circumcision," *The Telegraph*, February 16, 2007, accessed at http://www.telegraph. co.uk/news/uknews/1542803/Police-investigate-babys-death-after-circumcision.html.

127 *In 2009, seven-month-old:* "Baby Death Mystery," October 3, 2009, WABC-TV, accessed at http://abclocal.go.com/wabc/story?section=news/local&id=7046153.

127 *as many as 117 deaths:* "Lost Boys: An Estimate of U.S. Circumcision-Related Infant Deaths," *THYMOS: Journal of Boyhood Studies* 4, no. 1 (Spring 2010): 78–90.

128 *"Now that's illogical":* This and subsequent quotations: David Llewellyn, lawyer, in an interview with the author, July 25, 2011.

128 *20 percent of the patients he sees:* Christine S. Moyer, "Will male circumcision guidance reverse trends in the procedure?" amednews.com, September 3, 2012, accessed at http:// www.ama-assn.org/amednews/2012/09/03/hlsa0903.htm.

128 *approximately 1,250,000 boys:* Approximately 2,279,000 boys are currently born in the United States every year; of these, 54.7 percent are circumcised.

129 *Meatal stenosis:* AUA Foundation, "Meatal Stenosis," accessed at http://www.urologyhealth. org/urology/index.cfm?article=121.

129 *24 out of 329 circumcised boys:* Robert S. Van Howe, M.D., M.S., F.A.A.P., "Incidence of Meatal Stenosis Following Neonatal Circumcision in a Primary Care Setting," *Clinical Pediatrics* 45, no. 1 (January–February 2006): 49–54.

129 *had her oldest son circumcised:* Vicki Usagi, parent, in an interview with the author, March 30, 2012.

129 *buried penis:* E. Eroğlu, "Buried Penis After Newborn Circumcision," *Journal of Urology* 181, no. 4 (April 2009): 1841–1843, accessed at http://www.ncbi.nlm.nih.gov/pubmed/19233400.

129 *the problems are mostly easy to repair:* Aseem Shukla, M.D., pediatric urologist, in an interview with the author, July 29, 2011.

130 *injectable wrinkle treatments:* "Products: Vavelta," Consulting Room, accessed at http://www. consultingroom.com/Treatments/Vavelta.

130 *artificial skin:* Corey S. Maas, M.D., F.A.C.S., and David Lewis, M.D., "Soft Tissue Fillers and Facial Plastic Surgery Practice," Maas Clinic, accessed at http://www.maasclinic.com/soft-tissue-fillers-and-facial-plastic-surgery-practice/.

130 *wound dressings:* Andrew Heenan R.G.N., R.M.N., B.A. (Hons.), "Frequently Asked Questions: Alginate Dressings," last modified October 30, 2003, World Wide Wounds, accessed at http:// www.worldwidewounds.com/1998/june/Alginates-FAQ/alginates-questions.html.

130 *spa products:* Danik MedSpa, "Medical Skin Care," accessed at http://www.danikmedspa. com/skin-care-home.php.

130 *"Medical skin care products":* Ibid.

130 *"Hospital systems make money"*: Kris Ghosh, gynecologist, in an interview with the author, March 2, 2012.

130 *one petri dish of neonatal:* Genlantis, "Human Fibroblasts," accessed at http://www.genlantis. com/human-fibroblasts.html.

131 *more square footage than:* This warehouse is 126,000 square feet, according to ATCC's website, http://www.atcc.org/.

131 *Manassas, Virginia:* "Who We Are," ATCC, accessed at http://www.atcc.org/About/WhoWeAre/tabid/139/Default.aspx.

131 *over $85 million:* Financial information for ATCC is available at Guidestar.org.

131 *among the top-ten best-selling products:* "These Are Our Top Ten Product and Information Pages for Quick Reference," Cell Applications, Inc., accessed at http://www.cellapplications. com/sitemap.php.

131 *their customers include:* "Welcome to Cell Applications, Inc.," Cell Applications, Inc., accessed at http://www.cellapplications.com/content.php?id=2.

131 *cost is $730:* "Human Dermal Fibroblasts," Cell Applications, Inc., accessed at http://www. cellapplications.com/product_desc.php?id=78&category_id=.

131 *Lonza:* "Human Dermal Fibroblasts," Lonza, accessed at http://www.lonza.com/products-services/bio-research/primary-and-stem-cells/human-cells-and-media/fibroblasts-and-media/human-dermal-fibroblasts.aspx.

131 *Cell N Tech Advanced Cell Systems:* "Precision Media and Models," "Primary Dermal Fibroblasts," Cell N Tec, accessed at http://www.cellntec.com/node/337.

131 *Lifeline Cell Technology:* "Human Dermal Fibroblasts," Lifeline Cell Technology, accessed at http://www.lifelinecelltech.com/product-specs/SPC-Fibroblast-family.html.
    "About Our Company," Lifeline Cell Technology, accessed at http://www.lifelinecelltech. com/about.htm.

131 *Zen Bio, Inc.: Human Dermal Fibroblast Manual* (Research Triangle Park, N.C.: ZenBio, 2011), accessed at http://www.zen-bio.com/pdf/ZBM%20Human%20Adult%20Dermal%20 Fibroblast%20Manual.pdf.

131 *Life Technologies Corporation:* "Fibroblast Culture Systems," Life Technologies Corporation, accessed at http://www.invitrogen.com/site/us/en/home/Products-and-Services/Applications/Cell-Culture/Primary-Cell-Culture/Dermal_Fibroblast_Culture_Systems.html.

131 *PromoCell:* PromoCell, "Normal Human Dermal Fibroblasts," accessed at http://www. promocell.com/products/human-primary-cells/fibroblasts/normal-human-dermal-fibroblasts-nhdf/.

131 *System Biosciences: Human Foreskin Fibroblast Cell Lines (Neonatal and Pooled)* (Mountain View, Calif.: System Biosciences, 2009), accessed at http://www.systembio.com/downloads/Manual_HFFn_booklet_WEB.pdf.

131 *Applied Biological Materials:* "Human Primary Neonatal Fibroblasts," Applied Biological Materials (ABM), accessed at http://www.abmgood.com/Neonatal-Fibroblasts-T4104.html.

131 *AllCells:* "Normal Human Neonatal Dermal Fibroblast 80% Confluent, 0.5 Million/vial," AllCells, accessed at http://www.allcells.com/product_info.php/products_id/466/T/8.

131 *Millipore:* "Human Epidermal Keratinocytes, Neonatal," Millipore, accessed at http://www. millipore.com/catalogue/item/scce020.

131 *Allele Biotechnology:* Allele Biotechnology, "Feeder Cells," accessed at http://www. allelebiotech.com/feeder-cells/?gclid=CPapsIu6xK4CFUcHRQodNSApWg.

131 *ScienCell Research Laboratories:* "Human Epidermal Keratinocytes-neonatal [HEK-n]," ScienCell, accessed at http://www.sciencellonline.com/site/productInformation. php?keyword=2100.

131 *"If you want to study"*: David Bermudes, Ph.D., assistant professor, California State University at Northridge, in an interview with the author, January 17, 2012.

131 *"So if you want to study people"*: From 1991 to 2006, Bermudes worked at Yale University as a

research scientist and then as an adjunct assistant professor of clinical medicine. His lab team used cells cultured from foreskins to study *Toxoplasma gondii*, a microorganism excreted in cat feces. "Yale had a facility where they would get the trimmings from across the street from Yale–New Haven Hospital," Bermudes told me. Instead of the commercial price of about $400 a dish, Bermudez's lab bought the foreskin cells for only about $15. The cost of growing the cells was subsidized by government grant money: "To grow human cells is very expensive. It's technically demanding, it requires special facilities, highly trained people, and expensive reagents (the things that are used to make the cells grow). There was grant money from the U.S. government to pay someone full time to grow those cells."

132 *how much a circumcision . . . would cost:* Adam Ragusea, "CommonHealth: A $23,000 Circumcision?" RadioBoston, accessed at http://radioboston.wbur.org/2011/08/30/commonhealth-circumcision.

132 *The young mom was told the total charge:* Rachel Zimmerman, "The Saga of the $23,000 Circumcision," CommonHealth (blog), August 26, 2011, accessed at http://commonhealth.wbur.org/2011/08/the-saga-of-the-23000-circumcision-3.

132 *"[a] complex case":* Ragusea, "CommonHealth: A $23,000 Circumcision?"

132 *one bill for a circumcision:* Pat Palmer, founder, Medical Billing Advocates of America, in an interview with the author, March 5, 2012.

132 *"I see that happen all the time":* Kris Ghosh, M.D., in an interview with the author, March 2, 2012.

133 *Merrick Matthew Eagles, was born:* Robyn Eagles, parent, email communication with the author, March 26, 2012.

133 *"a lot of places don't use anything":* Anonymous labor and delivery nurse, Henry County Medical Center, in an interview with Brandeis research assistant, October 25, 2012.

134 *"All you have to do":* Tora Spigner, registered nurse, email communication with the author, September 28, 2012.

134 *"It's a completely cosmetic procedure":* Beth Hardiman, M.D., in an interview with the author, December 13, 2010.

134 *Cost of circumcision:* http://www.healthcarefees.com/2012/02/circumcision-procedure/.

134 *Circumstraint Newborn Immobilizer:* http://www.quickmedical.com/olympicmedical/circumstraint/immobolizer.html.

134 *Surgical Gamco Circumcision Clamp:* http://www.4mdmedical.com/gomco-circumcision-clamp-16mm.html#.UGPOHY5J9-I.

134 *Mogen Clamp:* http://www.israelect.com/come-and-hear/editor/br-clamps/index.html.

134 *Disposable PlastiBell Circumcision Device:* http://www.medexsupply.com/labor-delivery-circumcision-briggs-plastibell-circumcision-device-1-1-cm-x_pid-23705.html.

134 *Twelve-by-eight-inch tray:* "Human Dermal Fibroblasts," Cell Applications, Inc., accessed at http://www.cellapplications.com/product_desc.php?id=78&category_id=.

134 *Award by Fulton County jury to parents:* http://www.foxnews.com/story/0,2933,511809,00.html.

135 *Skadi Hatfield:* As told to the author on March 14, 2012.

Chapter 7 Bottled Profits: How Formula Manufacturers Manipulate Moms

138 *"the most amazing snuggle":* Claudine Jalajas, parent, in an email communication with the author, June 1, 2011.

138 *"If I'd been apart":* Jennifer Fink, parent, in an email communication with the author, June 3, 2011.

138 *oxytocin . . . beneficial in increasing feelings of trust:* Michael Kosfeld et al., "Oxytocin Increases Trust in Humans," *Nature* 435 (June 2, 2005): 673–676, accessed at http://www.nature.com/nature/journal/v435/n7042/full/nature03701.html.

138 *peace, well-being, and bonding:* H. J. Lee, "Oxytocin: The Great Facilitator of Life," *Progress*

in Neurobiology 88, no. 2 (June 2009): 127–151, accessed at http://www.ncbi.nlm.nih.gov/pubmed/19482229.

138 *less likely to develop breast cancer:* A. M. Stuebe et al., "Lactation and Incidence of Premenopausal Breast Cancer: A Longitudinal Study," *Archives of Internal Medicine* 169, no. 15 (August 10, 2009): 1364–1371; H. Jernström et al., "Breast-feeding and the Risk of Breast Cancer in BRCA1 and BRCA2 Mutation Carriers," *Journal of the National Cancer Institute* 96, no. 14 (May 2004): 1094–1098. Collaborative Group on Hormonal Factors in Breast Cancer, "Breast Cancer and Breastfeeding: Collaborative Reanalysis of Individual Data from 47 Epidemiological Studies in 30 Countries, Including 50,302 Women With Breast Cancer and 96,973 Women Without the Disease," *Lancet* 360, no. 9328 (July 20, 2002): 187–195.

138 *ovarian cancer:* K. N. Danforth et al., "Breastfeeding and Risk of Ovarian Cancer in Two Prospective Cohorts," *Cancer Causes Control* 18, no. 5 (June 2007): 517–523.

138 *endometrial cancer:* C. Okamura et al., "Lactation and Risk of Endometrial Cancer in Japan: A Case-Control Study," *Tohoku Journal of Experimental Medicine* 208, no. 2 (February 2006): 109–115, accessed at http://www.ncbi.nlm.nih.gov/pubmed/16434833.

138 *rheumatoid arthritis:* E. W. Karlson, "Do Breast-feeding and Other Reproductive Factors Influence Future Risk of Rheumatoid Arthritis? Results from the Nurses' Health Study," *Arthritis and Rheumatism* 50, no. 11 (November 2004): 3458–3467, accessed at http://www.ncbi.nlm.nih.gov/pubmed?term=Arthritis%20Rheum%2050%20(2004)%3A%203458-3467.

138 *heart disease:* A. M. Stuebe, "Duration of Lactation and Incidence of Myocardial Infarction in Middle to Late Adulthood," *American Journal of Obstetrics and Gynecology* 200, no. 2 (February 2009): 138, accessed at http://www.ncbi.nlm.nih.gov/pubmed?term=Am%20J%20Obstet%20Gynecol%202009%3B%20200%3A138; E. B. Schwarz, "Duration of Lactation and Risk Factors for Maternal Cardiovascular Disease," *Obstetrics and Gynecology* 113, no. 5 (May 2009): 974–982, accessed at http://www.ncbi.nlm.nih.gov/pubmed?term=Obstet%20Gynecol%202009%3B%20113%3A974-82.

138 *adult-onset diabetes:* A. M. Stuebe, "Duration of Lactation and Incidence of Type 2 Diabetes," *Journal of the American Medical Association* 294, no. 20 (November 23, 2005): 2601–2610; R. Villegas, "Duration of Breast-feeding and the Incidence of Type 2 Diabetes Mellitus in the Shanghai Women's Health Study," *Diabetologia* 51, no. 2 (February 2008): 258–266.

138 *natural method for family planning:* Mizanur Rahman and Shamima Akter, "Duration of postpartum amenorrhoea associated with breastfeeding pattern in Bangladesh," *International Journal of Sociology and Anthropology* 2, no. 1 (January 2010): 11–18, accessed at http://www.academicjournals.org/ijsa/PDF/pdf2010/January/Rahman%20and%20Akter.pdf.

138 *lose their pregnancy weight:* J. L. Baker et al., "Breastfeeding Reduces Postpartum Weight Retention," *American Journal of Clinical Nutrition* 88, no. 6 (December 2008): 1543–1551.

139 *women who do not breastfeed at all undergo:* G. G. Gallup Jr. et al., "Bottle Feeding Simulates Child Loss: Postpartum Depression and Evolutionary Medicine," *Medical Hypotheses* (2009): accessed at http://www.albany.edu/news/images/GGallupbottlefeeding.pdf.

139 *ear infections:* Burris Duncan et al., "Exclusive Breast-Feeding for at Least 4 Months Protects Against Otitis Media," *Pediatrics* 91, no. 5 (May 1, 1993): 867–872; ESPGHAN Committee on Nutrition, "Breast-feeding: A Commentary by the ESPGHAN Committee on Nutrition," *Journal of Pediatric Gastroenterology and Nutrition* 49, no. 1 (July 2009): 112–125, accessed at http://www.ncbi.nlm.nih.gov/pubmed?term=J%20Pediatr%20Gastroenterol%20Nutr.%202009%20Jul%3B49(1)%3A112-25. S. Ip et al., *Breastfeeding and Maternal and Infant Health Outcomes in Developed Countries*, Evidence Report/Technology Assessment no. 153, AHRQ Publication No. 07-E007 (Rockville, Md.: Agency for Healthcare Research and Quality, April 2007), accessed at http://www.breastfeedingmadesimple.com/ahrq_bf_mat_inf_health.pdf.

     ESPGHAN Committee on Nutrition, "Breast-feeding: A Commentary by the ESPGHAN Committee on Nutrition."

139 *flatulence, halitosis, digestive problems:* Alan M. Lake, M.D., "Food Protein-induced Proctitis/

colitis, Enteropathy, and Enterocolitis of Infancy," http://www.uptodate.com/contents/food-protein-induced-proctitis-colitis-enteropathy-and-enterocolitis-of-infancy. See also J. A. Vanderhoof, N. D. Murray, S. S. Kaufman, et al., "Intolerance to Protein Hydrolysate Infant Formulas: An Underrecognized Cause of Gastrointestinal Symptoms in Infants," *Journal of Pediatrics* 131, no. 5 (1997): 741.

139 *gastrointestinal illnesses:* P. W. Howie et al., "Protective Effect of Breast Feeding Against Infection," *British Medical Journal* 300, no. 6716 (January 6, 1990): 11, accessed at http://www.ncbi.nlm.nih.gov/pubmed/2105113. S. Ip et al., *Breastfeeding and Maternal and Infant Health Outcomes in Developed Countries*, Evidence Report/Technology Assessment no. 153, AHRQ Publication No. 07-E007 (Rockville, MD: Agency for Healthcare Research and Quality, April 2007), accessed at http://www.breastfeedingmadesimple.com/ahrq_bf_mat_inf_health.pdf.

139 *respiratory tract infections:* L. Duijts et al., "Prolonged and Exclusive Breastfeeding Reduces the Risk of Infectious Diseases in Infancy" *Pediatrics* 126, no. 1 (July 2010): e18–e25; C. J. Chantry et al., "Full Breastfeeding Duration and Associated Decrease in Respiratory Tract Infection in U.S. Children," *Pediatrics* 117, no. 2 (February 2006): 425–432; P. Nafstad et al., "Breastfeeding, Maternal Smoking and Lower Respiratory Tract Infections," *European Respiratory Journal* 9, no. 12 (December 1996): 2623–2629.

139 *bouts of bad behavior:* K. Heikkilä, "Breast feeding and Child Behaviour in the Millennium Cohort Study," *Archives of Disease in Childhood* 96, no. 7 (July 2011): 635–642, accessed at http://www.ncbi.nlm.nih.gov/pubmed?term=Arch%20Dis%20Child%2096%20no.%207%3A%20635.

139 *70 percent less likely to die of SIDS:* F. R. Hauck et al., "Breastfeeding and Reduced Risk of Sudden Infant Death Syndrome: A Meta-analysis," *Pediatrics* 128, no. 1 (July 2011): 103–110.

139 *necrotizing enterocolitis:* S. Ip et al., *Breastfeeding and Maternal and Infant Health Outcomes in Developed Countries*.

139 *lining of the intestinal wall: Necros* means "dead" in ancient Greek and *ize* is from the Greek root "to become."

139 *"among the most common and devastating":* Josef Neu, M.D., and W. Allan Walker, M.D., "Necrotizing Enterocolitis," *New England Journal Medicine* 364 (January 20, 2011): 255–264.

139 *15 to 25 percent of the infants:* Marion C. W. Henry and R. Lawrence Moss, "Necrotizing Enterocolitis," *Annual Review of Medicine* 60 (February 2009): 111–124.

139 *reduce significantly the risk of late-onset septicemia:* A. Rønnestad et al., "Late-Onset Septicemia in a Norwegian National Cohort of Extremely Premature Infants Receiving Very Early Full Human Milk Feeding," *Pediatrics* 115, no. 3 (March 1, 2005): e269–e276, accessed at http://pediatrics.aappublications.org/content/115/3/e269.abstract?sid=cb271168-06e9-4887-88ec-40866a1058ee.

139 *less likely to become obese:* Thomas Harder et al., "Duration of Breastfeeding and Risk of Overweight: A Meta-Analysis," *American Journal of Epidemiology* 162, no. 5 (September 1, 2005): 397–403; C. G. Owen et al., "Effect of Infant Feeding on the Risk of Obesity Across the Life Course: A Quantitative Review of Published Evidence," *Pediatrics* 115, no. 5 (May 2005): 1367–1377; Bernardo L. Horta, *Evidence on the Long-Term Effects of Breastfeeding: Systematic Review and Meta-Analyses* (Geneva, Switzerland: World Health Organization, 2007), accessed at http://whqlibdoc.who.int/publications/2007/9789241595230_eng.pdf.

S. Ip et al., *Breastfeeding and Maternal and Infant Health Outcomes in Developed Countries*.

139 *develop diabetes:* S. M. Virtanen and M. Knip, "Nutritional Risk Predictors of Beta Cell Autoimmunity and Type 1 Diabetes at a Young Age," *American Journal of Clinical Nutrition* 78, no. 6 (December 2003): 1053–1067; J. Karjalainen et al., "A Bovine Albumin Peptide as a Possible Trigger of Insulin-Dependent Diabetes Mellitus," *New England Journal of Medicine* 327, no. 5 (July 30, 1992): 302–307; H. C. Gerstein, "Cow's Milk Exposure and Type I Diabetes Mellitus. A Critical Overview of the Clinical Literature," *Diabetes Care* 17, no. 1 (January 1994): 13–19; J. M. Norris and F. W. Scott, "A Meta-analysis of Infant Diet

and Insulin-dependent Diabetes Mellitus: Do Biases Play a Role?" *Epidemiology* 7, no. 1 (January 1996): 87–92; Patricia A. McKinney et al., "Perinatal and Neonatal Determinants of Childhood Type 1 Diabetes. A Case-Control Study in Yorkshire U. K.," *Diabetes Care* 22, no. 6 (June 1999): 928–932.

139 *have asthma:* "Our analysis showed that breastfeeding for at least 3 months was associated with a 27 percent (95% CI 8% to 41%) reduction in the risk of asthma in those subjects without a family history of asthma compared with those who were not breastfed. For those with a family history of asthma, there was a 40 percent (95% CI 18% to 57%) reduction in the risk of asthma in children less than 10 years of age who were breastfed for at least 3 months compared with those who were not breastfed." S. Ip et al., *Breastfeeding and Maternal and Infant Health Outcomes in Developed Countries,* 3–4.

139 *other allergies:* A. Lucas et al., "Early Diet of Preterm Infants and Development of Allergic or Atopic Disease: Randomised Prospective Study," *British Medical Journal* 300, no. 6728 (March 31, 1990): 837–840.

139 *childhood leukemia:* "We found breastfeeding of at least 6 months duration was associated with a 19 percent (95% CI 9% to 29%) reduction in the risk of childhood ALL [acute lymphocytic leukemia]. The previous meta-analysis also reported an association between breastfeeding of at least 6 months duration and a 15 percent reduction (95% CI 2% to 27%) in the risk of acute myelogenous leukemia (AML). Overall there is an association between a history of breastfeeding for at least 6 months duration and a reduction in the risk of both leukemias (ALL and AML)." S. Ip et al., *Breastfeeding and Maternal and Infant Health Outcomes in Developed Countries,* 5.

139 *"breastfeeding may have":* C. G. Owen, "Infant Feeding and Blood Cholesterol: A Study in Adolescents and a Systematic Review," *Pediatrics* 110, no. 3 (September 2002): 597–608.

139 *less likely to be abused or neglected:* Lane Strathearn et al., "Does Breastfeeding Protect Against Substantiated Child Abuse and Neglect? A 15-Year Cohort Study," *Pediatrics* 123, no. 2 (February 1, 2009): 483–493.

139 *prompts the development of white matter:* E. B. Isaacs, "Impact of Breast Milk on Intelligence Quotient, Brain Size, and White Matter Development," *Pediatric Research* 67, no. 4 (April 2010): 357–362.

139 *more intelligent:* S. Kramer, "Breastfeeding and Child Cognitive Development: New Evidence from a Large Randomized Trial," *Archives of General Psychiatry* 65, no. 5 (May 2008): 578–584, accessed at http://www.ncbi.nlm.nih.gov/pubmed/18458209; Bernardo L. Horta, *Evidence on the Long-Term Effects of Breastfeeding: Systematic Review and Meta-Analyses* (Geneva, Switzerland: World Health Organization, 2007), accessed at http://whqlibdoc.who.int/publications/2007/9789241595230_eng.pdf.

139 *scoring higher on tests of mental development:* Lise Eliot, Ph.D., *What's Going On in There?* (New York: Bantam, 1999), 184.

140 *recommending babies be breastfed:* American Academy of Pediatrics, "Breastfeeding and the Use of Human Milk," *Pediatrics* 115, no. 2 (February 1, 2005): 496–506, accessed at http://aappolicy.aappublications.org/cgi/content/full/pediatrics;115/2/496.

140 *supplementation with other fluids should not:* In the 1960s it was popular to supplement breastfeeding with bottles of water, a practice that continues in many parts of the world (to the great detriment of infants in developing countries who often die needlessly as a result of drinking contaminated water). My three older brothers and I were given bottles of water because breast milk was thought not to be adequately hydrating.

140 *"Breastfeeding is an unequalled way":* World Health Organization/UNICEF, *Global Strategy for Infant and Young Child Feeding* (Geneva, Switzerland: World Health Organization, 2003), 7–8, accessed at http://whqlibdoc.who.int/publications/2003/9241562218.pdf.

140 *our breastfeeding rates are among the lowest:* Save the Children, *Nutrition in the First 1,000 Days: Save the World's Mothers 2012* (Westport, Conn.: Save the Children, May 2012), 39.

140 *only 77 percent of American women:* "Breastfeeding Report Card 2012, United States: Outcome

Indicators," Centers for Disease Control and Prevention, accessed at http://www.cdc.gov/breastfeeding/data/reportcard2.htm. Statistics on how many American women breastfeed vary depending on who is collecting the data. The numbers have one thing in common: They are all much lower than they should be to promote optimal child and maternal health.

140 *only 36 percent are exclusively:* "Breastfeeding Report Card 2012, United States: Outcome Indicators," Centers for Disease Control and Prevention, http://www.cdc.gov/breastfeeding/data/reportcard2.htm.

140 *4.3 million babies born:* CIA, *World Factbook,* estimated 2011 statistics (based on a population of 313,847,465 and a live birth rate of 13.68 births/1,000 population), accessed at https://www.cia.gov/library/publications/the-world-factbook/geos/us.html.

140 *nearly 3 million babies:* Many women who identify themselves as breastfeeding moms give their infants formula as well. Because of this, the number of exclusively nursed American infants is probably much lower than these statistics indicate.

141 *"I just assumed that breastfeeding":* Annie Urban, parent, in an interview with the author, May 18, 2011.

142 *"tongue-tie":* D. M. B. Hall and M. J. Renfrew, "Perspectives: Tongue Tie," *Archives of Disease in Childhood* 90 (2005): 1211–1215, accessed at http://adc.bmj.com/content/90/12/1211.1.full.

142 *does not mandate paid leave:* Human Rights Watch, *Failing Its Families: Lack of Paid Leave and Work-Family Supports in the U.S.* (New York: Human Rights Watch, February 2011), 1, accessed at http://mediavoicesforchildren.org/?p=8085.

142 *industry-sponsored:* Infant Feeding and Nutrition, http://www.infantformula.org/about-ifc.

142 *"Infant Formula: A Safe and Nutritious Feeding Choice":* Infant Feeding and Nutrition, http://www.infantformula.org/.

142 *corporate-sponsored medical societies:* American Academy of Pediatrics, Honor Roll of Giving, *AAP News* 31 (2010): 26. Abbott Nutrition and Mead Johnson Nutrition, manufacturers of the two leading formula brands in the United States, each gave in the $750,000 and above category.

142 *America ranked* last: Save the Children, *Nutrition in the First 1,000 Days*, 42.

142 *Breastfeeding was much more difficult:* Leslie Ott, parent, in an interview with the author, May 17, 2011.

144 *"She said she was really hungry":* Ruby Wentz, parent, in an interview with the author, May 30, 2011. Many hospitals have a nonevidence-based hypoglycemia protocol. Falling blood sugar, or hypoglycemia, is a normal occurrence after birth. Though most, if not all, babies have a drop in blood sugar immediately following birth, their blood sugar will start to rise by three hours of age or so. Large babies are not necessarily at higher risk of hypoglycemia unless the mother was diabetic (Ruby was not). Then they need to be watched very carefully because extreme low blood sugar that lasts for too long can result in brain damage, since the brain needs glucose for fuel. But breastfed babies have a compensatory mechanism whereby ketone bodies elevate during the couple of hours after birth that it takes to normalize blood glucose values. The brain uses ketone bodies as an alternative fuel source until the baby starts eating more. For more information see N. Wight, K. A. Marinelli, and the Academy of Breastfeeding Medicine Protocol Committee, "ABM Clinical Protocol #1: Guidelines for Glucose Monitoring and Treatment of Hypoglycemia in Breastfed Neonates," Revision June 2006, *Breastfeeding Medicine* 1, no. 3 (2006): 178–184.

144 *When she gave birth at a hospital in Downey, California:* Lana Wahlquist, parent, in discussion with the author, April 4, 2012.

144 *the hospital pediatrician was concerned:* Melissa Bartick, M.D., parent, her birth story and subsequent quotations from an interview with the author, March 9, 2011.

145 *importance of avoiding pacifiers:* Cynthia R. Howard, M.D., M.P.H., Ruth A. Lawrence, M.D., et al., "Randomized Clinical Trial of Pacifier Use and Bottle-Feeding or Cupfeeding and Their Effect on Breastfeeding," *Pediatrics* 111, no. 3 (March 1, 2003): 511–518; A. M. Vogel

et al., "The Impact of Pacifier Use on Breastfeeding: A Prospective Cohort Study," *Journal of Paediatrics and Child Health* 37, no. 1 (February 2001): 58–63; Jack Newman, "Breastfeeding Problems Associated with the Early Introduction of Bottles and Pacifiers," *Journal of Human Lactation* 6, no. 2 (June 1990): 59–63; C. G. Victora et al., "Use of Pacifiers and Breastfeeding Duration," *Lancet* 341, no. 8842 (February 13, 1993): 404–406.

145 *optimal infant-mother bonding:* Marshall H. Klaus and John H. Kennell, *Maternal-Infant Bonding: The Impact of Early Separation or Loss on Family Development* (St. Louis, Mo.: Mosby, 1976); P. de Château et al., "A Study of Factors Promoting and Inhibiting Lactation," *Developmental Medicine and Child Neurology* 19, no. 5 (October 1977): 575–584; P. de Chateau and B. Wiberg, "Long-Term Effect on Mother-Infant Behaviour of Extra Contact During the First Hour Post Partum. III. Follow-Up at One Year," *Scandinavian Journal of Social Medicine* 12, no. 2 (1984): 91–103; S. K. McGrath and J. H. Kennell, "Extended Mother-Infant Skin-to-Skin Contact and Prospect of Breastfeeding," *Acta Paediatrica* 91, no. 12 (December 2002): 1288–1289; P. de Château, "The Interaction Between the Infant and the Environment: The Importance of Mother-Child Contact After Delivery," *Acta Paediatrica Scandinavica Supplement* 344 (1988): 21–30.

146 *"I work at Cedars-Sinai":* Jay Gordon, pediatrician, in an interview with the author, April 2, 2012.

146 *"Every mother is told":* Marsha Walker, lactation consultant, in an interview with the author, May 24, 2011.

146 Selling Out Mothers and Babies: Marsha Walker, *Selling Out Mothers and Babies: Marketing of Breast Milk Substitutes in the USA* (Weston, Mass.: National Alliance for Breastfeeding Advocacy, 2001). Available from the National Alliance for Breastfeeding Advocacy, Research, Education, and Legal Branch.

147 *first took a job at Narragansett Bay Pediatrics:* She started working for that practice in 1997.

147 *Musial liked the practice because:* This and subsequent quotations: Sandra Musial, M.D., pediatrician, in an interview with the author, March 30, 2011.

147 *"Most pediatricians don't receive adequate":* Nancy Mohrbacher, lactation consultant, in an interview with the author, August 1, 2011.

148 *"Compared to other practices":* Maggie Kozel, M.D., pediatrician, in an interview with the author, April 6, 2011.

148 *"The formula reps would hand-deliver it":* Maggie Kozel, M.D., pediatrician, in an interview with the author, February 21, 2011.

148 *Abbott Laboratories . . . employs 91,000 people:* "Fast Facts," Abbott, 2012, accessed at http://www.abbott.com/global/url/content/en_US/10.17:17/general_content/General_Content_00054.htm.

148 *$38.9 billion in sales: 2011 Annual Report* (Abbott Park, Ill.: Abbott, 2011), 2, accessed at http://media.corporate-ir.net/media_files/irol/94/94004/Proxy_Page/AR2011.pdf.

148 *Abbott made record earnings selling:* Ibid., 53.

148 *reported earnings of $3.1 billion:* Mead Johnson Nutrition, *Growing Stronger Every Day: Mead Johnson Nutrition Annual Report 2010* (Glenville, Ill.: Mead Johnson Nutrition Company, 2010), 6.

148 *sales . . . grew 17 percent:* Ibid., 7.

149 *in the highest donation bracket:* American Academy of Pediatrics, "Honor Roll of Giving," *AAP News* 33, no. 20 (2012).

149 *together donating $1.5 million:* The AAP puts them in the $750,000 and above category but does not disclose the actual dollar amount.

149 *voted to sell the company:* David Woodruff et al., "Strained Peas, Strained Profits?" *BusinessWeek*, June 6, 1994, accessed at http://www.businessweek.com/archives/1994/b337543.arc.htm.

149 *Nestlé bought Gerber from Novartis:* "Novartis Completes Its Business Portfolio Restructuring,

Divesting Gerber for USD 5.5 Billion to Nestlé," press release, Novartis, April 12, 2007, accessed at http://www.evaluatepharma.com/Universal/View.aspx?type=Story&id=122986. See also, Hugo Miller and Eva von Schaper, "Nestle Buys Gerber for $5.5 Billion," *Bloomberg News,* April 13, 2007, accessed at http://www.washingtonpost.com/wp-dyn/content/article/2007/04/12/AR2007041200372.html.

149 *Nestlé, the largest food company in the world:* "Nestlé Is . . . ," Nestlé, 2012, accessed at http://www.nestle.com/AboutUs/Pages/AboutUs.aspx.

149 *the smallest share in the formula market:* Victor Oliveria, Elizabeth Frazão, and David Smallwood, "The Infant Formula Market: Consequences of a Change in the WID Contract Brand," U.S. Department of Agriculture, accessed at http://www.ers.usda.gov/media/121286/err124.pdf.

149 *Its donations to the AAP:* American Academy of Pediatrics, "Honor Roll of Giving," *AAP News* 31, no. 26 (2010).

149 *endorsement of exclusive breastfeeding:* L. M. Gartner, "Breastfeeding and the Use of Human Milk," *Pediatrics* 115, no. 2 (February 2005): 496–506, accessed at http://pediatrics.aappublications.org/content/115/2/496.long.

149 *donated more than $6.7 million:* Numbers totaled from the *AAP News,* Honor Roll of Giving (2005–2006 to 2009–2010), which is published each fall for the previous fiscal year and made available to members. Publication courtesy of the AAP.

149 *Formula manufacturers also contributed:* Peggy O'Mara, "The Dastardly Deeds of the AAP," *Mothering* 123, accessed at http://mothering.com/the-dastardly-deeds-of-the-aap.

149 *"something came up":* Deborah Jacobson, media relations manager, AAP, in email communication with the author, July 6, 2012.

149 *"The AAP accepts corporate funding":* Ibid.

150 *"I'm sorry, but that is":* Deborah Jacobson, media relations manager, AAP, in email communication with Melissa Chianta (researcher and fact checker for the author), September 27, 2012.

150 *Steven M. Altschuler, M.D., the current chief:* Mead Johnson Nutrition, "Board of Directors," accessed on September 27, 2012, at http://www.meadjohnson.com/Company/Pages/Board-of-Directors.aspx#17.

150 *"The vast majority of doctors":* This and subsequent quotations: Stefan Topolski, M.D., assistant professor of Family and Community Health at the University of Massachusetts, in discussion with the author, June 21, 2012.

151 *"benefited significantly from":* Mead Johnson Nutrition, *Growing Stronger Every Day,* 2.

151 *"nourish the world's children":* Ibid., inside front cover.

151 *its advertising and promotional campaigns have been so duplicitous:* Jim Edwards, "Strike Five: Why Mead Johnson Keeps Airing Misleading Baby-Formula Ads," *CBS Money Watch,* May 20, 2010, accessed at http://www.cbsnews.com/8301-505123_162-42744908/strike-five-why-mead-johnson-keeps-airing-misleading-baby-formula-ads/.

151 *found guilty five times:* Ibid.

151 *In another, a television actor playing a pediatrician:* "NAD Recommends Mead Johnson Modify, Discontinue Certain Claims for 'Enfamil Premium,' Following Challenge by Abbott," National Advertising Division, news release, May 20, 2010, accessed at http://www.nadreview.org/DocView.aspx?DocumentID=8046&DocType=1.

152 *"You caught me off guard":* Christopher Perille, spokesperson, Mead Johnson, in an encounter with the author, August 16, 2011.

152 *"She wasn't helpful at all":* Margaret Pemberton, parent, in an interview with the author, June 20, 2011.

153 *a service to new moms offering support:* Similac, Feeding Expert, June 20, 2011, accessed at https://similac.com/feeding-nutrition/baby-feeding-expert.

153 *They must then pass a rigorous exam:* Breastfeeding-education.com, a website maintained by IBCLC Gini Baker, explains the current requirements for becoming an Internationally Board

Certified Lactation Consultant (accessed on June 21, 2011). The IBCLC's website also details the requirements for certification: http://www.americas.iblce.org/how-to-qualify (accessed on June 21, 2011).

153 *They agree to "disclose":* "Code of Professional Conduct for IBCLCs," International Board of Lactation Consultant Examiners, November 1, 2011, accessed at http://www.iblce.org/upload/downloads/CodeOfEthics.pdf.

154 *New moms are advised to:* "Breastfeeding Basics," Similac, accessed at http://similac.com/feeding-nutrition/breast-feeding/getting-started.

154 *shocked when she received a package:* Erin Kotecki Vest, patient, in an email communication with the author, August 25, 2011.

154 *Erin, who was too sick:* Ibid. When Erin called the hospital to complain, they denied having given out her contact information. Recovering from major surgery, she did not have the energy to pursue the matter further.

154 *more than 95 percent of mothers given free formula:* Marsha Walker, *Selling Out Mothers and Babies: Marketing of Breast Milk Substitutes in the USA* (Weston, Mass.: National Alliance for Breastfeeding Advocacy, 2001), 13.

154 *In almost every hospital:* As of May 2012, there were 143 U.S. Baby Friendly hospitals and birth centers, according to Baby Friendly Hospital Initiative USA, accessed on June 30, 2012, at http://www.babyfriendlyusa.org/eng/03.html.

154 *about 2 percent of all American hospitals:* According to the American Hospital Association, there are 5,754 hospitals in the United States ("Fast Facts on U.S. Hospitals," American Hospital Association, updated, January 3, 2012, http://www.aha.org/aha/resource-center/Statistics-and-Studies/fast-facts.html). I counted up the number of birth centers listed by the American Association of Birth Centers, http://www.birthcenters.org/find-a-birth-center/index.php, which is 113.

154 *formula sales representatives have often:* Marsha Walker, *Selling Out Mothers and Babies,* 27.

155 *there were reports of nurses:* Ibid.

155 *succumbing to pressure from corporate interests:* Gabrielle Palmer, *The Politics of Breastfeeding: When Breasts Are Bad for Business,* 3rd revised ed. (London: Pinter & Martin, 2009), 258.

155 *"Health workers should not":* World Health Organization, *International Code of Marketing of Breast-milk Substitutes* (Geneva: World Health Organization, 1981), 12, accessed at http://www.who.int/nutrition/publications/code_english.pdf.

155 *"The people who work":* Sylvia Fine, obstetrician, in an interview with the author, December 15, 2010.

155 *On June 16, 2010, Mead Johnson Nutrition:* Description of this event was accessed online on June 7, 2011, at www.regonline.com/register/checkin.aspx?eventid=853294.

156 *John H. Stroger Jr. Hospital:* I toured John H. Stroger Jr. Hospital on the South Side of Chicago on August 24, 2011.

156 *"There's a bag we get":* Rosemarie Mamei Tamba, nurse, head of the Maternal Child Nursing Division, John H. Stroger Jr. Hospital, in discussion with the author, August 24, 2011.

156 *According to the procurement office:* Sonja Vogel, communications and marketing coordinator, John H. Stroger Jr. Hospital and ACHN, Cook County Health & Hospitals System, in an email communication with the author, October 27, 2011.

156 *"the interests of the nation's":* American Nurses Association, "About ANA," accessed on June 16, 2011, at http://www.nursingworld.org/FunctionalMenuCategories/AboutANA.aspx.

156 *"The nurse's primary commitment":* American Nurses Association, *Code of Ethics for Nurses with Interpretive Statements* (Silver Spring, Md.: American Nurses Association, November 15, 2010), 1, accessed at http://www.nursingworld.org/MainMenuCategories/EthicsStandards/CodeofEthicsforNurses/Code-of-Ethics.aspx.

156 *"one of the key things":* This and subsequent quotations: Katie Brewer, senior policy analyst, American Nurses Association, in an interview with the author, June 21, 2011.

156 *"Nurses are often the front line"*: Katie Brewer, e-mail communication with the author, June 15, 2011.

157 *"The nurses come to think of formula"*: Marsha Walker, lactation consultant, in an interview with the author, May 24, 2011.

157 *received a phone call from Similac*: Margaret Cividino, parent, in an interview with the author, June 7, 2011.

158 *"No one disputes the association"*: Lise Eliot, *What's Going On in There?*, 184.

158 *"There's the social sensory interaction"*: This and subsequent quotations: Lise Eliot, neuroscientist and author, *What's Going On in There?*, in discussion with the author, August 22, 2011.

159 *Formula is a highly processed*: Since 2002 formula manufacturers have been adding DHA and RHA—and heavily promoting these new ingredients in advertising campaigns—because these fatty acids, naturally found in breast milk, are crucial to brain and eye development. See Charlotte Vallaeys, *Replacing Mother—Imitating Human Breast Milk in the Laboratory* (Cornucopia, Wisc.: Cornucopia Institute, 2008), 1, accessed at http://cornucopia.org/DHA/DHA_Executive_Summary_web.pdf.

159 *Breast milk, which contains*: Florence Williams, *Breasts: An Unnatural and Natural History* (New York: Norton, 2012), excerpted in the *Guardian*, June 15, 2012: http://www.guardian.co.uk/lifeandstyle/2012/jun/16/breasts-breastfeeding-milk-florence-williams. See also Naomi Baumslag, M.D., M.P.H., and Dia L. Michels, *Milk, Money, and Madness: The Culture and Politics of Breastfeeding* (Westport, Conn.: Bergin & Garvey, 1995), 67.

159 *"The features that make human milk"*: J. Bruce German, as cited in Judy Dutton, "Liquid Gold: The Booming Market for Human Breast Milk," *Wired* (June 2011).

160 *peers at breast milk under the microscope*: Carl Morten Laane, professor of molecular biology, University of Oslo, in discussion with the author, September 16, 2011.

160 *"The cow milk has some resemblance"*: Carl Morten Laane in an email communication with the author, September 27, 2011.

160 *Norway's outstanding maternal and infant outcomes:* "Immigrants and Norwegian-born to Immigrant Parents: Most New Immigrants from the New EU Countries," *Statistics Norway*, April 26, 2012, http://www.ssb.no/innvbef_en/main.html.

161 *about 80 percent*: Save the Children, *Nutrition in the First 1,000 Days: Save the World's Mothers 2012*, "Breastfeeding Policy Scorecard for Developed Countries," 43.

161 *"Now I know that everything"*: This and subsequent quotations: Gro Nylander, M.D., Ph.D., obstetrician, in an interview with the author, September 18, 2011.

161 *Babies were supplemented*: Anne Hagen Grøvslien and Morten Grønn, "Donor Milk Banking and Breastfeeding in Norway," *Journal of Human Lactation* 25 (2009): 206.

161 *Nylander found that babies exclusively nursed*: Gro Nylander et al., "Unsupplemented Breastfeeding in the Maternity Ward. Positive Long-Term Effects," *Acta Obstetricia et Gynecologica Scandinavica* 70, no. 3 (1991): 205–209.

162 *A baby in the United States:* "Country Comparison: Infant Mortality Rate," CIA, *World Factbook,* accessed at https://www.cia.gov/library/publications/the-world-factbook/rankorder/2091rank.html.

162 *escape or delay death if they were breastfed*: A. Chen and W. J. Rogan, "Breastfeeding and the Risk of Postneonatal Death in the United States," *Pediatrics* 5, no. 113 (May 2004): 435–439.

162 *more than $13 billion in medical costs saved*: M. Bartick and A. Reinhold, "The Burden of Suboptimal Breastfeeding in the United States: A Pediatric Cost Analysis," *Pediatrics* 125, no. 5 (May 2010): 1048–1056.

162 *the single most important factor in saving children's lives*: R. E. Black et al., "Where and Why Are 10 Million Children Dying Every Year?" *Lancet* 361, no. 9376 (June 28, 2003): 2226–2234, accessed at http://www.thelancet.com/journals/lancet/article/PIIS0140-6736(03)13779-8/abstract.

162 *In Norway . . . manufacturers are forbidden:* "Sales in Europe accounted for a small percentage

of our global business and are heavily concentrated in specialty formulas . . . that are primarily distributed through pharmacies," Mead Johnson notes in its 2010 annual report. Mead Johnson Nutrition, *Growing Stronger Every Day,* 2.

162 *contaminated by warehouse insects:* "Abbott Recalls Certain Similac® Brand Powder Infant Formulas," U.S. Food and Drug Administration, October 26, 2010, accessed at http://www.fda.gov/Food/FoodSafety/Product-SpecificInformation/InfantFormula/ AlertsSafetyInformation/ucm227039.htm?utm_campaign=Google2&utm_ source=fdaSearch&utm_medium=website&utm_term=infant%20formula%20recall&utm_ content=3.

162 *recalled for its unpleasant smell:* "Enforcement Report for January 12, 2005," U.S. Food and Drug Administration, January 12, 2005, accessed at http://www.fda.gov/Safety/ Recalls/EnforcementReports/2005/ucm120332.htm?utm_source=fdaSearch&utm_ medium=website&utm_term=Enfamil%20recall&utm_content=1.

162 *causing gastrointestinal complaints:* "Gerber Good Start Gentle Powdered Infant Formula: Recall—Off-Odor," U.S. Food and Drug Administration, March 9, 2012, accessed at http:// www.fda.gov/Safety/MedWatch/SafetyInformation/SafetyAlertsforHumanMedicalProducts/ ucm295435.htm?utm_source=fdaSearch&utm_medium=website&utm_term=Good%20 Start%20recall&utm_content=1.

163 *Amount of health care savings:* M. Bartick and A. Reinhold, "The Burden of Suboptimal Breastfeeding in the United States: A Pediatric Cost Analysis," *Pediatrics* 125, no. 5 (May 2010): 1048–1056.

163 *Number of infant deaths that would be avoided:* Ibid.

163 *Amount donated by Abbott:* Numbers calculated for 2004–2012 from the AAP's "Honor Roll of Giving," published each September in the *AAP News.*

163 *Formula industry donations to the AAP:* Ibid.

163 *Net profits of Abbott: Abbott 2011 Annual Report* (Abbott Park, Ill.: Abbott, 2012), accessed at http://www.abbott.com/static/content/microsite/annual_report/2011/downloads/Abbott_ AR2011_Full.pdf.

163 *Net profits of Mead Johnson: Hitting Our Stride: Mead Johnson Nutrition Annual Report 2011* (Glenview, Ill.: Mead Johnson Nutrition, 2012), accessed at http://www.annualreports.com/ HostedData/AnnualReports/PDF/MJN2011.pdf.

163 *Net profits of Nestlé:* Nestlé Annual Report 2011 (Vevey, Switzerland: Nestlé S. A., 2012), accessed at http://www.nestle.com/Common/NestleDocuments/Documents/Library/ Documents/Annual_Reports/2011-Annual-Report-EN.pdf.

163 *Cost of formula for an infant for 12 months:* A 23.4-ounce tub of Enfamil premium costs around $26. Each tub has twenty-one servings (one serving equals one bottle). If a baby has an average of five bottles a day, then one tub will last around four days. There are niney-one four-day segments in a year. Ninety-one times twenty-six equals $2,366.

163 *Maria:* As told to the author by Maria on June 2, 2011.

Chapter 8 Diaper Deals: How Corporate Profits Shape the Way We Potty

167 *"I'd tell her":* This and subsequent quotations: Angela Akins, parent, in discussion with the author, July 15, 2011.

167 *90 percent of American children were potty trained:* Erica Goode, "Two Experts Do Battle over Potty Training," *New York Times,* January 15, 1999.

167 *By 2001 the average age of potty training:* T. R. Schum et al., "Factors Associated with Toilet Training in the 1990s," *Ambulatory Pediatrics* 1, no. 2 (March–April 2001): 79–86.

167 *"toileting troubles are epidemic":* Steve J. Hodges, M.D., with Suzanne Schlosberg, *It's No Accident: Breakthrough Solutions to Your Child's Wetting, Constipation, UTIs, and Other Potty Problems* (Guilford, Conn.: Lyons Press, 2012), 3.

167 *up to 30 percent of children:* M. M. van den Berg et al., "Epidemiology of Childhood Constipation: A Systematic Review," *American Journal of Gastroenterology* 101, no. 10 (October 2006): 2401–2409.

168 *was told by her son's preschool teachers:* Tiffany Vandeweghe, parent, in an interview with the author, February 27, 2012.

168 *"I began to realize if you allow":* T. Berry Brazelton, M.D., pediatrician and founder, Child Development Unit, Boston Children's Hospital, in an interview with the author, December 13, 2010.

169 *An ingenious, energetic, and creative inventor:* Malcolm Gladwell, "Annals of Technology: The Disposable Diaper and the Meaning of Progress," *New Yorker*, November 26, 2001, accessed at http://www.gladwell.com/2001/2001_11_26_a_diaper.htm.

169 *These diapers were bulky, expensive:* Heather McNamara, executive director, Real Diaper Association, in an interview with the author, June 30, 2011.

169 *replace the cellulose in the diaper's core:* Gladwell, "Annals of Technology."

169 *in this case superabsorbent synthetic:* Ibid., 132.

169 *died of toxic shock syndrome after:* Alecia Swasy, *Soap Opera: The Inside Story of Procter & Gamble* (New York: Touchstone, 1993), 130–131.

170 *Despite the fact that the company knew:* Ibid., 133.

170 *sixty million sample packets to American households:* Ibid., 134.

170 *Procter & Gamble could no longer ignore:* Ibid., 130.

170 *their babies' bottoms were breaking out:* Ibid., 152.

170 *making Ultra Pampers with carboxymethylcellulose:* Ibid., 154–155.

171 *"Pediatricians are most valued":* Procter & Gamble, "Benchmark Survey January 1998. Wave II Survey July 1998," Brazelton Papers, Box 78, Folder 162. Francis A. Countway Library of Medicine Center for the History of Medicine, Boston.

171 *Personable, polite, and enthusiastic, Brazelton:* Brazelton Papers. Countway Library of Medicine, Boston.

171 *"I had to watch them":* This and subsequent quotations: T. Berry Brazelton, M.D., pediatrician and author, in an interview with the author, December 13, 2010.

172 *Brazelton graduated from Princeton:* T. Berry Brazelton, M.D., curriculum vitae, prepared May 2010. Courtesy of Suzanne Otcasek, executive assistant to T. Berry Brazelton, Brazelton Touchpoints Center, Boston.

172 *He wrote three more books:* Ibid.

172 *"Our market capitalization":* "Purpose & People: The Power of Purpose," Procter & Gamble, accessed on July 19, 2011, at http://www.pg.com/en_US/company/purpose_people/index.shtml.

172 *"I was pretty surprised":* As quoted in Michael Lasalandra, "Bottom Line in Potty Dispute Is Proper Time for Training," *Boston Herald*, January 13, 1999.

172 *"a fairly blatant conflict of interest":* As quoted in Goode, "Two Experts Do Battle Over Potty Training."

172 *"This child and all the others":* Beverly Beckham, "Flush Harvard Baby Doc's Diaper Pitch," *Boston Herald*, January 15, 1999.

172 *From 1983 until the last one:* According to the show's producer, there were 221 episodes of *What Every Baby Knows* plus several one-hour specials and 26 episodes of *Brazelton on Parenting*, for a total of more than 247 episodes. They started producing the shows in 1983 and mastered the last one in May 2000.

173 *"During the 1980s you had":* Henry O'Karma, producer and founder, New Screen Concepts, in an interview with the author, July 18, 2011.

173 *an online parenting resource:* "Overview Vision/Ideal PPI," Brazelton Papers. Countway Library of Medicine, Center for the History of Medicine.

173 *"Procter & Gamble came to me":* T. Berry Brazelton, M.D., pediatrician and founder, Child

Development Unit, Boston Children's Hospital, in an interview with the author, December 13, 2010.

173 *"It took me a long time"*: As quoted in Goode, "Two Experts Do Battle Over Potty Training."

173 *Brazelton worried that accepting:* This and subsequent quotations: T. Berry Brazelton, M.D., in an interview with the author, December 13, 2010.

174 *"I don't like that"*: When my research assistant followed up with Brazelton's executive assistant, Suzanne Otcasek, via email, to ensure that what I had written was accurate, Suzanne responded on Brazelton's behalf that the way I framed our conversation, "gives the impression that he holds himself to a more lenient standard than the one to which he holds other physicians. His position is that physicians need detailed knowledge about whatever they recommend, and their recommendations should never be motivated or influenced by their own financial interests. Again, the services for which Procter & Gamble paid Dr. Brazelton allowed him to express his carefully researched conclusions that he had arrived at long before engaging with that company and that did not change in any way as a result of that engagement." (Suzanne Otcasek, executive assistant to Dr. T. Berry Brazelton, in email communication with Melissa Chianta, July 13, 2012.)

174 *starting potty training late:* Sonna, *Early-Start Potty Training*, 29.

174 *One child who experienced social ostracism:* Brigid Schulte, "Girl's Suspension a Sign of the Times for Potty Training," *Washington Post*, January 30, 2011, accessed at http://www.washingtonpost.com/wp-dyn/content/article/2011/01/29/AR2011012904520.html.

174 *Zoe was suspended:* Ibid.

174 *"urge incontinence"*: Joseph G. Barone et al., "Later Toilet Training Is Associated with Urge Incontinence in Children," *Journal of Pediatric Urology* 5, no. 6 (December 2009): 458–461.

175 *started toilet training at a* later *age:* See E. Bakker and J. J. Wyndaele, "Changes in the Toilet Training of Children During the Last 60 Years: The Cause of an Increase in Lower Urinary Tract Dysfunction?" *British Journal of Urology International* 86, no. 3 (August 2000): 248–252; Wilhelmina Bakker, *Research into the Influence of Potty-Training on Lower Urinary Tract Dysfunctions* (Belgium: University of Antwerp, 2002); and E. Bakker et al., "Results of a Questionnaire Evaluating the Effects of Different Methods of Toilet Training on Achieving Bladder Control," *British Journal of Urology International* 90, no. 4 (September 2002): 456-461.

175 *"I have always been a proponent"*: As quoted in Scott Tennant, "Toilet Training More Beneficial When Started Early: Incontinence Rates Increase in Children Who Begin Training Later, Data Show," *Urology Times*, April 1, 2010.

175 *"Potty training was not such"*: This and subsequent quotations: Jean-Jacques Wyndaele, M.D., urologist, in an interview with the author, July 18, 2011.

176 *an average of $27 million* per day: Approximately 4.29 million children are born each year and they stay in diapers for approximately three years. I took the number of American children born per year multiplied by the average toilet training age of three years, times average number of diapers per day (between six and eleven), times the cost of the diapers.

177 *Leslie and I sit:* This and subsequent quotations: Leslie Becknell Marx, former assistant brand manager, Procter & Gamble, in discussion with the author, April 7, 2011.

178 *"We don't give away Pampers"*: This and subsequent quotations: Kai Abelkis, sustainability coordinator, Boulder Community Foothills Hospital, in an interview with the author, January 27, 2010. A version of this discussion about the environmental harm of plastic diapers originally appeared in *Mothering* magazine.

179 *"I know toddlers"*: Shawna Cummings, parent, in an interview with the author, November 23, 2009.

179 *also contain trace amounts of dioxin:* Michelle Allsopp, *Achieving Zero Dioxin: An Emergency Strategy for Dioxin Elimination* (Amsterdam: Greenpeace International, 1994), accessed at http://archive.greenpeace.org/toxics/reports/azd/azd.html.

179 *highlighted the harmful effects of dioxin:* Simina Mistreaneu, "MU's Frederick vom Saal Wants FDA to Ban BPA, Endocrine Disruptors," *Missourian*, January 31, 2012, accessed at http://www.columbiamissourian.com/stories/2012/01/31/mus-frederick-vom-saal-wants-fda-ban-bpa-endocrine-disruptors/.

179 *"Dioxins can be toxic":* This and subsequent quotations: Jay Bolus, vice president of technical operations, MBDC, in an interview with the author, January 15, 2010.

179 *the most vulnerable to dioxins:* "Dioxins and Their Effects on Human Health," World Health Organization Media Centre Fact Sheet, no. 225 (May 2010), accessed at www.who.int/mediacentre/factsheets/fs225/en/.

180 *"more than a good idea":* Gladwell, "Annals of Technology."

180 *Instead, the spokesperson will become:* Procter & Gamble customer-service representative in an interview with Tara Crist (Southern Oregon University student research assistant), November 11, 2009.

181 *"RECALL PAMPERS DRY MAX":* Mandy Fonck, Rebecca Boxer, Jenniffer Brown, June and July 2011, comments on "RECALL PAMPERS DRY MAX DIAPERS!" Facebook page.

181 *Proctor & Gamble denied any problem existed:* Martinne Geller, "P&G Dismisses Dry Max Pampers Rash Rumors," *Reuters*, May 6, 2010.

" 'Pampers Bring Back the Old CRUISERS/SWADDLERS': Our Side . . . Finally," Life360, June 9, 2010, accessed at http://www.life360.com/blog/pampers-dry-max-criticism/.

181 *"We've been accused of many things":* Ibid.

181 *Ultimately the Consumer Product Safety Commission did not find:* "United States District Court for the Southern District of Ohio Western Division, Judge Timothy S. Black, Re: Dry Max Pampers Litigation Case No. 1: Lo-Cv-00301-Tsb Doc No. 54. Joint Motion for Certification of Settlement Class, Preliminary Approval of Settlement, Approval of Notice Plan and Notice Administrator and Appointment of Lead Class Counsel." Filed: May 27, 2011, accessed at http://www.diaperclassactionsettlement.com/docs/jointmo.pdf.

182 *scrotal skin temperatures were significantly higher:* C.-J. Partsch et al., "Scrotal Temperature Is Increased in Disposable Plastic Lined Nappies," *Archives of Disease in Childhood* 83, no. 4 (October 2000): 364–368, accessed at http://adc.bmj.com/content/83/4/364.abstract.

182 *linked to asthma:* According to the CDC, 18.7 million adults and 7.0 million children currently suffer from asthma. FastStats, "Asthma," Centers for Disease Control and Prevention, page updated, June 19, 2012, accessed at www.cdc.gov/nchs/FASTATS/asthma.htm.

182 *eye, nose, and throat irritation:* R. C. Anderson and J. H. Anderson, "Acute Respiratory Effects of Diaper Emissions," *Archives of Environmental Health* 54, no. 5 (September–October 1999): 353–358, accessed at www.ncbi.nlm.nih.gov/pubmed/10501153.

182 *"You want to think":* Deborah Gordon, M.D., family physician, in an interview with the author, January 22, 2010.

182 *"We'd rather put":* Kai Abelkis, sustainability coordinator, Boulder Community Foothills Hospital, in an interview with the author, January 27, 2010.

182 *three hundred times their weight in water:* Pfizer, "Super Absorbent Polymers," Pfizer Education Initiative, Ala Kazoo Ala Kazam, accessed at http://www.docstoc.com/docs/83118768/Superabsorbent-Polymers.

183 *"It just was the way":* This and subsequent quotations: Heather McNamara, parent, in an interview with the author, June 30, 2011.

184 *more than 50 percent of the world's children:* Tina Kelley, "A Fast Track to Toilet Training for Those at the Crawling Stage," *New York Times*, October 9, 2005, accessed at http://www.nytimes.com/2005/10/09/nyregion/09diapers.html.

184 *In Niger . . . plastic diapers are so uncommon:* I lived in Niger, West Africa, in 1992–1993 and again in 2006–2007.

184 *A traveler to Ghana, Togo, Benin:* These are all countries that I have visited and this is an experience I have had firsthand.

185 *keep their infants dry and diaper-free:* Christine Gross-Loh, *The Diaper-Free Baby: The Natural Toilet Training Alternative* (New York: HarperCollins), 4.

185 *a recent luxury:* Christine Gross-Loh, author, *The Diaper-Free Baby*, in an interview with the author, April 13, 2012.

185 *"I thought I, a hip":* Christine Gross-Loh, *The Diaper-Free Baby*, 8.

186 *"I began to realize":* Ibid.

186 *"Seasoned grandmothers would tell me":* Christine Gross-Loh, in an interview with the author, April 13, 2012.

186 *remembers sitting in her living room:* This and subsequent quotations: Melinda Rothstein, parent, in an interview with the author, July 11, 2011.

186 *Bauer sleeps her unclad infant:* Ingrid Bauer, *Diaper Free: The Gentle Wisdom of Natural Infant Hygiene* (New York: Plume, 2006), 154.

187 *wearing svelte German-made baby underwear:* Kelley, "A Fast Track to Toilet Training for Those at the Crawling Stage."

187 *DiaperFreeBaby:* In 2011 DiaperFreeBaby had mentors—moms who help other women interested in learning EC—in more than thirty-five states and eleven countries. "Connect Online: About DiaperFreeBaby," updated February 22, 2011, accessed at http://www.diaperfreebaby.org/aboutdfb.htm.

187 *"I'll never forget that moment":* Tiffany Vandeweghe, parent, in an interview with the author, February 27, 2012.

189 *Cost of cloth diapers:* Twenty-four diapers times $20 per diaper. Many cloth diapers cost much less but many families buy more than two dozen diapers.

189 *Errol Matherne:* Errol Matherne, parent, as told to the author, February 10, 2012.

Chapter 9 Boost Your Bottom Line: Vaccinating for Health or Profit?

192 *diphtheria, pertussis, tetanus:* Until 1996 whole cells were used in diphtheria-tetanus-pertussis vaccine (DTP). One of the ingredients of this vaccine was the pertussis toxin, which is used in laboratory research on animals to induce severe brain swelling. Now babies are given the vaccine with pertussis in an attenuated (weakened) form: DTaP. The FDA keeps a complete list of vaccines licensed for use in the United States, which can be found at U.S. Food and Drug Administration, Vaccines, Blood & Biologics, "Complete List of Vaccines Licensed for Immunization and Distribution in the US," accessed at http://www.fda.gov/BiologicsBloodVaccines/Vaccines/ApprovedProducts/UCM093833. A version of this discussion about childhood vaccines first appeared in *Mothering* magazine.

192 *HiB, PCV, poliovirus:* Department of Health and Human Services, Centers for Disease Control and Prevention, "FIGURE 1: Recommended Immunization Schedule for Persons Aged 0 through 6 Years—United States, 2012," accessed at http://www.cdc.gov/vaccines/recs/schedules/downloads/child/0-6yrs-schedule-pr.pdf.

192 *At four months of age a baby receives:* Ibid.

192 *children receive no fewer than:* Department of Health and Human Services, Centers for Disease Control and Prevention, "Recommended Immunization Schedules for Persons Aged 0 Through 18 Years, United States, 2012," http://www.cdc.gov/vaccines/recs/schedules/downloads/child/0-18yrs-11x17-fold-pr.pdf.

192 *more than four times as many injections:* According to Barbara Loe Fisher, executive director of the National Vaccine Information Center (NVIC), in the late 1970s, most children received five DPT shots (fifteen doses of three vaccines) and five doses of oral polio virus at two, four, six, and eighteen months of age, and between four and six years, plus one dose of MMR

between twelve and fifteen months. More information about vaccine licensure dates can be found at www.immunize.org/timeline, a website operated by the Immunization Action Coalition and funded by the CDC. See also Paul A. Offit, M.D., and Louis M. Bell, M.D., *Vaccines: What You Should Know* (Hoboken, N.J.: John Wiley & Sons, 2003), 99.

192 double *the number . . . in Norway:* "Adolescent Health Programme and its contributions to the success of vaccination, Norway," Hanne Nøkleby, Norwegian Institute of Public Health. See Norway: http://www.vhpb.org/files/html/Meetings_and_publications/Presentations/LJUS48S4Nokleby.pdf.

192 *Now the CDC is considering whether to add:* U.S. Food and Drug Administration, "FDA Approves the First Vaccine to Prevent Meningococcal Disease in Infants and Toddlers," news release, April 22, 2011, accessed at http://www.fda.gov/NewsEvents/Newsroom/PressAnnouncements/2011/ucm252392.htm.

192 *"I don't know any rational person":* Centers for Disease Control and Prevention community meeting, Ashland Middle School, Ashland, Oregon (January 10, 2009).

193 *Martin G. Myers, M.D.:* Martin G. Myers and Diego Pineda, *Do Vaccines Cause That?! A Guide for Evaluating Vaccine Safety Concern* (Galveston, Tex.: Immunizations for Public Health, 2008).

193 *98 percent of the world:* I got this number by the following calculation: The U.S. State Department recognizes 195 independent countries in the world. Six of them, according to the WHO, still have some cases of wild polio. U.S. Department of State, "Independent States in the World," Fact Sheet, Bureau of Intelligence and Research, January 3, 2012, accessed at http://www.state.gov/s/inr/rls/4250.htm.

194 *"One instructor didn't vaccinate":* Ann Miller, registered nurse, in an interview with the author, January 24, 2011.

194 *vaccines are neither as safe nor as effective:* International Medical Council on Vaccination, "About," January 4, 2011, accessed at http://www.vaccinationcouncil.org/about/.

194 *Bernadine Healy . . . publicly critiqued the current schedule:* Bernadine Healy signaled this fact on *Larry King Live*: "I think there is so much more to learn. Simple things like a comparison of children who have and have not been vaccinated. This is something that we have talked about doing for many years. It has not been done. It can be done through various models, through case control model models. It can be done retrospectively. It has to be done." CNN *Larry King Live*, "Jenny McCarthy and Jim Carrey Discuss Autism; Medical Experts Weigh In," April 3, 2009, accessed at http://transcripts.cnn.com/TRANSCRIPTS/0904/03/lkl.01.html.

194 *"How can these vaccinations":* Michele Pereira, registered nurse, in an interview with the author, June 28, 2012.

195 *Hepatitis B is not common:* "Vital Hepatitis Statistics & Surveillance: Table 3.1 Reported Cases of Acute, Hepatitis B, by State—United States, 2006–2010," Centers for Disease Control and Prevention, page last updated June 6, 2012, last accessed at http://www.cdc.gov/hepatitis/Statistics/2010Surveillance/Table3.1.htm.

196 *can also contract the disease:* "Hepatitis B FAQs for Health Professionals," Centers for Disease Control and Prevention, page last updated January 31, 2012, accessed at http://www.cdc.gov/hepatitis/HBV/HBVfaq.htm#overview.

196 *their infant has little, if any, chance of getting the disease:* The only plausible way an American infant in a hepatitis B–negative home could get it is if he needed a blood transfusion and was exposed to hepatitis B–tainted blood. While this can be common in developing countries that lack strict controls, donated blood in the United States is carefully screened and is tainted with hepatitis B only 1 in every 65,000 to 500,000 blood units (Robert W. Sears, *The Vaccine Book: Making the Right Decision for Your Child,* Completely Revised and Updated [New York: Little, Brown & Company, 2007], 47). An older child could theoretically contract the disease from another child if he were bitten or had sexual contact.

Only thirty infants a year become infected with hepatitis B, with virtually all of these cases contracted from their mothers. Sears, *The Vaccine Book*, 50.

196 *"The hepatitis B vaccine"*: This and subsequent quotations: Larry Palevsky, M.D., pediatrician, in personal communication with the author, March 1, 2011.

196 *"If I'm a rational person"*: The hepatitis B vaccine was licensed in 1981 and recommended for people in known high-risk groups. In 1991 the recommendation was extended to include all infants.

197 *the most common adverse reactions:* Merck & Co., *Recombivax HB* (Whitehouse Station, N.J.: Merck & Co., 1998), 1–13, accessed at http://www.merck.com/product/usa/pi_circulars/r/recombivax_hb/recombivax_pi.pdf.

197 *no increase in infectious causes of the fevers:* N. Linder et al., "Unexplained Fever in Neonates May Be Associated with Hepatitis B Vaccine," *Arch Dis Child Fetal Neonatal Ed* 81 (1999): F206–F207, doi:10.1136/fn.81.3.F206, accessed at http://fn.bmj.com/content/81/3/F206.full.

197 *revealed myriad problems with the hepatitis B vaccine:* Burton A. Waisbren, M.D., "Universal Hepatitis B Vaccination: Is It a Sword of Damocles Hanging Over the Head of the American People?," New Yorkers for Vaccination Information and Choice, accessed at http://ffitz.com/nyvic/health/hep-b/sword.htm.

197 *lupus:* P. Tudela et al., "Systemic Lupus Erythematosus and Vaccination Against Hepatitis B," *Nephron* 62, no. 2 (1992): 236, accessed at http://www.ncbi.nlm.nih.gov/pubmed/1436323.

197 *fatal form of inflammation:* E. Sindern, J. M. Schroder, M. Krismann, and J. P. Malin, "Inflammatory Polyradiculoneuropathy with Spinal Cord Involvement and Lethal Outcome After Hepatitis B Vaccination," Neurologische Klinik, BG-Kliniken Bergmannsheil, Ruhr-Universitat, Burkle-de-la-Camp-Platz 1, 44789, Bochum, Germany, *Journal of the Neurological Sciences* 186, nos. 1–2 (May 2001): 81–85.

197 *more risk for juvenile diabetes:* J. B. Classen, "The Diabetes Epidemic and the Hepatitis B Vaccines," *New Zealand Medical Journal* 109, no. 1030 (September 27, 1996): 366, accessed at http://www.ncbi.nlm.nih.gov/pubmed/8890866

Michael Devitt, "Hepatitis B Vaccine May Be Linked to Juvenile Diabetes," *Dynamic Chiropractic* 18, no. 15 (July 10, 2000), accessed online at http://www.dynamicchiropractic.com/mpacms/dc/article.php?id=31783.

John Barthelow Classen, "Clustering of Cases of IDDM 2 to 4 Years after Hepatitis B Immunization Is Consistent with Clustering after Infections and Progression to IDDM in Autoantibody Positive Individuals," *Open Pediatric Medicine Journal* 2 (2008): 1–6, accessed at http://www.benthamscience.com/open/topedj/articles/V002/1TOPEDJ.pdf.

P. Pozzilli et al., "Hepatitis B Vaccine Associated with an Increased Type I Diabetes in Italy." Presented at the annual meeting of the American Diabetes Association, San Antonio, Texas, June 13, 2000.

197 *other chronic autoimmune disorders:* J. Toft et al., "Subacute Thyroiditis After Hepatitis B Vaccination," *Endocrine Journal* 45, no. 1 (February 1998): 135. No abstract available. PMID: 9625459; UI: 98287141. L. M. Tartaglino et al., "MR Imaging in a Case of Postvaccination Myelitis," *American Journal of Neuroradiology* 16, no. 3 (March 1995): 581–582. PMID: 7793384; UI: 95313683. R. Treves et al., "Erosive Nodular Rheumatoid Athritis Triggered by Hepatitis B Vaccination," *Presse Médicale* 26, no. 14 (April 26, 1997): 670. F. Trevisani et al., "Transverse Myelitis Following Hepatitis B Vaccination," *Journal of Hepatology* 19, no. 2 (September 1993): 317–318.

197 *cell death and mitochondrial disorders:* H. Hamza, J. Cao, X. Li, et al., "Hepatitis B Vaccine Induces Apoptotic Death in Hepa 1-6 Cells," *Apoptosis* 17, no. 5 (May 2012): 516–527. doi: 10.1007/s10495-011-0690-1. Accessed at http://www.ncbi.nlm.nih.gov/pubmed/22249285.

198 *abnormal neurodevelopmental responses:* L. Hewitson, "Delayed Acquisition of Neonatal Reflexes in Newborn Primates Receiving a Thimerosal-Containing Hepatitis B Vaccine:

Influence of Gestational Age and Birth Weight," *Journal of Toxicology and Environmental Health, Part A,* 73, no. 19 (August 12, 2010): 1298–1313.

198 *Rotavirus is most severe in the first year:* Sears, *The Vaccine Book,* 63.

198 *the disease was responsible for:* Centers for Disease Control and Prevention, "Reduction in Rotavirus After Vaccine Introduction—United States, 2000–2009," *MMWR* 58, no. 41 (October 23, 2009): 1146–1149, accessed at http://www.cdc.gov/mmwr/preview/mmwrhtml/mm5841a2.htm.

198 *1 in 400,000 children under five:* The total number of children under 5 in 2005 in the United States (approximately 20.5 million according to the U.S. Census Bureau) divided by 50, the average number of deaths from rotavirus.

198 *the chance of an American baby:* The average number of deaths from rotavirus divided by the total number of children under 5 in 2005 in the United States.

198 *"Brittany lost five pounds":* Karen Driscoll, parent, in discussion with the author, May 17, 2011.

199 *Wyeth's RotaShield, was taken off the market:* Department of Health and Human Services, Centers for Disease Control and Prevention, "Vaccines and Preventable Diseases: Rotavirus Vaccine (RotaShield) and Intussusception," accessed at phttp://www.cdc.gov/vaccines/vpd-vac/rotavirus/vac-rotashield-historical.htm.

199 *During the first months when Rotateq:* Sears, *The Vaccine Book,* 68.

199 *About 20 percent of the infants:* Ibid., 66.

199 *bloody stools:* Ibid., 67.

199 *rotavirus infection from vaccinated child to nonvaccinated person:* Merck & Co., Inc., *Patient Information: RotaTeq* (Whitehouse Station, NJ: Merck & Co., Inc., 2012), 1–3, accessed at http://www.merck.com/product/usa/pi_circulars/r/rotateq/rotateq_ppi.pdf.

199 *hives:* Ibid.

199 *During safety trials, 1 in 1,000:* Sears, *The Vaccine Book,* 66.

199 *Kawasaki syndrome, a poorly understood:* "Kawasaki Disease," *A. D. A. M. Medical Encyclopedia,* PubMed Health, last reviewed, June 20, 2011, accessed at http://www.ncbi.nlm.nih.gov/pubmedhealth/PMH0001984.

199 *fivefold increase . . . in children in Mexico:* "Intussusception Risk and Health Benefits of Rotavirus Vaccination in Mexico and Brazil," *New England Journal of Medicine* 364, no. 24 (June 2011): 2283–2292, accessed at http://www.nejm.org/doi/full/10.1056/NEJMoa1012952.

199 *"The pediatrician said it was a fluke":* John E. Trainer III, M.D., family physician, in an interview with the author, January 23, 2009.

199 *"I did research at Children's Hospital":* Lyn Redwood, in an interview with the author, May 17, 2011.

200 *"During my training years":* Sears, *The Vaccine Book,* 69.

200 *doctors . . . in Italy, France, The Netherlands:* World Health Organization, WHO Vaccine Preventable Disease Monitoring System, last updated May 18, 2012, accessed on June 3, 2012, at http://apps.who.int/immunization_monitoring/en/globalsummary/ScheduleSelect.cfm. For infant mortality rates, see both the CIA *Factbook* and the United Nations World Population Prospects (http://esa.un.org/unpd/wpp/Excel-Data/mortality.htm).

200 *not found . . . in much of Europe:* World Health Organization, WHO Vaccine Preventable Disease Monitoring System, last updated May 18, 2012, accessed on June 3, 2012, at http://apps.who.int/immunization_monitoring/en/globalsummary/ScheduleSelect.cfm.

Some countries, like Spain, recommend a child get vaccinated against chicken pox at age ten if he has not already been exposed to the disease. Germany recommends universal varicella vaccination—see http://www.ncbi.nlm.nih.gov/pubmed/20600490—as do Canada, Australia, and Japan.

200 *about a hundred deaths each year:* Department of Health and Human Services, Centers for Disease Control and Prevention, Vaccine Information Statement (Interim): Varicella Vaccine, "Chickenpox Vaccine: What You Need to Know," 42 U.S.C. (March 13, 2008): §300aa-26, accessed at www.cdc.gov/vaccines/pubs/vis/downloads/vis-varicella.pdf.

For an extended discussion of how the CDC numbers may be misleading, see Brian Wimer, Jacquelyn L. Emm, and Deren Bader, "Chickenpox Party: Developing Natural Varicella Immunity," *Mothering* 122 (January–February 2004): 30–37, accessed at www.mothering.com/articles/growing_child/child_health/chickenpox_party.html.

200 *the vaccine was first designed:* Barbara Loe Fisher, director, National Vaccine Information Center, in personal communication with the author, January 15, 2009.

201 *"If working mothers could":* Susan Market, M.D., with Linda F. Palmer, *What Your Pediatrician Doesn't Know Can Hurt Your Child* (Dallas, Tex.: BenBella Books), 253.

201 *"We now have an epidemic":* Barbara Loe Fisher, vaccine safety advocate, in personal communication with the author, January 15, 2009.

201 *one million people a year get shingles:* Department of Health and Human Services, Centers for Disease Control and Prevention, "Vaccines and Preventable Diseases: Shingles Disease— Questions and Answers (Herpes Zoster)" (October 19, 2006), accessed at www.cdc.gov/vaccines/vpd-vac/shingles/dis-faqs.htm.

201 *90 percent increase in the number of adults infected:* W. K. Yih et al., "The Incidence of Varicella and Herpes Zoster in Massachusetts as Measured by the Behavioral Risk Factor Surveillance System (BRFSS) During a Period of Increasing Varicella Vaccine Coverage, 1998–2003," *BMC Public Health* 5 (June 16, 2005): 68.

201 *Other studies have estimated increases as well:* G. S. Goldman, "Cost-Benefit Analysis of Universal Varicella Vaccination in the U.S. Taking into Account the Closely Related Herpes-Zoster Epidemiology," *Vaccine* 23, no. 25 (May 9, 2005): 3349–3355.

G. S. Goldman, "The Case Against Universal Varicella Vaccination," *International Journal of Toxicology* 25, no. 5 (September–October 2006): 313–317.

201 *concern over rising rates of shingles:* Clare Murphy, "Why Don't We Vaccinate Against Chickenpox?" *BBC News,* last updated March 10, 2010, http://news.bbc.co.uk/2/hi/8557236.stm.

201 *found in their shingles prevention study:* M. N. Oxman et al., "A Vaccine to Prevent Herpes Zoster and Postherpetic Neuralgia in Older Adults," *New England Journal of Medicine* 352, no. 22 (June 2, 2005): 2271–2284, accessed at http://content.nejm.org/cgi/content/full/352/22/2271.

201 *"Shingles was with us":* This and subsequent quotations: John Grabenstein, senior medical director, Merck & Co., personal communication with the author, April 13, 2009.

202 *"It's a coin toss of efficacy":* John E. Trainer III, M.D., family physician, in an interview with the author, January 23, 2009.

202 *Guillin-Barre syndrome, encephalitis, and seizures:* Department of Health and Human Services, Centers for Disease Control and Prevention, Vaccine Information Statement (Interim): Varicella Vaccine, "Chickenpox Vaccine: What You Need to Know."

Merck & Co., Inc., "Varivax: Varicella Virus Vaccine Live" (Whitehouse Station, N.J.: November 2008), accessed at www.merck.com/product/usa/pi_circulars/v/varivax/varivax_pi.pdf.

202 *the CDC recommends getting two shots: Chickenpox* (Washington, D.C.: Centers for Disease Control and Prevention, 2012): http://www.cdc.gov/VACCINES/vpd-vac/varicella/downloads/PL-dis-chickenpox-color-office.pdf.

202 *Two children in Dr. Sears's practice:* Robert Sears, M.D., pediatrician, in personal communication with the author, February 10, 2009.

203 *"I think Joey was":* Sarah Lipoff, parent, in personal communication with the author, April 25, 2011.

203 *Her husband is the director:* William Redwood was included as one of two top emergency room doctors in a peer-nominated list of Atlanta's top twenty doctors in *Lifestyle Magazine's* "2011 Top Docs."

204 *"I used to preach vaccinations":* Lyn Redwood, member, Board of Health for Fayette County, Georgia, in an interview with the author, February 1, 2011.

204 *"Nobody used the A-word"*: Lyn Redwood, member, Board of Health for Fayette County, Georgia, in an interview with the author, May 17, 2011.

204 *On a test that was like a baby IQ:* Ibid.

204 *"action level" for exposure to mercury in water:* "Environmental Geochemistry of Mercury Mines in Alaska," Fact Sheet 94-072, U.S. Geological Survey, last modified September 30, 2005, accessed at http://pubs.usgs.gov/fs/fs-0072-94/.

205 *an exposure to mercury . . . approximately 125 times:* Lyn Redwood, member, Board of Health for Fayette County, Georgia, in an interview with the author, May 17, 2011.

Lyn Redwood, "Poison in Our Vaccines: Investigating Mercury, Thimerosal, and Neurodevelopmental Delay," *Mothering*, no. 115 (November–December 2002): 36–39.

David Kirby, *Evidence of Harm: Mercury in Vaccines and the Autism Epidemic* (New York: St. Martin's Press, 2005), 55.

205 *In 1981, Kimberley was nine weeks old:* Press conference, May 10, 2011, 12:00 p.m., U.S. Court of Claims, 717 Madison Place, NW, Washington, D.C.

205 *including death:* Though death is a rare side effect of vaccines, every year infants and toddlers still die after being vaccinated. One confidential government report in England revealed that eighteen babies and toddlers died in a four-year period in Great Britain: Beezy Marsh, "Secret Report Reveals 18 Child Deaths Following Vaccinations," *Telegraph*, February 13, 2006, accessed at http://www.telegraph.co.uk/news/uknews/3336455/Secret-report-reveals-18-child-deaths-following-vaccinations.html.

205 *changes in how we define or identify autism:* Mark F. Blaxill, "What's Going On? The Question of Time Trends in Autism," *Public Health Reports* 119 (November–December 2004), accessed at http://www.ncbi.nlm.nih.gov/pmc/articles/PMC1497666/pdf/15504445.pdf. The synopsis of this report reads: "Increases in the reported prevalence of autism and autistic spectrum disorders in recent years have fueled concern over possible environmental causes. The author reviews the available survey literature and finds evidence of large increases in prevalence in both the United States and the United Kingdom that cannot be explained by changes in diagnostic criteria or improvements in case ascertainment. Incomplete ascertainment of autism cases in young child populations is the largest source of predictable bias in prevalence surveys; however, this bias has, if anything, worked against the detection of an upward trend in recent surveys. Comparison of autism rates by year of birth for specific geographies provides the strongest basis for trend assessment. Such comparisons show large recent increases in rates of autism and autistic spectrum disorders in both the U.S. and the U.K. Reported rates of autism in the United States increased from 3 per 10,000 children in the 1970s to 30 per 10,000 children in the 1990s, a 10-fold increase. In the United Kingdom, autism rates rose from 10 per 10,000 in the 1980s to roughly 30 per 10,000 in the 1990s. Reported rates for the full spectrum of autistic disorders rose from the 5 to 10 per 10,000 range to the 50 to 80 per 10,000 range in the two countries. A precautionary approach suggests that the rising incidence of autism should be a matter of urgent public concern."

206 *1 in every 150 children:* In 2009 the Centers for Disease Control and Prevention's Autism and Developmental Disabilities Monitoring Network estimated that 1 in every 150 eight-year-olds in the United States has an autism spectrum disorder. (Centers for Disease Control and Prevention, Autism Information Center, "Frequently Asked Questions—Prevalence.")

206 *1 in every 88 American children:* Centers for Disease Control and Prevention, "Autism Spectrum Disorders: Data & Statistics," last updated March 29, 2012, accessed at http://www. cdc.gov/ncbddd/autism/data.html.

206 *adjuvant aluminum, which is a neurotoxin:* Robert A. Yokel and Mari S. Golub, *Research Issues in Aluminum Toxicity* (Washington, D.C.: Taylor & Francis, 1996).

206 *can cause seizure in primates:* H. M. Wisniewski, R. C. Moretz, J. A. Sturman, et al., "Aluminum Neurotoxicity in Mammals," *Environmental Geochemistry and Health* 12, nos. 1–2 (1990): 115–120, doi: 10.1007/BF01734060.

206 *In June 2012, the Italian Healthy Ministry:* Sue Reid, "MMR: A Mother's Victory," *Mail Online*, June 15, 2012, accessed at http://www.dailymail.co.uk/news/article-2160054/MMR-A-mothers-victory-The-vast-majority-doctors-say-link-triple-jab-autism-Italian-court-case-reignite-controversial-debate.html.

206 *the U.S. courts awarded $1.5 million in damages:* Sharyl Attkisson, "Family to Receive $1.5M+ in First-Ever Vaccine-Autism Court Award," *CBS News*, September 9, 2010, accessed at http://www.cbsnews.com/8301-31727_162-20015982-10391695.html.

206 *the nine shots Hannah received:* Claudia Wallis, "Case Study: Autism and Vaccines," *Time*, March 10, 2008, accessed at http://www.time.com/time/health/article/0,8599,1721109,00.html.

206 *"As a parent I researched":* Lyn Redwood, interview with the author, February 1, 2011.

208 *if vaccines could cause long-term immune dysfunction:* Centers for Disease Control Community Meeting, Ashland Middle School, Ashland, Oregon (January 10, 2009).

208 *"[T]he technology used to make":* "Vaccine Technology Outpacing Ability to Predict Adverse Events, FDAer Says," *Pink Sheet,* November 29, 1999, 8.

208 *"in every office visited there were":* Ibid., 9.

208 *"increasing the risk of mistakenly administering":* Daniel R. Levinson, *Vaccines for Children Program: Vulnerabilities in Vaccine Management* (Washington, D.C.: Department of Health and Human Services, 2012), 1–48, accessed at http://oig.hhs.gov/oei/reports/oei-04-10-00430.pdf.

208 *"Our national vaccine program":* Lyn Redwood, in an interview with the author, May 17, 2011.

209 *an infant can be vaccinated against ten thousand illnesses:* Paul A. Offit, M.D., "Addressing Parents' Concerns: Do Multiple Vaccines Overwhelm or Weaken the Infant's Immune System?" *Pediatrics* 109, no. 1 (January 2002): 124–129, accessed at http://pediatrics.aappublications.org/cgi/content/full/109/1/124.

210 *"There's a benefit from vaccination":* Heather Zwickey, professor of immunology, National College of Natural Medicine, in an interview with the author, November 11, 2010.

210 *Michele Pereira's instructors informed students that:* Michele Pereira, registered nurse, in an interview with the author, June 28, 2012.

212 *the children developed allergies at significantly lower rates:* K. L. McDonald, "Delay in Diphtheria, Pertussis, Tetanus Vaccination Is Associated with a Reduced Risk of Childhood Asthma," *Journal of Allergy and Clinical Immunology* 121, no. 3 (March 2008): 626–631.

212 *wild measles virus have lower rates of allergies:* Helen Rosenlund et al., "Allergic Disease and Atopic Sensitization in Children in Relation to Measles Vaccination and Measles Infection," *Pediatrics* 123, no. 3 (March 2009): 771–778.

212 *cases of type 1 diabetes among children:* Dan Hurley, *Diabetes Rising: How a Rare Disease Became a Modern Pandemic, and What to Do About It* (New York: Kaplan Publishing, 2011).

212 *asthma, allergies, Crohn's disease:* In March 2005, the National Institutes of Health issued a lengthy report to Congress stating that autoimmune diseases, which include "more than 80 chronic, and often disabling, illnesses that develop when underlying defects in the immune system lead the body to attack its own organs, tissues, and cells," and "affect 14.7 to 23.5 million people [in the United States], and—for reasons unknown—their prevalence is rising." National Institutes of Health, "Progress in Autoimmune Diseases Research," March 2005, accessed at http://www.niaid.nih.gov/topics/autoimmune/Documents/adccfinal.pdf. According to the CDC, 7.1 million children currently suffer from asthma in 2009 (CDC, FastStats, "Asthma," www.cdc.gov/nchs/FASTATS/asthma.htm). According to a comprehensive CDC report, "Asthma prevalence rates among children remain at historically high levels following dramatic increases from 1980 until the late 1990s" (http://www.cdc.gov/nchs/data/ad/ad381.pdf).

212 *statistically significant increase in type 1 diabetes:* John Barthelow Classen, "Risk of Vaccine Induced Diabetes in Children with a Family History of Type 1 Diabetes," *Open Pediatric*

Medicine Journal 2 (2008): 7–10, accessed at http://www.benthamscience.com/open/topedj/articles/V002/7TOPEDJ.pdf.

212 *responsible for the exponential rise in autoimmune disorders:* Jean-François Bach, "The Effect of Infections on Susceptibility to Autoimmune and Allergic Diseases," *New England Journal of Medicine* 347 (September 19, 2002): 911–920.

212 *"Fatal allergies and autoimmune diseases":* This and subsequent quotations: Larry Palevsky, M.D., pediatrician, in an interview with the author, March 1, 2011.

213 *more than $50 billion in sales in 2010: Novartis Group Annual Report 2010* (Basel, Switzerland: Novartis International AG, 2010), accessed at http://www.novartis.com/downloads/newsroom/corporate-publications/novartis-annual-report-2010-en.pdf.

213 *net sales of $2 billion:* Novartis, *Vaccines and Diagnostics at a Glance* (Cambridge, Mass.: Novartis AG, 2012), 1–2, accessed at http://www.novartis.com/downloads/newsroom/corporate-fact-sheet/5b_Vaccines_Diagnostics_at_a_glance_EN.pdf.

　　Novartis Group Annual Report 2011 (Basel, Switzerland: Novartis International AG, 2011), 1–284, accessed at http://www.novartis.com/downloads/investors/reports/novartis-annual-report-2011-en.pdf.

214 *"That's why it angers":* Kenneth Saul, M.D., pediatrician, in discussion with the author, May 5, 2011.

214 *"are the bread and butter of [the] specialty":* Robert S. Mendelsohn, M.D., *How to Raise a Healthy Child . . . in Spite of Your Doctor* (New York: Ballantine Books, 1987): 233.

215 *Merck saw their sales increase to $46 billion:* In 2003 Merck made just over $1 billion in vaccine sales.

215 *Merck's sales of vaccines have more than tripled:* Merck & Co., *Form 10-K* (Merck, Whitehouse Station, N.J.: 2006), accessed at http://media.corporate-ir.net/media_files/irol/73/73184/10k/031306_MERCKCOINC10K.pdf.

215 *conditions . . . linked to early vaccination:* K. L. McDonald, S. I. Huq, L. M. Lix, et al., "Delay in Diphtheria, Pertussis, Tetanus Vaccination Is Associated with a Reduced Risk of Childhood Asthma," *Journal of Allergy and Clinical Immunology*, 121, no. 3 (2008): 626–631.

　　T. Kemp, N. Pearce, P. Fitzharris, et al., "Is Infant Immunization a Risk Factor for Childhood Asthma or Allergy?" *Epidemiology* 8, no. 6 (1997): 678–680.

215 *profits of 3.5 billion pounds:* GlaxoSmithKline, *Do More, Feel Better, Live Longer: Annual Report for Shareholders 2011* (Brentford, Middlesex, United Kingdom: GlaxoSmithKline, 2011), 1–249, accessed at http://www.gsk.com/investors/reps11/GSK-Annual-Report-2011.pdf.

215 *688 million pounds:* Ibid.

215 *$2.595 billion in revenue in 2010:* "Key Facts," Sanofi Pasteur, updated April 29, 2011, accessed at http://www.sanofipasteur.us/front/index.jsp?siteCode=SP_CORP&codeRubrique=8&lang=EN.

215 *part of the persuasiveness of a drug pitch:* Anonymous drug representative, in an interview with the author, March 17, 2011.

216 *Nick is an "executive immunization specialist":* Nick Servies, LinkedIn profile, accessed at http://www.linkedin.com/in/nickservies.

216 *I would have to submit a list of questions in writing:* This telephone exchange (it would be too much of a stretch to call it an "interview," "noninterview" might be a more appropriate term) took place between Nick Servies (GlaxoSmithKline representative) and the author on April 22, 2011.

216 *almost $7.4 million:* The exact figure was $7,412,685.

216 *donated at least $500,000:* American Academy of Pediatrics, "Honor Roll of Giving," *AAP News* 33, no. 20 (2012).

216 *since bought by U.S. drug giant Pfizer:* Josephine Moulds, "Pfizer Buys Wyeth for $68bn," *Telegraph*, January 26, 2009, accessed at http://www.telegraph.co.uk/finance/newsbysector/pharmaceuticalsandchemicals/4345182/Pfizer-buys-Wyeth-for-68bn.html.

217 *There have been no cases of wild polio:* http://www.cdc.gov/vaccines/vpd-vac/polio/dis-faqs. htm.

217 *eight cases of paralysis a year:* Department of Health and Human Services, Centers for Disease Control and Prevention, "Vaccines and Preventable Diseases: Polio Disease—Questions and Answers" (April 6, 2007), accessed at www.cdc.gov/vaccines/vpd-vac/polio/dis-faqs.htm. Sears, *The Vaccine Book,* 74–75.

217 *Nigeria, Afghanistan, Pakistan:* As well, twelve other countries in Africa have been affected by poliovirus linked to Nigeria, including Benin, Botswana, Burkina-Faso, Cameroon, Central African Republic, Chad, Côte d'Ivoire, Ethiopia, Ghana, Guinea, Mali, and Togo. By February 2012, only three countries in the world, Afghanistan, Nigeria, and Pakistan, are still experiencing outbreaks of wild polio. (World Health Organization, "Poliomyelitis," Fact Sheet 114, February 2012, accessed at http://www.who.int/mediacentre/factsheets/fs114/en/index.html.)

217 *fewer than 1 percent of polio infections result in paralysis:* WHO African Region: Nigeria, "Polio Eradication in Nigeria, Frequently Asked Questions," accessed at http://www.who. int/countries/nga/areas/polio/faq/en/index.html. According to the most recent statistics, only 0.5 percent of people (1 in 200) who get the disease will become irreversibly paralyzed. (World Health Organization, "Poliomyelitis," Fact Sheet 114, February 2012. Accessed at http://www.who.int/mediacentre/factsheets/fs114/en/index.html.)

218 *about 20 percent of his two thousand patients:* This and subsequent quotations: Robert W. Sears, M.D., pediatrician, in an interview with the author, February 10, 2009.

218 *wealthier, better-educated parents . . . more vaccine concerns:* Susan Leib, M.D., M.P.H.; Penny Liberatos, Ph.D.; and Karen Edwards, M.D., M.P.H., "Pediatricians' Experience with and Response to Parental Vaccine Safety Concerns and Vaccine Refusals: A Survey of Connecticut Pediatricians," *Public Health Reports* 126, Suppl. 2 (2011): 13–23, accessed at http://www.ncbi. nlm.nih.gov/pmc/articles/PMC3113426/.

218 *Five children became sick from Hib:* Department of Health and Human Services, Centers for Disease Control and Prevention, "Invasive *Haemophilus influenzae* Type B Disease in Five Young Children—Minnesota, 2008," *Morbidity and Mortality Weekly Report* 58 (Early Release, January 23, 2009): 1–3, accessed at www.cdc.gov/mmwr/preview/mmwrhtml/ mm58e0123a1.htm.

218 *known side effects . . . include high-pitched crying:* Merck & Co., "Comvax® [Haemophilus B Conjugate (Meningococcal Protein Conjugate) and Hepatitis B (Recombinant) Vaccine]" (Merck & Co.: Whitehouse Station, N.J.: 2010), 9, http://www.merck.com/product/usa/pi_ circulars/c/comvax/comvax_pi.pdf.

218 *Guillain-Barre syndrome:* For a detailed description of the debilitating effects of Guillain-Barre syndrome, as related by a mom who got it from the flu vaccine, see Lisa Marks Smith, "Get Your Affairs in Order," in *Vaccine Epidemic: How Corporate Greed, Biased Science, and Coercive Government Threaten Our Human Rights, Our Health, and Our Children,* edited by Louise Kuo Habakus and Mary Holland (New York: Skyhorse Publishing, 2011), 127–131.

218 *serious Hib infections:* Sears, *The Vaccine Book,* 8.

218 *associated with food poisoning and diarrhea:* "Questions and Answers About Hib Recall," see http://www.lincoln.ne.gov/city/health/data/cdc/HIB.pdf.

218 *Hib bacteria is actually common in the human body:* Sears, *The Vaccine Book,* 1.

218 *complications like pneumonia from Hib are very rare:* Ibid., 3.

219 *she thought about vaccinating:* Rebecca Mehta, parent, in an interview with the author, January 19, 2011.

219 *In November 2008 parents in Maryland:* "Md. Parents Face Jail if Kids Skip Shots," CBS News, February 11, 2009, accessed at http://www.cbsnews.com/stories/2007/11/17/health/ main3517722.shtml.

219 *61 percent of pediatricians reported supporting:* Aaron Wightman, M.D., "Washington State

Pediatricians' Attitudes Toward Alternative Childhood Immunization Schedules," *Pediatrics* 128, no. 6 (December 1, 2011): 1094–1099, accessed at http://pediatrics.aappublications.org/content/128/6/1094.abstract.

219 *accepting families who chose to forgo some vaccines:* Shirley S. Wang, "More Doctors 'Fire' Vaccine Refusers," *Wall Street Journal*, February 15, 2012, accessed at http://online.wsj.com/article/SB10001424052970203315804577209230884246636.html.

219 *"I think he might":* Rebecca Mehta, parent, in an interview with the author, January 19, 2011.

220 *"Continued refusal after adequate discussion":* Douglas S. Diekema, M.D., M.P.H., and the Committee on Bioethics, American Academy of Pediatrics, "Responding to Parental Refusals of Immunization of Children," *Pediatrics* 115, no. 5 (May 2005), accessed at http://aappolicy.aappublications.org/cgi/reprint/pediatrics;115/5/1428.pdf. (Statement reaffirmed in 2009, see http://pediatrics.aappublications.org/content/123/5/1421.full.)

220 *"They've never had anything":* Tasha Pittser, parent, in an email communication with the author, September 24, 2012.

220 *One pediatrician responded with open belligerence:* Jake Marcus, parent, in an interview with the author, May 18, 2011.

221 *"Some doctors are very adamant":* Kenneth Saul, M.D., pediatrician, in an interview with the author, May 29, 2011.

221 *"I have asked if I have been":* Kenneth Saul, M.D., pediatrician, email communication with the author, May 19, 2012.

221 *A 2011 study:* Neil Z. Miller and Gary S. Goldman, "Infant Mortality Rates Regressed Against Number of Vaccine Doses Routinely Given: Is There a Biochemical or Synergistic Toxicity?" *Human & Experimental Toxicology* 30, no. 9 (September 2011): 1420–1428, accessed at http://het.sagepub.com/content/early/2011/05/04/0960327111407644.full.pdf+html.

221 *"These findings demonstrate":* Ibid., 1427.

222 *"the government has made":* Claudia Wallis, "Case Study: Autism and Vaccines," *Time*, March 10, 2008, accessed at http://www.time.com/time/health/article/0,8599,1721109,00.html#ixzz1zb3VHNJE.

222 *In December 2009, Gerberding:* "Dr. Julie Gerberding Named President of Merck Vaccines," Merck & Co., news release, December 21, 2009. Drugs.com, accessed at http://www.drugs.com/news/dr-julie-gerberding-named-president-merck-vaccines-21744.html.

222 *The same holds true in Norway:* Children who attend Waldorf (Steiner) schools are sometimes undervaccinated or unvaccinated.

223 *Merck & Co.'s revenue for 2012:* Merck Annual Report 2011 (Whitehouse, N.J.: Merck, 2012), accessed at http://merck.online-report.eu/2011/ar/servicepages/downloads/files/entire_merck_ar11.pdf.

223 *Amount Merck & Co. gave to AAP in 2010:* "Honor Roll of Giving," *AAP News* 31, no. 26 (2010).

223 *Money paid by the government to parents:* "National Vaccination Injury Compensation Program: Statistics Reports," U.S. Department of Health and Human Services, Health Resources and Services Administration, September 4, 2012, accessed at http://www.hrsa.gov/vaccinecompensation/statisticsreports.html.

223 *Money paid by British government to parents:* "£3.5m Paid Out in Vaccine Damages," BBC News, March 16, 2005, accessed at http://news.bbc.co.uk/2/hi/health/4356027.stm.

223 *Money awarded by the court on August 28, 2010:* Martin Delgado, "Family Win 18 Year Fight Over MMR Damage to Son: £90,000 Payout Is First Since Concerns Over Vaccine Surfaced," MailOnline, August 28, 2010, accessed at http://www.dailymail.co.uk/news/article-1307095/Family-win-18-year-fight-MMR-damage-son--90-000-payout-concerns-vaccine-surfaced.html?ito=feeds-newsxml.

223 *Michelle Maher Ford:* As told to the author by Michelle Maher Ford, parent, April 22, 2011.

Chapter 10 Sick Is the New Well: The Business of Well-Baby Care

228 *ear infections are one of the most overdiagnosed:* Susan Markel, M.D., and Linda F. Palmer, *What Your Pediatrician Doesn't Know Can Hurt Your Child: A More Natural Approach to Parenting* (Dallas, Tex.: BenBella Books, 2010): 203. See also Robert S. Mendelsohn, M.D., *How to Raise a Healthy Child . . . in Spite of Your Doctor* (New York: Ballantine Books, 1987), 140.

228 *more than thirty million courses of antibiotics:* Markel and Palmer, *What Your Pediatrician Doesn't Know Can Hurt Your Child,* 203.

228 *well-child visits:* "Well-child Visits" *MedlinePlus,* updated January 17, 2011, accessed at http:// www.nlm.nih.gov/medlineplus/ency/article/001928.htm.

228 *18.3 minutes with a child under three:* Edward L. Schor, M.D., "Rethinking Well-Child Care," *Pediatrics* 114, no. 1 (July 1, 2004): 210–216, accessed at http://pediatrics.aappublications. org/cgi/content/full/114/1/210.

228 *well-child care accounts for 57 percent:* Ibid.

228 *backbone of pediatrics:* Mendelsohn, *How to Raise a Healthy Child . . . in Spite of Your Doctor,* 230.

229 *written about the systematic problems with well-child visits:* These article include: "Rethinking Well-Child Care," *Pediatrics* 114, no. 1 (July 1, 2004): 210–216, accessed at http://pediatrics. aappublications.org/cgi/content/full/114/1/210; Edward L. Schor et al., "Rethinking Well-Child Care in the United States: An International Comparison," *Pediatrics* 118, no. 4 (October 1, 2006): 1692–1702; Edward Schor, "Quality of Child Health Care: Expanding the Scope and Flexibility of Measurement Approaches," *The Commonwealth Fund* 54 (May 22, 2009): 1–10; and Edward L. Schor, "Improving Pediatric Preventive Care," *Academic Pediatrics* 9, no. 3 (May–June 2009): 133–135.

229 *to help make sense of what's going on:* This and subsequent quotations: Edward Schor, M.D., policy analyst and pediatrician, in an interview with the author, February 14, 2011.

229 *$168,650 a year:* Total mean annual income for pediatricians is $168,650 according to the U.S. Department of Labor, Bureau of Labor Statistics, Occupational Employment Statistics, "Occupational Employment and Wages, May 2011, 29-1065 Pediatricians, General," last modified March 27, 2012, accessed at http://www.bls.gov/oes/current/oes291065.htm. Obstetricians and gynecologists have a mean annual wage of $218,610, according to the U.S. Department of Labor, Bureau of Labor Statistics, Occupational Employment Statistics, "Occupational Employment and Wages, May 2011, 29-1064 Obstetricians and Gynecologists," last modified March 27, 2012, accessed at http://www.bls.gov/oes/current/oes291064.htm.

229 *While pediatricians are on the low-end:* Leslie Kane, "Medscape Physician Compensation Report 2011 Results," *Medscape News Today,* slide 2, accessed at http://www.medscape.com/ features/slideshow/compensation/2011/. Bureau of Labor Statistics, U.S. Department of Labor, *Occupational Outlook Handbook, 2012–2013 Edition,* Physicians and Surgeons, accessed on July 6, 2012, at http://www.bls.gov/ooh/healthcare/physicians-and-surgeons.htm.

229 *more than three times the salary of the average American:* "National Average Wage Index," Automatic Determinations, Social Security Online, accessed on July 6, 2012, at http://www. ssa.gov/oact/cola/AWI.html. The average salary in America is about $42,000 a year. The average salary of a taxi driver in America is about $32,000 ("U.S. National Averages: Taxi Driver," Salary.com, accessed on April 5, 2012, at http://swz.salary.com/SalaryWizard/Taxi-Driver-Salary-Details.aspx); the average salary of a high school teacher is about $54,500 ("U.S. National Averages: Teacher High School," Salary.com, accessed on April 5, 2012, at http://www1.salary.com/High-School-Teacher-salary.html). The average salary of a college professor, who has as many years of higher education as a doctor, is about $80,600 ("U.S. National Averages: Professor Liberal Arts," Salary.com, accessed on April 5, 2012, at http:// www1.salary.com/Professor-Liberal-Arts-salary.html).

229 *debt of at least $100,000*: "Medical Student Debt: Background," American Medical Association, accessed on April 5, 2012, at http://www.ama-assn.org/ama/pub/about-ama/our-people/ member-groups-sections/medical-student-section/advocacy-policy/medical-student-debt/ background.shtml.

230 *"Anywhere between sixty and sixty-five percent"*: Brandon Betancourt, pediatric practice administrator, in an interview with the author, July 27, 2012.

230 *It's so lengthy and complicated*: The 832-page spiral-bound book *Current Procedural Terminology* retails for $109.95 (American Medical Association, *CPT 2012* [*Current Procedural Terminology*], Professional Edition. Chicago: American Medical Association, 2011).

231 *"Sometimes we would pick up"*: This and subsequent quotations: Maggie Kozel, M.D., physician and author, in an interview with the author, February 21, 2011. Kozel describes in detail why she chose to leave pediatrics in her memoir, *The Color of Atmosphere: One Doctor's Journey In and Out of Medicine* (White River Junction, Vt.: Chelsea Green Publishing, 2011), which also informed this chapter.

231 *The AAP publishes guidelines for infant care*: Bright Futures, American Academy of Pediatrics, http://brightfutures.aap.org/.

232 *"Pediatricians look at me"*: Sharon Rising, certified nurse midwife and founder, Centering Healthcare Institute, in an interview with the author, March 18, 2012.

232 *These days Kozel does a mean imitation*: This and subsequent quotations: personal communication, February 21, 2011. For the longer version of why Dr. Kozel chose to leave pediatrics, see also her memoir, *The Color of Atmosphere: One Doctor's Journey In and Out of Medicine*, which also informed this chapter.

232 *"falling off her growth curve"*: Jennifer Rosner, parent, in an interview with the author, March 1, 2011, and March 2, 2011.

233 *almost exclusively white, bottle-fed babies*: L. A. Nommsen-Rivers and K. G. Dewey, "Growth of Breastfed Infants," *Breastfeeding Medicine* 4, Suppl. 1 (October 2009): S45–S49.

233 *used predominantly bottle-fed babies to plot*: C. Garza and M. de Onis, "Rationale for Developing a New International Growth Reference," *Food and Nutrition Bulletin* 25, Suppl. 1 (March 2004), accessed at http://www.ncbi.nlm.nih.gov/pubmed/15069915.

234 *"Women who were breastfeeding"*: This and subsequent quotations: Jay Gordon, M.D., pediatrician and author, in an interview with the author, April 2, 2012.

234 *new information . . . released . . . in 2006*: "The WHO Child Growth Standards," World Health Organization, accessed at http://www.who.int/childgrowth/standards/en.

234 *"For my clinical practice"*: This and subsequent quotations: Jeffrey Brosco, M.D., professor of clinical pediatrics, University of Miami, in an interview with the author, April 2, 2012.

234 *Today Sophia is eleven*: Jennifer Rosner, parent, in an email communication with the author, April 4, 2012.

235 *The Oyakawas and their five children*: Janie Oyakawa, parent, in an interview with the author, March 30, 2012.

235 *"The use of standard growth charts"*: Mendelsohn, *How to Raise a Child . . . In Spite of Your Doctor*, 27.

235 *use of growth charts has led pediatricians to*: Ibid., 28–29.

235 *estrogen therapy to prevent girls from growing too tall*: J. M. Lee and J. D. Howell, "Tall Girls: The Social Shaping of a Medical Therapy," *Archives of Pediatrics and Adolescent Medicine* 160, no. 10 (October 2006): 1035–1039.

235 *prescribing growth hormones*: L. Cuttler and J. B. Silvers, "Growth Hormone Treatment for Idiopathic Short Stature: Implications for Practice and Policy," *Archives of Pediatrics and Adolescent Medicine* 158, no. 2 (February 2004): 108–110, accessed at http://archpedi. jamanetwork.com/article.aspx?articleid=485610.

236 *"no harsh side effects"*: "Why MiraLAX?," MiraLAX, accessed on April 5, 2012, at http://www. miralax.com/miralax/consumer/aboutmiralax.jsp.

236 *long list of side effects reported by adults:* Ask a Patient, "Drug Ratings for MiraLAX," accessed on April 5, 2012, at http://www.askapatient.com/viewrating.asp?drug=20698 &name=MIRALAX.

236 *nausea, bloating . . . diarrhea and hives:* "Polyethylene Glycol 3350," MedlinePlus, accessed at http://www.nlm.nih.gov/medlineplus/druginfo/meds/a603032.html. Last reviewed February 1, 2009.

236 *also an ingredient in antifreeze:* Low molecular weight polyethylene glycol is the key ingredient in automotive antifreeze. In this form it is lethal to ingest.

236 *"a treasure trove of free stuff":* Vincent Iannelli, M.D., "Pediatric Freebies: Free Health Stuff for Your Kids," About.com: Pediatrics, updated January 9, 2010, accessed at http://pediatrics. about.com/cs/pediatrics101/a/free_hlth_stuff.htm.

237 *In 2006 Walt Disney advertised its:* Louise Story, "Anywhere the Eye Can See, It's Likely to See an Ad," *New York Times,* January 15, 2007, accessed at http://www.nytimes.com/2007/01/15/ business/media/15everywhere.html?pagewanted=all.

237 *"Other samples I have recently received":* Vincent Iannelli, "Pediatric Freebies—Free Health Stuff for Your Kids," About.com: Pediatrics, updated January 9, 2010, accessed at http:// pediatrics.about.com/cs/pediatrics101/a/free_hlth_stuff.htm.

237 *acne medication . . . for teens:* Vincent Iannelli, "Acne Treatments for Children," About. com: Pediatrics, updated November 20, 2003, accessed at http://pediatrics.about.com/cs/ conditions/a/acne_treatments.htm.

237 *allergy medication for young children:* Vincent Iannelli, "Allergy Treatments for Kids," About.com: Pediatrics, updated March 5, 2012, accessed at http://pediatrics.about.com/cs/ conditions/a/allergies.htm.

237 *$16 million worth of free samples:* M-A Gagnon and J. Lexchin, "The Cost of Pushing Pills: A New Estimate of Pharmaceutical Promotion Expenditures in the United States," *PLoS Medicine* 5, no.1 (January 3, 2008): e1, doi:10.1371/journal.pmed.0050001.

238 *"The rep for Zithromax":* Zithromax is a broad-spectrum antibiotic available by prescription.

239 *Pedialyte:* An electrolyte drink, like Gatorade, made by Abbott Laboratories, the global health care and medical research company that also makes Similac infant formula.

239 *grape-flavored Pedialyte:* Pedialyte also comes in bright orange (fruit), bright red (strawberry), bright pink (bubble gum), and murky white (unflavored).

239 *acesulfame (another artificial sweetener):* Center for Science in the Public Interest Reports, "Sample Quotes from Cancer Experts' Letters on Acesulfame Testing," Center for Science in the Public Interest, accessed on April 5, 2012, at http://www.cspinet.org/reports/asekquot.html.

239 *FD&C Blue #1:* Sarah Kobylewski and Michael F. Jacobson, Ph.D., *Food Dyes: A Rainbow of Risks* (Washington, D.C.: Center for Science in the Public Interest, June 2010), v.

239 *Red #40:* Ibid.

239 *dehydration is a leading cause of death:* "Diarrhoeal Disease," Fact Sheet No. 330, August 2009, World Health Organization, accessed at http://www.who.int/mediacentre/factsheets/fs330/ en/index.html.

239 *breastfeed as often as possible:* Department of Child and Adolescent Health and Development, *The Treatment of Diarrhoea: A Manual for Physicians and Other Senior Health Workers* (Geneva: World Health Organization, 2005), 9. "Food should *never* be withheld and the child's usual foods should *not* be diluted. Breastfeeding should *always* be continued" (p. 10).

240 *Abbott . . . over $750,000:* American Academy of Pediatrics, "Honor Roll of Giving," *AAP News* 33, no. 20 (2012): doi:10.1542/aapnews.2010319-26.

240 *"She will come to my house":* Elizabeth Hunter, parent, in email communication, June 27, 2012.

241 *"The mom and baby in the first year":* This and subsequent quotations: Laura Wise, M.D., family physician, in an interview with the author, March 21, 2012.

242 *there are now thirteen health centers:* Sharon Rising, founder, CenteringParenting and Centering Healthcare Institute, in an interview with the author, March 31, 2012.

242 *reported higher satisfaction than with traditional care:* Ada Fenick et al., "Health Care Utilization in Infants Receiving Group Pediatric Care," Academic Pediatric Association Presidential Plenary, Denver, May 2, 2011.

242 *"These residents are learning":* Ada Fenick, M.D., pediatrician and assistant professor of pediatrics, Yale School of Medicine, in an interview with the author, April 2, 2012.

243 *Salary of Average American:* U.S. Social Security Administration 2012 Average Wage Index.

243 *Salary of average pediatrician:* "Occupational Employment and Wages, May 2011: 29-1065 Pediatricians, General," Bureau of Labor Statistics, http://www.bls.gov/oes/current/oes291065.htm.

243 *Amount drug companies gave in free samples:* M-A Gagnon and J. Lexchin, "The Cost of Pushing Pills."

243 *Stephanie Precourt:* Stephanie Precourt, parent, as told to the author, February 8, 2012.

Chapter 11 So Where Do We Go from Here?

247 *Robbie Goodrich's wife, Susan, died:* Jahnke, Krista, "Moms for Moses: 2 Dozen Women Give Infant a Nurturing, Nutritional Start," *Deseret News,* August 3, 2009, accessed at http://www.deseretnews.com/article/705320848/Moms-for-Moses-2-dozen-women-give-infant-a-nurturing-nutritional-start.html.

247 *"He's a healthy, happy":* As quoted in Jessica Ravitz, CNN, December 1, 2009, "Nursing Moms: Moms Step in After Infant's Mother Dies," accessed at http://articles.cnn.com/2009-12-01/living/marquette.moms.nursing.moses_1_goodrich-family-nurse-amniotic/3?_s=PM:LIVING.

249 *Norway's family-friendly policies:* The information about Norwegian leave policies comes from Kirsti Bergstø, state secretary, in an interview with the author, September 15, 2011.

250 *In Korean culture a baby's first birthday:* Christine Gross-Loh, "Celebrating Milestones," All Things Mothering (blog), November 19, 2010.

251 *"Read your baby, not the books":* Claire Niala, "Why African Babies Don't Cry: An African Perspective," accessed at http://www.naturalchild.org/guest/claire_niala.html.

Glossary of Terms

260 *There are no federal:* "Radiation-Emitting Products: Ultrasound Imaging," FDA, updated 06/06/2012, http://www.fda.gov/Radiation-EmittingProducts/RadiationEmittingProduct-sandProcedures/MedicalImaging/ucm115357.htm.

Acknowledgments

If it takes a village to raise a child, it took an international metropolis to write this book. I owe a debt of gratitude to so many people, especially the men and women on three continents who openly shared their birth stories, parenting struggles, small triumphs, and bewildering defeats with me. Though I was not able to include every story, I learned something from each of the hundreds of interviews I conducted. Many of the parents I spoke to are named on these pages, others preferred to tell their stories anonymously, but all of you know who you are and you are greatly appreciated. Thank you.

Gillian MacKenzie, the best agent in the universe, was still willing to talk to me after I wasted a ridiculous amount of time searching for a place that served coffee in ceramic mugs instead of product-placement paper cups when we first met at the Roosevelt Hotel. Gillian has been a tireless champion, hand-holder, editor, and friend throughout the conception, gestation, and birth of this book. No one could wish for a better agent—or a better editor. Alexis Gargagliano is both a thoughtful reader and an insightful, smart, and spot-on critic. If this book reads well, it is because of her. Alexis helped me tease out the thornier issues and fixed mistakes on almost every page. I feel honored to have her as an editor, and even more honored to have been able to share a tiny bit in the joy of her first pregnancy (and advise her on cloth diapering and good books to read). Thanks to Alexis's lightning-quick and completely fearless assistant Kelsey Smith, as well as to all the other hardworking, diligent, and intelligent folks at Scribner (the best publishing house in the world), including Samantha Martin, who first saw the potential in this book; Susan Moldow, executive vice president; and vice president and editor in chief, Nan Graham. Scribner's attorney, Elisa Rivlin, as both a lawyer and a reader, has helped make this a better book, and I am grateful to my publicist, Sophie Vershbow, for her tireless support of this project.

This project would not have been possible without Melissa Chianta, a.k.a. the Most Diligent Fact Checker in the World, whom I first met when I was a contributing editor at *Mothering* magazine. Melissa has proved as indefatigable as she is persistent. Hanging in over the long haul and even canceling a trip so we could finish on time, Melissa has meticulously checked every fact and fixed countless errors. Any remaining mistakes are mine.

My in-real-life writing group, Debra Murphy, Rachel Murphy, and Debbie Zaslow, patiently read and reread drafts, caught countless mistakes, and provided me with invaluable feedback. My goal buddy, Marina Krakovsky, helped keep me on track. My friend and best-selling author extraordinaire, Alisa Bowman, proved a master at crafting abstruse concepts into readable prose and generously lent me a sympathetic ear when I needed it most. My research assistant, Caitlin Simmons, transcribed hours of interviews. My thirteen-year-old Waldorf mentee Lucy Neubeck organized my files. Friend, colleague, and diaper expert Christine

Gross-Loh provided a ready ear when we were both panicking in the home stretch. Medical student Sara Hopkins generously introduced me to her advisers. Laura Jessup made me the best sourdough bread in the world and took me on much-needed walks. Our babysitter, friend, and honorary daughter, Hannah Sayles, was happy to take our baby to the potty ("She did it!") and helped in a thousand other ways until she abandoned us to attend Colorado College. I am also grateful to my daughter's friend, Alex Westrick (and her whole wonderful family), who was always so enthusiastic when I biked her and Athena to school in the mornings, delighted to hear another "lecture" about everything from baby wash to baby food, inspired by the research for this book.

Thanks, also, to Shayna Perkinson and all the other wise moms on MamasMedicineWheel, who were an invaluable source of inspiration, stories, and thoughtful parenting advice. And an especially huge thanks to Sue Gries, I mean Susan Langston, my best friend in the whole world, for bringing us homemade strawberry jam, formatting footnotes, taking the kids on adventures, and making my daughter's birthday present while I remained squirreled away in my office.

Thank you to my writer friends on- and off-line: Tangren Alexander, Mark Anderson, Stephanie Auteri, Casey Barber, Edwin Battistella, Andrew Scot Bolsinger, Kris Bordessa, Jane Boursaw, Kerri Fivecoat Campbell, Scott Carney, Kimberly Ford Chisholm, Angela Decker, Kerry Dexter, Karen Driscoll, Hope Edelman, Meagan Francis, Mona Gable, Kristen J. Gough, Alexandra Grabbe, Melanie Haiken, Shu-Huei Henrickson, Sarah Henry, Donna Hull, Claudine Jalajas, Susan Johnson, Debbie Koenig, Sheryl Kraft, Cindy LaFerle, Richard Lehnert, Harriet Lerner, Maryn McKenna, Melanie McMinn, Virginia Morell (who first introduced me to Gillian), Theo Nestor, Charles Ornstein, Brett Paesel, Ruth Pennebaker, Meredith Resnick, Gretchen Rubin, Brette Sember, Jen Singer, Holly Smith, Stephanie Stiavetti, Judith Stock, Candace Walsh, Samantha Ducloux Waltz, Lauren Ware, Michele Warrence-Schreiber, Steve Weinberg, and the *USA Today* investigative journalist who also served as my mentor, Alison Young.

I am awed by how many tremendously talented birth photographers there are working today, some of whose work you will see on these pages. Thank you to everyone who provided me photographs to consider for the book. Harald Birkevold, an investigative reporter in Norway, helped facilitate my visit and generously shared his contacts. David Vanderlip, a digital photography whiz and trainer, has given me invaluable instruction, as have Christopher Briscoe and Sean Bagshaw, two Ashland-based photographers whose talents are internationally known.

A special thanks to founder and editor of *Mothering* magazine, Peggy O'Mara, a tireless champion of safe birth, safe medicine, breastfeeding, and empowering women. I'm grateful to Mothering.com's Web editor, Melanie Mayo, who loves Oregon as much as I do (though she lives in Minnesota). Stephanie Von Hirschberg at *More* magazine; Laurie Grossman, Julie Hogenboum, Laura Lambert, and Shannon Peavey at the Walt Disney Internet Group; Alex Pulaski, George Rede, and Cornelius Swart at the *Oregonian*; Dan Salzstein and Susan Ellingwood at the *New York Times*; Robin Doussard at *Oregon Business Magazine*; Laura Helmuth at *Smithsonian*; Paige Parvin and Mary Loftus at Emory University's alumni magazine; my editor and now good friend Abigail Kraft at the *Jefferson Monthly*; Peggy O'Mara at Mothering.com; Mona Gable at BlogHer.com; Kimberly Ford Chisholm; and the good folks at *FamilyFun Magazine* all gave me other writing projects that helped keep us afloat during the years it took to complete this book.

I am grateful to the dozens of medical professionals who generously shared their time and expertise. Many of them are quoted on these pages, but others stayed behind the scenes. A special thanks to my father's best friend and poker buddy Richard Sullivan, M.D., who made time to unpack the minutiae of medicine for me. A shout-out goes also to Robert Sears, M.D.; Jay Gordon, M.D.; Felicia Cohen, M.D.; Linda Hopkins, M.D.; Stephanie Koontz, M.D.; Stuart Fischbein, M.D.; Kenneth Saul, M.D.; Gro Nylander, M.D.; and midwives Dagny Zoega, Colleen Forbes, and Augustine Colebrook.

Florence George Graves, Melissa Ludtke, Sophie Eisner, and all the students and staff at the Schuster Institute for Investigative Journalism at Brandeis University, where I was appointed senior fellow, have given me both research assistance and moral support. A special thanks to Sandy Bergo, executive director of the Fund for Investigative Journalism, and the board of the Fund for Investigative Journalism, for giving me a grant that helped finance this project.

Nine-year-old Delia and her friends Sam and Kate were playing in Ordway Park in Newton, Massachusetts, when they found my black backpack that had been stolen. The camera was gone but the notes from a week of interviews were still there. Thanks to them, and their mom, Joanne Mead, for sending back what was invaluable to me but worthless to anyone else.

Thank you to my loving family: Judy Margulis; Jeffrey Kessel; Laurie Olsen; Michael Margulis; my cousins Jesse, Josh, Hannah, and Jacob; my brother Zachary Margulis-Ohnuma; Mary Margulis-Ohnuma; as well as Miranda, Atticus, and Maddie; Dorion Sagan; Tonio Sagan; Jeremy Sagan; Robin Kolnicki and my niece Sarah; Sarah Propis, a loving, attentive nurse who helped with resources, ideas, and a ready ear; Matthew, Marya, Donna, and John Propis; my mother-in-law Susan Selfridge; my father-in-law Jim Propis; Carol Propis; Jan Young; Roy Young; my husband's cousin Kristin Mannoni, who generously gave us a place to stay in New York City and carried the baby in a front pack while I was meeting with my agent and presenting at the American Society of Journalists and Authors; Great-Grandma Propis; my mother's *compañero*, Ricardo Guerrero, who has shared both his science and his love and support with me over the years; my dear father, Thomas N. Margulis; and my little sister, who is also a wonderful babysitter, caretaker, and friend, Katherine Margulis.

"You can't *not* write this book, Mom," my oldest daughters, Hesperus and Athena, scolded when I despaired that I had taken on too big a task, I was too upset by how women of childbearing age were being mistreated, and too many people would be as disturbed as I was when they read the book. "People need to read it! You have to tell moms! You have to save their lives." All four of my children, Hesperus, Athena, Etani, and baby Leone (who traveled with me on almost every book-related trip I took; much to Dr. Ken Saul's horror, Leone was picking up that goose poop on the path along the Deschutes), have provided me invaluable support, love, and encouragement.

I can't find words meaningful enough to thank my husband, James, who makes me decaf cappuccinos, edits my drafts, writes glossary definitions, and patiently listens to me despair over the mistreatment of new moms at three o'clock in the morning. If our children turn out well, it's because they have the best father in the world.

But my biggest debt of gratitude goes to my mother, Lynn Margulis (March 5, 1938–November 22, 2011), who died unexpectedly of a catastrophic brain hemorrhage five months before the manuscript was due. Her love, support, advice, keen intellect, and open-mindedness helped see me through every day of writing this book. I wasn't sure I could finish it without her. Mom, I've done my best. I hope I've done you proud.

Index

ABOUT THE AUTHOR

Jennifer Margulis, Ph.D., senior fellow at the Schuster Institute for Investigative Journalism at Brandeis University, is an award-winning journalist and the mother of four. Her writing has appeared in the *New York Times; O: The Oprah Magazine*; the *Washington Post*; and on the cover of *Smithsonian Magazine*. A Boston native, Jennifer lives in Ashland, Oregon.